MULTICRITERION DECISION IN MANAGEMENT

Principles and Practice

INTERNATIONAL SERIES IN
OPERATIONS RESEARCH & MANAGEMENT SCIENCE

Frederick S. Hillier, Series Editor
Stanford University

CONTENTS

4 Weighting methods and associated problems

5 Ordinal multicriterion methods

6 Additive utility functions and associated methods

7 Outranking methods

8 Other multicriterion decision methods

9 Computers, Artificial Intelligence, Interactivity and Multicriterion Decision

10 Software for discrete multicriterion decision

11 Multicriterion decision in practice

12 Multicriterion methods: features and comparisons

References

Author Index

Subject Index

FOREWORD

Why another book on multicriterion decision making?

Several answers can be given to this question. The first is that, as far as we know, there exists at present no book devoted exclusively to discrete multicriterion decision making. Within companies and organizations, multicriterion analysis is used in two different ways: there is discrete multicriterion decision making, which is concerned with choice among a finite number of possible alternatives such as projects, investments, decisions etc., and it is this domain which is the subject of the book. Then there is the other main area, a subject which we shall not be treating here: multiple criteria linear programming. Here, the approach is to extend the results of linear programming and the associated algorithms (the most well known of which are simplex, gradient and Karmarkar) to multiple criteria.

The first answer to our question, then, is that a book on one of the two components of multicriterion analysis which can be of practical help in company decision making and management should be welcome, and this book is foremost written for those who want to apply the methods of multicriterion analysis. *Thus, starting from the main scientific results which are the foundation of the domain, we will show the principle methods of multicriterion analysis, their advantages and their shortcomings.* Although the presentation is rigorous, it should be accessible to all readers who have given some thought to decision problems; it is designed to be put into practice by managers and decision makers in the course of their daily professional life.

The second answer to the question is that, despite the large body of scientific papers on multicriterion decision making, the tools, methods and thinking behind multicriterion analysis remain virtually unknown to managers and engineers at all levels. There are several reasons for this. The main one would appear to be cultural; at the present time – in Europe, at least, where this book was written – the scientific culture still favors the notion of the 'best decision', whereas, as we shall see, in multicriterion analysis any optimum in the strict sense of the term does not exist. Herein lies the original sin which is at the root of the almost total absence of the discipline in academic curricula and therefore the understandable ignorance of the subject by engineers and managers. It is true that multicriterion analysis lacks the huge body of mathematical results which make optimization so attractive, but it can nevertheless offer several very interesting properties, ignorance of which leads decision makers to re-invent the wheel every day. More seriously, in companies this ignorance results in an impoverishment of thinking, a kind of self-censoring where all too often, analysis halts at the first convenient optimization which is quite clearly a pale shadow of the real complexity of the situation.

We believe that that is the main reason for using multicriterion analysis; it is no longer possible to ignore the fact that each real decision is the result of a compromise between several solutions which all have their advantages and disadvantages, depending on one's point of view. In organizations in future, it will become harder and harder to disregard the complexity of points of view, motivations and objectives. The day of the single objective (profit, social, environment etc.) is over, and the wishes of all those involved in all their diversity must be taken into account; and to do this, a minimum knowledge of multicriterion analysis is necessary. One of the objectives of this book is to supply that knowledge and enable it to be applied.

Acknowledgements

This book incorporates material from lectures and research work carried out by the authors in the last twenty years in their respective universities.

At the Université P. et M. Curie in Paris, the lectures were partly to fourth year undergraduates, partly to post-graduates in the specialty 'Decision methods and algorithms' under J.-Y. Jaffray. Large amounts of material have been borrowed from this teaching, which began in 1975 and continues today. Chapters 2, 3 and 6 in particular re-state the demonstrations and ideas coming directly from J.-Y. Jaffray's lectures on decision theory; we would like to express our gratitude to him for his permission to use the material and for sharing with us his extremely erudite knowledge of the subject.

In 1987, inside the venerable walls of the Universidad de Alcalá de Hénares, the authors held a summer school on multicriterion decision making. This agreeable experience was, through the interest shown by the participants and the support of the University authorities, at the origin of this book. Every year a doctoral course on multicriterion decision making is given at Alcalá by S. Barba-Romero, who would like to express his appreciation for the financial support from the Spanish National Research Plan (CAICYT project 499/84) given to the multicriterion decision making research team which he directs in the University of Alcalá. Among the members of this team, we would especially like to thank J. Pérez Navarro, who checked several chapters of the book.

In the field of multicriterion analysis research, the work of B. Roy is paramount. The LAMSADE laboratory (Université de Paris-Dauphine), of which he is Director, is the leading research center in the domain in France; we cannot stress too greatly how invaluable his help and support have been. The quality of the LAMSADE documentation center and the intensity of research activities taking place there make a visit obligatory for all who are interested in multicriterion analysis. We are extremely grateful to B. Roy for the warm welcome we have always received at LAMSADE, and we would particularly like to thank D. Champ-Brunet for his cooperation in the bibliographic search which was vital to the achievement of this book. Finally, we thank M.-J. Pomerol for her help in the final stages.

THE MAIN SYMBOLS USED IN THIS WORK

$\mathcal{A}$	choice set	p. 18
P or $\succ$	strictly preferred to	p. 21
I or $\approx$	indifferent to	p. 21
Q or $\succeq$	preferred or indifferent to	p. 21
NC	not comparable to	p. 21
$\mathcal{R}$	related pairs	p. 21
$\overline{\mathfrak{R}}$	non-related pairs	p. 21
R	binary relation	p. 21
$\mathbb{R}$	set of real numbers	p. 27
P_f or $\succ_f$	weak preference	p. 46
C	complete set	p. 60
O	set of efficient points	p. 61
A^c	convex envelope of A	p. 84
$\mathcal{PR}$	set of preorders	p. 124
x_{-i}	excluding coordinate i	p. 156
ϕ	flow in a graph	p. 197

1 WHAT IS MULTICRITERION DECISION MAKING[1]?

1.1 Choice in the presence of multiple criteria

In both everyday life and in organizations we are very often confronted with difficult choices where we are unable to decide between a number of imperatives. Here are a few examples of this kind of situation. Suppose there are available p different models N_1, N_2, ... N_P of a product N; everybody wants to buy model N_i which is the cheapest, looks best, is easiest to look after and is the most robust (the list is non-exhaustive). These various features are all purchasing criteria. Consider for example price and strength; experience tells us that the cheapest product is not the strongest: price and strength are two conflicting criteria. If we use price as a criterion for choice we may end up buying a product which is not among the strongest. On the other hand, if we buy the strongest product we may also be buying the most expensive. As the saying goes, you can't have your cake and eat it, and it is obvious that conflicting desires will always lead to a compromise. This is the type of situation we shall be interested in.

The buyer's situation we have just described is almost identical to the case of the investor. Suppose a company has p investment plans P_1, P_2, ... P_p, and wants to choose which one actually to apply. Among the criteria for decision it will almost certainly consider the forecast yield, the total cost, the usefulness in terms of

---

[1] As B. Roy points out in his preface to Vincke (1989), the term 'multicriterion decision' is not a very good one since, strictly speaking, decisions are neither monocriterion nor multicriterion, but are a choice, often an action and almost always an intention – in other words, a host of things which cannot be qualified as mono or multicriterion. What really is multicriterion is the decision support model, which is why a better name would be 'multicriterion modeling of choice', or else the term advocated by Roy and adopted by Vincke (1989): 'multicriterion decision aid'. The reader must forgive us for retaining the terms 'multicriterion decision' and 'multicriterion analysis' to denote the multicriterion modeling of choice and/or decision situations. Note finally that in this book, the terms 'monocriterion' and 'multicriterion' will be used as adjectives.

strategy and/or image, and will probably also be forced to consider the social and environmental impacts of the various plans. Clearly, the best plan for the environment is not necessarily going to be the cheapest. Here again, criteria clash.

The problem of multicriterion decision making sketched out above is also met in administrations, where decisions on public amenities rarely enjoy universal approval. Take for instance the route of a new highway; it is rare indeed that the cheapest solution is the one that serves the most people and respects the environment most: that would be too good to be true! As we know, it is cost which so often comes into conflict with the other criteria—the eternal problem of limited resources and unlimited needs!

In many situations, the risk factor, too, refuses to act in the same direction as the other criteria. We can illustrate this by the choice of a financial investment. If you have money to invest you will realize that National Savings do not pay the highest interest rates; on the other hand, there is practically no risk. A bigger yield can be had by buying shares, but as you stand to lose a fair amount if the shares go down, you conclude that the risk is greater. The rule of thumb soon emerges: the highest yields carry the greatest risk. This risk minimization criterion is also present in the company investment problem described above; in fact it is involved in nearly all decisions in as much as a decision has consequences in the future; and the future is not, as we know, given to man.

As these examples show, the situation of having to make a choice in the presence of multiple criteria is extremely common if not universal. We can summarize the facts as follows. The decision maker is faced with making one choice among several possibilities which we call alternative*s*, the set of these alternatives forming what is called the *choice set*. To make his choice from this set, the decision maker adopts several points of view, often contradictory, which we call *criteria* (these points will be formalized in the next chapter). These criteria are at least partially contradictory in that, if the decision maker adopts one of the points of view, for example risk minimization, he will not choose the same alternative as he would from the standpoint of another criterion, for instance best yield.

The reader will by now have realized that multicriterion analysis is relevant to many diverse situations; yet most analyses carried out in professional circles take no account of multicriterion decision making, for reasons that that will become clear as we proceed. The first reason is that multicriterion decision making has only become established as a scientific discipline relatively recently, as we shall see in the next section.

To conclude this introduction, we should point out that this book is devoted exclusively to discrete multicriterion decisions (*i.e.* there is a finite number of choices available), and is concerned neither with the continuous domain (where there is an infinite number of choices—see section 1.5), nor with multicriterion choice in databases; the latter is generally dealt with in works on databases, and in any case comes back to discrete multicriterion in the special case where different alternatives are evaluated on the basis of a simple 'yes' or 'no'.

1.2 Historical background

The idea of contradictory criteria has existed in popular culture since time immemorial, generally appearing in the form of proverbs and fables of which the miller, his son and the ass is perhaps the most well-known. The conflict between two equally desirable criteria (usually love and duty) is the source of innumerable dramatic plots which existed long before the story of Chimene. Then there is the tale of Buridan's ass who, equally hungry and thirsty, was unable to decide on which criterion to base his decision, and died. If only he had studied multicriterion analysis! However, such is not the stuff of scientific study, even if it is all based on common sense!

From the point of view of science, the research into economics which took place at the end of the nineteenth century and the beginning of the twentieth are one of the sources of inspiration for the domain we are interested in here. At that time economists were beginning to look for links between the behavior of economic agents and the economy itself. One of the basic factors governing behavior, applying both to producer and consumer, is the way choices are made in consumption and production. The formalization adopted at that time was to say that each sought to maximize their utility functions. We shall see that this utility function is a global way of expressing the choice of the consumer or the producer. In other words, the notion of distinct and more or less contradictory criteria was not taken into consideration at the time. Among those who first studied individual economic agents were Walras, Cournot and Pareto. The latter, who was both an economist and a sociologist, was a professor at the University of Lausanne. Like Walras and Cournot, he studied economic situations where several agents made different and sometimes conflicting choices. He showed that in such circumstances, not all agents can achieve maximum satisfaction simultaneously; since resources are finite, the gains of one agent are to the detriment of another. A situation in which agents cannot simultaneously improve their satisfaction is called a *Pareto optimum* (Lecture in political economy, 1896). If we consider a group made up of several agents, each with different preferences, as a single collective agent with several different choice criteria, we come back to the multicriterion problem and the attendant Pareto optimum, which we shall be discussing further in chapter 3.

The economists of the late nineteenth and early twentieth centuries may thus be considered as the precursors of multicriterion analysis as it stood after the Second World War; however we must also acknowledge an even earlier antecedent in the political thinking in France in the second half of the eighteenth century. For if we go back to the idea described above of a group of economic agents each having different wishes resembling a single agent possessing several different criteria for choice, we can say that two judges with differing opinions on the guilt of various suspects possess different choice criteria. Suppose now that those two judges have to deliver a single judgement; they will then find themselves in the situation of stating a single choice:—who is guilty—while they each have different criteria. They are performing multicriterion analysis without knowing it. This is precisely the problem that the Marquis de Caritat de Condorcet was studying around 1780 and which he published in 1785 in his book 'Essay on the application of analysis to the probability of decisions made by a plurality of voices'. Condorcet was a product of the Age of Enlightenment and a friend of Turgot, the economist who under Louis XVI fell out of favor when he tried to introduce ideas of economic freedom. Member of the Académie des Sciences in 1769 at the age of twenty six, he was a champion of reform and knew perfectly well that the legal context of judges meeting as a jury was strictly the same as the political context of voting where n electors, each with their own criteria or motives, must choose a single winner from m candidates; a society made up of electors is just like a jury, but with the problem of social choice (social in the sense 'from society'). We can understand why, in the climate of political reflection so characteristic of the end of the *Ancien Régime*, these problems came to be studied. During the years 1784-5, Condorcet's memoirs were discussed by the *Académie* along with the previous ideas on the same subject of the Chevalier de Borda (the memoir of 1781, first presented as a speech in 1770). Borda (1733-99), who had a less theoretical approach than Condorcet, proposed a simple method of social choice, a method less well known than that of Condorcet. Borda's method was adopted by the *Académie* for the election of new members (see Black, 1958, p. 178-80). We shall meet these two methods again in chapter 5. Before leaving the eighteenth century we must mention the first multicriterion choice method ever really presented in the form of a choice between contradictory criteria (or rather, arguments). This was the subject of a letter from Benjamin Franklin to Joseph Priestly in which the writer advised separating the criteria between those who are for the decision in question and those against. Franklin then suggests, using a sort of weighted sum (see chapter 4) to eliminate arguments pro and con that are of equal weight so that after this simplification it is easy to see where the balance hangs (*cf. e.g.* Zionts, 1992).

It was not until after the Second World War that political and economic thinking converged to become the theory of social choice, the theory of voting and multicriterion analysis, all sharing a common base. This fusion of thinking first came about in the domain of microeconomics, under the instigation of numerous economists including Hicks, Bergson and Samuelson, founders of the 'New welfare economy'. In 1944, von Neumann and Morgenstern's book alluded directly to the problem but offered no analysis. In 1951, Koopmans introduced the notion of the efficient vector, a new incarnation of the Pareto optimum. The problem of multiple criteria in linear programming was also tackled by Kuhn and Tucker (1951), and later by specialists in operational research such as Hitch (1953), Klahr (1958) and several others.

From microeconomics came a feeling that the link between the individual behavior of agents and the results observed in society should be deepened. In the early fifties, after the von Neumann-Morgenstern publication of 1944, discussion on the rationality of individual choice in the face of uncertainty became very animated with an initial polemic on the realism of the axioms used in obtaining a representation of 'simple' preferences (Allais, 1953a and b). From another point of view, the relation between choice and preference order became fundamental in the consumer theory. The next theory to be studied in depth was that of revealed preferences, pioneered by Samuelson (1938), taken up by Ville (1946) and later by the American school. Closer to multicriterion decision making is the problem of choice in a group of agents, or social choice; here, the American school lead the way with Arrow, whose well-known theorem (1951) we shall meet later. Arrow, followed by May (1954) explored this problem in depth.

Fundamental contributions to preference theory have been made by Savage (1954) and Debreu (1960). Psycho-mathematical aspects of individual decision have been treated by Luce (1956) and Raiffa (in their book published in 1957) and Tversky (1969), and the subject has been reviewed by Fishburn (1970) and Krantz, Luce and Tversky (1971).

By 1960, multicriterion analysis is acquiring its own vocabulary and problem formulations as outlined at the beginning of this chapter: the problem of choosing one alternative in the presence of multiple criteria. Between 1960 and 1970 there appeared a number of methods which have now become classical: the 'goal programming' of Charnes and Cooper (1961), who solved the problem of multicriterion choice in linear programming by a search for a solution at minimal distance from a multicriterion goal, generally non-achievable, set by the decision maker. The first book to deal with multicriterion decision by means of a weighted sum, Kepner and Tregoe, appeared in 1965. At the same time, at the SEMA, a French group was engaged in pioneering work in the field of applied mathematics, including work on multicriterion problems. From this came the notion of outranking (Roy, 1968a) and the associated method of multicriterion decision, ELECTRE (Roy, 1968a). Also at SEMA, Benayoun and Tergny (1969) designed an interactive method POP (followed by STEP) for linear programming with several criteria.

The first scientific meeting devoted to multicriterion analysis took place at the seventh congress on mathematical programming at The Hague (Netherlands) in 1970, where, among other things, a wide survey of the state of the art was given by Roy, and the first two interactive multicriterion methods were launched: the improved method of Benayoun and Tergny (Benayoun, de Montgolfier, Tergny and Larichev, 1971) and that of Geoffrion (published the following year in the paper by Geoffrion, Dyer and Feinberg, 1972). The next meeting, organized in 1972 at the University of Columbia in South Carolina by Cochrane and Zeleny (who first studied multicriterion analysis at the SEMA) was devoted entirely to 'Multiple Criteria Decision Making'. The congress report was to contain no fewer than sixty papers.

1972 was the year when multicriterion analysis came into its own, though the gradual emergence of ideas had started in the sixties. The process was not entirely painless since the dominant paradigm at that time was of operational research with the principle of 'search for an optimal decision maximizing an economic function'. As Roy (1987b) points out, this paradigm, largely inspired by physics, became a model for economists and is also to be found in several human science fields. The paradigm is a very powerful one and is still dominant in the work of many researchers; its influence extends to numerous university syllabuses. It was to be a long time before there was general agreement that human beings—and living things in general—was different from mechanical systems. After the first stirrings of doubt in the sixties followed by Simon's rejection of the old paradigm of maximizing expected utility, specialists in aids to decision, having to deal with real-life decision makers, were quick to reject in their turn the 'optimizer' paradigm. By 1968 Roy was able to write, 'Operational research must be de-optimized', an argument set out in detail by Roy in papers of 1976 and 1977. In 1977 Keen introduced his 'apprehensive man' (see section 11.1), an accurate pointer to the direction ideas were taking. It is difficult now, with our modern notions of complexity and the study of complex systems, to imagine how incongruous the multicriterion idea seemed during the sixties; indeed, the idea is not always perfectly understood even now and, as Roy (1987b) has pointed out, the single-criterion paradigm is one of those 'lost paradigms' which numerous methods of aggregation have as their sole aim to re-introduce as quickly as possible. Not the least merit of the multicriterion modeling which you will learn about in this book is to foster the idea that there is no optimization, but rather, compromise, equilibrium and a legitimate multiplicity of points of view.

In the United States in the seventies, thinking on multicriterion decision was dominated by discussions on the additivity of preferences, for which Leontief (1947). Debreu (1960), Fishburn (1965 and 1970) had paved the way; others were subsequently to supply the first really formal results (an excellent history of this subject is to be found in Wakker (1989)). In our field of interest, the most lasting result was the MAUT method, popularized by Keeney and Raiffa in their book (1976).

From 1975 onwards, numerous refinements were made in the various avenues explored up to then. The French school (Brans, Jacquet-Lagrèze, Roy, Roubens, Vansnick and Vincke, among others) thoroughly investigated discrete multicriterion, outranking relations and decision maker preference. American workers were divided between supporters of the additive utility of Keeney-Raiffa and pragmatists using other methods (Saaty, Yoon, Zeleny and Zionts being among the most active). Various other methods were introduced by other European workers including Rietveld, Paelinck and Wallenius. Continuous multicriterion decision making is still a very active field with numerous methods (*e.g.* Vincke, 1976; Steuer and Choo, 1983), improvements to existing methods such as 'goal programming' (*e.g.* Ignizio, 1976 and Spronk, 1981), and effective algorithms for Pareto point searching (Evans and Steuer, 1973; Ecker and Kouada, 1978; Isermann, 1977 and 1978; Zionts and Wallenius, 1883a). By 1985, multicriterion methods had come to enjoy world-wide recognition with many countries contributing. European and American schools, more active than ever, were joined by the Pacific school (Takeda, Seo, Sawaragi, Tabucanon, Changkong).

The most decisive element of the eighties was the introduction of computer methods into multicriterion decision making. By 1970, interactive methods were already being proposed (see above); the main novelty was the ease with which they could be installed and the fact that the tasks which computers—particularly microcomputers— were now capable of performing had a powerful influence on the design of methods. Visualizing and interacting became items for study in their own right (through the work of Angehrn, Belton, Korhonen and Vanderpooten). And finally, with informatics came computer-specific methods such as artificial intelligence (Bratko-Rajkovic and Lévine-Pomerol being among the first in the field).

Multicriterion decision making can now be considered as a field of activity in which practical application and informatics are dominant. Theoretical research is not of course devoid of interest, but it is now more concerned with giving depth to existing ideas than in innovating. On the other hand, the possibilities of informatics have not yet been wholly explored; we may even say that actual application of multicriterion methods in professional contexts has only just begun.

1.3 The role of multicriterion analysis in organizations

At this point the reader should be getting an idea of the theme of this book, but is probably not yet convinced of the utility of multicriterion formalism. The advantages of multicriterion modeling must be set against classical modeling: the aim in monocriterion modeling is to end up with a *maximizing problem free of constraints, whose optimal solution represents the best choice.* As an example, take the investment problem once again. Each decision maker wishes to maximize his profit expectancy with the constraint $\sum_i v_i \leq I$ where each v_i represents an investment and I is the total sum to invest (a limit will probably be required also on the number of investments that can be managed). Suppose now that the investments under consideration present different risks to the environment. One, for example, may involve increasing the capacity of a factory which has already attracted the wrath of ecologists, while others may involve a new plant to acceptable environmental standards. How will this be integrated into our model? The profit function will have to be changed by estimating cost increases or decreases and by evaluating the risk of a successful challenge by the environmentalists. This is clearly a difficult exercise. The formulation of the problem is exactly the same for the social impact of the various projects, or the effect on company image (and more criteria could be added to the list). The exercise of translating these elements into costs affecting the profit function is not, in the vast majority of cases, carried out, so that monetary costs alone are taken into consideration. Whatever example we take leads to the same conclusion: *certain aspects of choice are very hard to evaluate in terms of cost.*

   Another possibility in single-criterion modeling is to treat factors which are not immediately monetary as constraints. This is the solution generally recommended by advocates of classical operational research. In our example, suppose each investment project to have an image mark Im_i out of ten. If investment i has a mark $Im_i = 8$, this will tell the investment advisor that it will give the company a good image (*e.g.* building a production unit which fits in perfectly with the environment). If we want to introduce this into the constraints we shall have to state an image threshold, for example 7. The model will then contain a constraint of the form:

$$Im_i \geq 7$$

   Defining such a threshold is extremely difficult, especially outside the context of the decision. For the decision maker to define the threshold he has to know the marks given to other projects, and the strength of the constraint, whether active or not, in determining the solution. In other words, even though it is not realistic to try to define such constraints *a priori*, this is precisely what must be done in this type of modeling.

Notice also that it is only in defining constraints that we can avoid trivial solutions in optimization models. Suppose, for example, that a company wishes to supply distributors as cheaply as possible. No computer is required to determine the optimal solution: just leave the trucks in the garage! This is obviously not the best solution if other criteria such as customer satisfaction or survival of the company are taken into account![2]. From a single-criterion point of view, the only way of avoiding this trivial solution is to introduce constraints on minimal supplies. The drawback is that constraints are treated by the decision maker as just that: constraints, with the attendant negative connotations, when in fact they are simply one of the factors involved in the decision. Some breaks in supply may well be acceptable if this lowers costs. In any rational organization a compromise between these two criteria will be found, according to the aims in view and the type of product and activity.

From this last example, we see that multicriterion modeling gives a freedom of judgement to the decision maker which is obscured by single-criterion modeling. Multicriterion modeling is thus much more realistic, since it takes pseudo-constraints for what they really are: elements involved in the decision, *i.e.* criteria. In certain models, it may be left up to the decision maker to set the criteria and constraints (see Korhonen *et al.*, 1989). In multicriterion analysis, profit expectancy, the environment, image, quality of supply and any other factors in the decision process will, for as long as possible, be evaluated separately as criteria affecting the decision. The first result of this is that the model will at all times remain intelligible to the decision maker with the direct consequence of a high quality interactive analysis of the decision problem. *Including the criteria in the function to be maximized or in the constraints is not only an artifice, however conceptually admissible, but is also harmful to the decision process since it prevents intervention of the decision maker and makes choices highly rigid.*

Another difficulty arising from this rigidity appears in the very frequent context where the decision is a more or less foregone conclusion in the organization. It is obvious that the 'maximizer' of profit expectancy must pay attention not only to the unions and the environment but also to the members of the board who will not necessarily agree among themselves. A closed model in which the criteria of the various interested parties are hidden in the function to be optimized or in the constraints is of no use whatsoever in negotiations (or discussions). On the other hand, a model which leaves everybody's criteria visible can be used as a tool for consensus-seeking; this point is not the least advantage of multicriterion analysis, and we shall be returning to it.

The other drawback of 'optimization' modeling is its lack of realism from the human point of view. This criticism has been fully developed by Simon (1983) in his work on criticism of the utility expectancy maximization model. He points out that in human administrative management the practice as far as he had seen was to

---

[2] In a humorous vein, the Dean of the Columbia Business School remarked: 'As for conflicting objectives, quality *vs.* lower cost, better products *vs.* cheaper raw materials, for example, just about any idiot can maximize a simple function. Anybody can increase sales. After all, if nothing else matters, you can decrease the price to zero. In fact you don't have to stop there. If they won't take it at zero, you pay them to take it.' (Hermes, vol. 3, No. 2, 1975, quoted by O'Leary, 1986).

use different—and often conflicting— criteria at different times. Let us give an actual example. One day a chief executive in a group proposed the appointment of an administrative and financial director (AFD) ready to do some belt tightening, mainly on the side of computer costs; the first criterion for decision making: reduce costs. A little while later, the same AFD is required to submit various cost forecasts together with a financial viability evaluation for each operation; second set of criteria: risk evaluation and strategic choice. Given the number and complexity of operations, the AFD needs a computer decision support system (DSS) to answer all the questions, for which he requests the necessary resources ... and gets them. Net result: a substantial increase in computer costs! The story is perhaps banal but it does nicely demonstrate how decision criteria can vary with time. Simon makes the same point in his introduction to the multicriterion dimension in his modeling of limited rationality.

Multicriterion analysis has two key features: realism and legibility; these are vital today in organizations where the complex nature of decisions is recognized by most of those involved, even if they are not all equally sensitive to the same criteria. They know perfectly well that any decision, even a personal one, is a compromise between requirements that can never all be completely satisfied at once. Between the principles of pleasure and reality, the non-psychopathic individual has to arbitrate. This does not imply that desires vanish; they remain in the wings as criteria for choice, but multi-criterion analysis manages to model the problem without hiding this fact: it has the merit of frankness.

1.4 An example to introduce some basic notions

Before beginning a rigorous and formal presentation of the basic principles of multicriterion analysis, we shall introduce a few of the ideas through an example illustrating the various problems and notions involved. This will be helpful to the reader at a later stage when the problems are discussed.

Our illustration is based on the classical example of investment. Suppose a wine producer is thinking of diversifying into a new activity. Several plans are on the drawing-board.

Plan 1 (P_1) Purchase and develop an estate in Napa valley
Plan 2 (P_2) Set up a poultry unit in Kansas
Plan 3 (P_3) Set up a pig breeding unit in Georgia
Plan 4 (P_4) Take over a canning factory in California
Plan 5 (P_5) Invest in a pork abattoir and set up in the commercial production of pork products (in Illinois).

The company directors will draw up an estimate of the profit margin of each activity (we shall call this criterion *pr*). But this margin will depend on so many factors that it will be cloaked in uncertainty. However, the figure can be based on the average profit margin of other firms with comparable activities. Less difficult, perhaps, is the evaluation of market growth (criterion *gr*). Risk factors (criterion *ri*) must also be evaluated, and it is obvious that poultry breeding, where there are already many exporters in the Near and Middle Eastern countries, is a riskier undertaking than pork products where the consumers are for the most part Americans and Europeans.

The impact on the environment must also be evaluated; if the production unit is subsequently rejected by the neighborhood, expenses may be incurred. (criterion ev).

Finally, pay-back time, the time for the balance sheet to become positive, is a very important criterion, involving state aid and which affects later investment (criterion *pb)*.

At this stage of the analysis the decision maker gathers all the information together in a decision matrix (or performance table), one project per row, and evaluates them against each criterion.

	Profit margin *pr*	Growth *gr*	Risk *ri*	Environment *ev*	Pay-back time *pb*
P_1	14%	8%	low	good	7
P_2	16%	8%	high	bad	2
P_3	12%	9%	medium	very bad	4
P_4	13% .	10%	low	medium	4
P_5	20%	12%	medium	bad	5

Figure 1.1 Quantitative/qualitative decision matrix

There are two types of criterion. *Quantitative* criteria (profitability, growth, pay-back time) are criteria which can be expressed on a numerical scale (in percentage for profitability and growth and in years for the turn-around time).

The other criteria are *qualitative*. It is often useful to choose a scale on which they can be expressed as numbers so that they can be ordered (this is called a *utility function*). The choice of scale is often difficult and may have repercussions on the final choice made (see chapter 2). Here we shall choose a scale of 1 to 5 for risk and 1 to 10 for the environment. The riskiest project will be marked 5 and the best environment will be given 10. Note that, among the various criteria, we are seeking to *maximize* profitability, growth and the environmental mark while *minimizing* risk and pay-back time. The numerical decision matrix is shown in figure 1.2.

	Max *pr*	Max *gr*	Min *ri*	Max *ev*	Min *pb*
P_1	14	8	1	8	7
P_2	16	8	5	3	2
P_3	12	9	3	1	4
P_4	13	10	1	5	4
P_5	20	12	3	3	5

Figure 1.2 Numerical decision matrix

A glance at the matrix will show that, out of P_3 and P_4, nobody would choose project P_3: P_4 is better than P_3 on all counts. P_4 is said to *dominate* P_3 and P_3 is *not Pareto optimal*.

On the other hand, if we look at the other projects we find that no one project is dominated: these are all *Pareto optimal projects*. We now gather all these non-dominated projects into a new decision matrix (figure 1.3).

So how do we choose between the remaining projects? What properties is it desirable to have in a choice function? These are the two main questions that this book sets out to answer.

	Max *pr*	Max *gr*	Min *ri*	Max *ev*	Min *pb*
P_1	14	8	1	8	7
P_2	16	8	5	3	2
P_4	13	10	1	5	4
P_5	20	12	3	3	5

Figure 1.3 Decision matrix of non-dominated projects

1.5 Continuous multicriterion decision making

In the preceding section we showed a simple example of discrete multicriterion decision making, and we have limited the scope of this book to this field; however, we now give an example of what is called continuous multicriterion decision making, or more accurately, multiple criterion linear programming. We hope thereby to give the reader an idea of the type of model involved so that he can compare it with his own problem and see if the continuous model applies there; the reader will in addition be aware of what this book does not cover.

Suppose we wish to set up a production plan for a factory which manufactures two products A and B. The manufactured quantity of product A is x_1. We assume x_1 is a real number, so that product A is divisible (*e.g.* granulated sugar or gasoline, but not products like trucks unless the production unit – such as sugar lumps – is small compared with the number produced). Similarly, the quantity of product B produced is x_2.

Each manufactured unit of A fetches p_1 and of each one of B fetches p_2 (the units are immaterial). Products A and B use a number of common resources. The amount of machine time available is limited to 6600 hours; the maximum quantity of basic raw material used in the two products is 2100 tons. To give a concrete example, suppose the factory manufactures two types of cement: ordinary Portland and Prompt (early set cement), both from the same raw materials, limestone and clay. Only one furnace is available. Note, by the way, that in most actual cases constraints on resources are *real* constraints (*cf.* section 1.3) which are external to the model, at least at our level of analysis.

We have to draw up a monthly production plan which assumes that prices are external to the model and that all the products on offer are sold. (In practice these assumptions are too simplistic). With the processes used, each unit of product A leads to r_1 units of atmospheric pollution, each unit of product B leads to r_2 units.

Classically, the optimum plan is the result of solving the following program:

Maximize $p_1 x_1 + p_2 x_2$
subject to

$$a_1 x_1 + a_2 x_2 \leq 6600$$

$$b_1 x_1 + b_2 x_2 \leq 2100$$

$$x_1 \geq 0, x_2 \geq 0$$

The set of feasible productions (x_1, x_2) is a polyhedron in the plane. In order to visualize this polyhedron let $p_1 = 1$, $p_2 = 1.05$, $a_1 = 1$ (number of machine-hours to manufacture a unit of A), $a_2 = 1.1$ (number of machine-hours to manufacture a unit of B), $b_1 = 0.5$ (quantity of raw material required to produce one unit of A) and $b_2 = 0.25$ (quantity of raw material required to produce one unit of B). With these figures which are, of course, entirely imaginary, the program to be optimized becomes:

Maximize $x_1 + 1.05x_2$
subject to

$$x_1 + 1.1 x_2 \leq 6600$$

$$0.5 x_1 + 0.25 x_2 \leq 2100$$

$$x_1 \geq 0, x_2 \geq 0$$

The set of feasible productions is an infinite set consisting of the set of pairs of real numbers in the hatched part of figure 1.4. Mathematically, this set is said to be continuous. This is why in this case we refer to continuous as opposed to discrete programming. The program solution is $x_1 = 2200$ and $x_2 = 4000$, *i.e.* in one month 2200 units of A and 4000 units of B must be produced. The solution corresponds to the vertex Q of the polyhedron of feasible productions (figure 1.4).

Up to now we have not included pollution in our program. In fact, the product which produces the most profit happens also to be the one whose production causes the most pollution. Manufacturing one unit of B produces 1 volume of pollution while one unit of A produces 0.8 volumes of pollution. The pollution function to be minimized is thus $0.8 x_1 + x_2$. This is equivalent to maximizing $- (0.8 x_1 + x_2)$. There are three ways of incorporating pollution into our model (*cf.* section 1.3). First, we can penalize the production of x_2 and say that the associated profit is 0.9 instead of 1.05 (a tax of 0.15 per unit) and then solve the program:

Maximize $x_1 + 0.9 x_2$
subject to

$$x_1 + 1.1 x_2 \leq 6600$$

$$0.5 x_1 + 0.25 x_2 \leq 2100$$

$$x_1 \geq 0, x_2 \geq 0$$

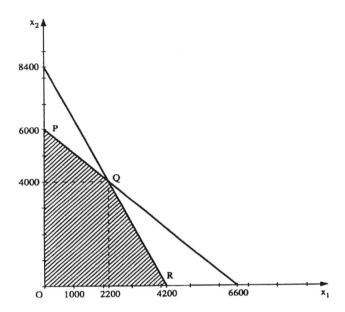

Figure 1.4 Feasible productions polyhedron

The result will be the same as before. The pollution tax has to reach 0.35 per unit produced for the solution to change (and it will then become $x_1 = 4200$ and $x_2 = 0$, giving logical preference to the least polluting product). What we have done in this first solution is to enter pollution into the function to be optimized (or economic function), *i.e.* to attribute a cost to the pollution criterion. Within the context we have chosen, this solution is far from absurd.

The second, much more debatable, solution is to set a pollution level – such as 5000 – which must not be exceeded (hence, a constraint). The program is then solved as follows:

Maximize $x_1 + 1.05 \, x_2$
subject to
$$x_1 + 1.1 x_2 \leq 6600$$
$$0.5 x_1 + 0.25 x_2 \leq 2100$$
$$0.8 x_1 + x_2 \leq 5000$$
$$x_1 \geq 0, x_2 \geq 0$$

The set of feasible productions (figure 1.5) is now different from that in figure 1.4; the solution to the program has also changed and now corresponds to the point Q' in figure 1.5 ($x_1 = 2833$ and $x_2 = 2733$). This solution involves a pollution level of 5000, this being a fixed threshold; in other words, the pollution constraint determines the solution *de facto*. This is an illustration of the case where the real decision maker is the person who sets the constraint.

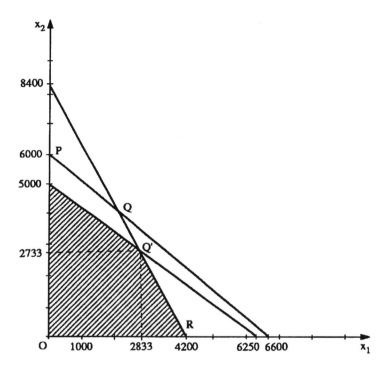

Figure 1.5 The new set of feasible productions under pollution level constraint

What we observe in the two methods above is the need to introduce either a price for pollution or a maximum level of pollution; and these are both difficult to justify. Unfortunately, once these more or less arbitrary values have been introduced the solution immediately pops out without any visible link for the non-specialist, nor even for the professional modeler when the program is complex and carries numerous values of the same type. This idea of replacing criteria by constraints is used systematically in continuous linear programming in so-called 'epsilon-constraint programming' (see Goicoechea *et al.* (1982)).

The final possibility is to analyze the multicriterion program:

Maximize $x_1 + 1.05\ x_2$ and Minimize $0.8\ x_1 + x_2$
subject to
$$x_1 + 1.1\ x_2 \leq 6600$$
$$0.5\ x_1 + 0.25\ x_2 \leq 2100$$
$$x_1 \geq 0, x_2 \geq 0$$

Looking at this program, we note that all the points on the polyhedron of figure 1.4 lying on the broken line QRO are possible production for which the pollution drops from 5760 to 0 when the profit decreases from 6400 to 0. Since the two criteria conflict, it is up to the decision maker to choose an acceptable compromise. The decision maker may also try to transform the two criteria into a single one. Here, he may decide to maximize $x_1 + 1.05\ x_2 - (0.8\ x_1 + x_2)$, which will give the program:

Maximize $0.2\ x_1 + 0.05\ x_2$
subject to
$$x_1 + 1.1 x_2 \leq 6600$$
$$0.5 x_1 + 0.25 x_2 \leq 2100$$
$$x_1 \geq 0, x_2 \geq 0$$

This program has the solution $x_1 = 4200$ and $x_2 = 0$ corresponding to the point R on the polyhedron, producing a pollution level of 3360 and a profit of 4200. The optimum profit, we remember, is 6400 associated with a pollution level of 5760, while the minimum pollution level is 0 for zero production (we obtain this obvious pollution-minimizing solution if we do not set a production minimum – see section 1.3).

We should stress once more that continuous programming, whether multicriterion or not, falls outside the scope of this book. This type of modeling obviously finds application in practice when dealing with big problems involving many variables. It is particularly interesting given that there exist highly effective algorithms for solving this type of program, at least in the single-criterion case; these algorithms can, however, be adapted to the multicriterion case as well.

One reason why we decided to restrict this work to the discrete case, *i.e.* to a finite number of alternatives, is that, paradoxically, it appears that this type of modeling, which is adaptable to many relatively simple practical situations, is even less well-known than continuous programming. The second reason for our decision is that the discrete case is entirely sufficient to illustrate and help understand the problems and possibilities of multicriterion analysis. We see popularization of the ideas of multicriterion analysis as an urgent task and the underlying force behind this book, which we wish to be accessible to a wide public of engineers and decision makers with no previous knowledge of the subject.

1.6 How to use the book

We wrote this book for two types of reader. The first was engineers, decision makers and negotiators: professionals engaged in modeling, discussion and decision making within a firm. It was with this type of reader in mind that we decided that the scientific knowledge required should be of high school level. The reader will learn that there exist numerous methods of multicriterion choice all of which have their advantages and disadvantages; a wide range of methods is included in the book, and the reader should find here a method suited to his particular problem. *It is our belief that the actual choice of method is rather less important than the initial multicriterion analysis followed by high quality discussion.* The other point we consider important is to make readers and future users aware of the pitfalls and limits of multicriterion methods.

Nothing could be more prejudicial, both to users and to the very discipline of multicriterion analysis, than to think of it as a panacea providing instant miracle solutions to all decision problems. Multicriterion analysis often has, as we have already said in section 1.4, the merits of legibility and straightforwardness; but decision itself is, by definition, subject to political – or perhaps entrepreneurial – choice (political in the widest sense of the term). It is therefore important to understand that multicriterion methods do not carry with them an inherent rationality, whatever many technocrats and modelers might say. A good knowledge of the field is required before the difference between what the models really say and what smart analysts can make them say can be appreciated.

The other type of reader we have in mind is the student or teacher who wants to know what multicriterion analysis can do. Needless to say, in the world of higher education the optimization paradigm is still dominant and multicriterion analysis is not – with rare exceptions – taught there; but our hope is that multicriterion analysis will attract more and more interest as workers in the field of economics become aware of the complexity of choice with which they are faced. There will then be a need for basic textbooks which cover the field without sacrificing scientific rigor. This is precisely why we have striven to retain scientific rigor in our treatment. We have not included every scientific result that exists, but we have tried to be as complete as possible. This is reflected in the bibliography which should be extensive enough for those looking for proofs or further reading.

2 BASIC PRINCIPLES AND TOOLS

2.1 The discrete multicriterion decision (DMD) paradigm

2.1.1 The decision maker and the analyst

In discrete multicriterion decision the existence of a decision maker is assumed; this decision maker will obviously feature as an element in the model and as such is an abstraction. In practice the term 'decision maker' corresponds either to a single person faced with a choice or to a group of individuals. The word 'decision maker' may also apply to people with the job of analyzing certain choices, even if they do not actually make the decision, or if the decision is made afterwards within terms of reference of which they are ignorant or disapprove.

Without discussing the validity of a model featuring an abstract decision maker, we shall always assume the existence of a real decision maker, even in actual situations where he or she is sometimes difficult to identify and we have to use an idealized abstraction. But even when hard to identify, we consider that the notion of a decision maker is still present as a limiting case which, though it may not be attained, is at least approximated satisfactorily. This is rather similar to the notion of perfect competition as an idealized form of real competition between companies. To get an idea of the wide variety of contexts in which the decision maker plays a role, see Roy (1985).

Nor shall we be discussing the questions, 'What is the essence of decision?' or 'Do decision makers exist?'; some people push the argument against the ideal decision maker to the limit where no decision maker or group ever makes a decision, and what we call decision is nothing more than social, economic, psychological, etc. determinism. We totally reject this point of view, simply because it would lead us into areas of little interest and leave workers out there in the real world with nothing on which to base their operational thinking. It is however clear that the notion of decision is at the same time an ideal notion. The important thing is not to make a mythology out of the instant of decision; the instant is only rarely identifiable and for most workers since Simon (1977) in the science of organization, the decision process within organizations is spread out in time and goes through at least four phases: information gathering (Simon's 'intelligence'), design, choice and review. Phases do not follow one after another in a fixed or linear order, but can switch in a more or less random fashion as the thinking of the decision maker(s) progresses.

In multicriterion analysis, the information gathering phase will apply to the whole of the problem at hand, involving a survey of the criteria and possible alternatives. The design phase consists of constructing the choice sets, *i.e.* the alternatives, and evaluating them for each of the chosen criteria. The choice phase is the moment when an alternative is finally selected. It is often the case that the definitive choice is made independently from the analysis and at times or in cases which are outside the field of multicriterion analysis as performed by specialists. As for formal decision reviews, these are very rarely carried out, whether in multicriterion analysis or other decision aids, even though the decision makers are almost certainly aware of them. Analysis review as an aid to multicriterion decision has not really been exploited up to now.

In addition to the decision maker, it is often convenient to introduce a second person – whom we shall call the *analyst* – into the proceedings. The analyst is the person who actually models the situation under study and who will make any recommendations as to the final choice. He will not express any personal preferences, but will simply gather those of the decision maker and treat them as objectively as possible

Decision makers nowadays often try replacing the analyst by a computer program. The decision maker can then feed the computer with information which is processed into a form which is meant to help decision making. The computer, however, is not able to create new models, only to propose pre-defined ones, so the initial analysis rests on the shoulders of the decision maker, with the computer remaining simply an aid to formalization, memorizing and reflection.

2.1.2 The choice set

The second element in our model is the set of choices or alternatives. Here we assume that the choice of the decision maker is to be made from a *finite number of alternatives*, and the set of these alternatives is called *the choice set*.

In practice these alternatives will be projects, candidates, investments, plans etc. among which a choice has to be made. In general we shall denote these alternatives by A_i, $i = 1,...,m$, and the choice set $\{A_1, A_2,..., A_m\}$ by $\mathcal{A}$.

Unless otherwise stated we shall adopt the classical modelling system where we assume that the alternatives are *different and that they* make *up the whole of the decision set*. The latter is quite a strong assumption and it can be useful to make it known, for its effect on the decision maker is to forbid the choice of a mixed solution, intermediate between A_i and A_k. It also forbids any alternative which does not belong to the choice set under consideration, even if such a choice is the result of mature reflection. If the decision maker introduces a new alternative then in principle the analysis process must be repeated with the newly formed choice set. We shall come back in chapter 5 to the question of the consequences of introducing new alternatives.

2.1.3 Attributes and criteria

When the decision maker is making his choice among the alternatives within the choice set we assume that he has at his disposal several (at least one) lines of evaluation; these may be, for example, price, quality, appearance and strength for a choice involving a product. These evaluation lines are known as the *attributes of the alternatives*; the terms 'evaluation axis' and 'characteristic' are also used. When a minimum amount of information about a decision maker 's preferences is added to these attributes, the latter become *criteria*. In other words, *a criterion* expresses more or less accurately the preferences of the decision maker on a given attribute. We shall see in section 2.3 which information the decision maker is supposed to bring to these attributes, and what alternatives must be selected if the concept of criterion is to be rigorously constructed. We make the permanent assumption that *there is a finite number of criteria $C_1, C_2,...,C_j,...,C_n$.*

In the most basic model we say nothing about independence of or, on the other hand, relations between these criteria. We do not, for example, ask whether strength is a component of quality. The model accepts both criteria as distinct and unequivocal. Similarly the basic paradigm assumes that all criteria are on an equal footing and is not concerned about whether some are more important than others. Methods do of course exist which take just such considerations into account, and we shall be dealing with these later.

Finally, we shall sometimes be making a distinction between two types of evaluation among these criteria; certain attributes such as price, speed, costs, and percentages (market share, profit margins) are inherently numerical, and we denote these as *quantitative criteria*. Then there are other evaluations which cannot be made on a numerical basis, including corporate image, social risk and quality for which no canonical unit of measurement exists; these we shall call *qualitative criteria*.

2.1.4 The decision matrix

Before going into any discussion on decision maker preferences, we shall make the fairly strong assumption that for each of the attributes considered, the decision maker can, for each alternative in the choice set, give a numerical or symbolic symbol a_{ij} (for the alternative i and the attribute j) which expresses an evaluation of A_i relative to the attribute j. Thus for example a_{ij} may be a mark, a price, or an assessment of the type 'good', 'bad', 'excellent', etc.

The matrix (a_{ij}) is called the *decision matrix or the performance table*. Each line of the matrix (figure 2.1) expresses the performances of alternatives i relative to the n attributes considered. Each column j expresses the evaluations of all the alternatives adopted by the decision maker, relative to the attribute j. Given the basic ingredients described above, it should be reasonable within the multicriterion decision paradigm to seek the answer to the question: what is the best decision? We shall see that this is actually far from being the case and, even worse, the above question is not very well posed.

Attributes or criteria

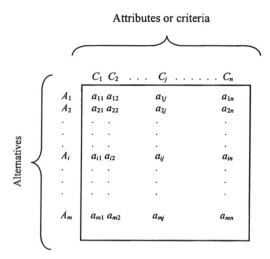

Figure 2.1 The decision matrix

2.1.5 Related models

In the introduction we mentioned the fact that other models exist that are related to multicriterion decision. Historically, the first of these was the vote. Imagine a society with several members; these members have to elect a candidate (or choose a victim in the case of Condorcet) from a set of candidates $\mathcal{A}$. Each member of the society has his own classification of the candidates or knows in his heart of hearts whom he wants elected. The reader will have noted that the set of candidates $\mathcal{A}$ is the choice set and that the decision maker is the society. The criteria can be found by identifying each member's classification to one criterion. In other words, each member of the society expresses his own criterion. If we identify 'candidate' = 'action', 'decision maker' = 'society' and 'preference of each elector' = 'vote', the two problems of vote and multicriterion decision become almost the same except that in voting, a single candidate is in general elected, while in multicriterion decision we often try to establish a ranking of all the alternatives; however, many of the results in the voting problem are transferable to multicriterion decision, and we shall be returning to this in chapter 5.

The other related model is that of social choice; again we have a society with n agents and a choice set (*e.g.* collective assets), and the agents require a choice procedure for reaching agreement on a classification of the assets contained in the choice set. If there is consensus on the choice, the latter is known as the social choice. Here too, each agent is a criterion in that he expresses his own classification of the choice set. Since in this context we require classification of the set of assets both by the agents and by the society, we are in a situation identical to that of multicriterion decision of the most classical sort.

2.2 The decision maker's preferences and order relations

2.2.1 Decision maker's preferences

Since the decision maker as defined above is the person who chooses, it is reasonable to suppose that, on considering two alternatives a and b within his choice set, he is able to say that he prefers a to b or vice-versa. We shall in fact be a little more indulgent with this decision maker and allow him the possibility of being indifferent to the two alternatives concerned, *i.e.* of considering them as equivalent.

Formally we shall say that:

– the decision maker strictly prefers a to b if he chooses a unambiguously, and we write:

$$a\,P\,b \text{ or } a \succ b$$

$\succ$ thus means '*is strictly preferred to*';

– the decision maker is indifferent to a or b if he accepts indifferently one or other alternative, and we denote:

$$a\,I\,b \text{ or } a \approx b$$

$\approx$ thus means '*is indifferent to*';

– when the decision maker does not know whether he strictly prefers a to b or if he is indifferent between the two we shall say that a is preferred or indifferent to b and we shall denote:

$$a\,Q\,b \text{ or } a \succcurlyeq b$$

$\succcurlyeq$ thus means '*is preferred or indifferent to*'.

In practice it may be that the decision maker is incapable of choosing or refuses to choose between two alternatives; for the analysts this means that the alternatives are not comparable, *i.e.* that we have neither a equal to or greater than b nor b equal to or greater than a, which we denote:

not $(a \succcurlyeq b$ or $b \succcurlyeq a)$ or $a\,NC\,b$, which is also equivalent to not $(a \succcurlyeq b)$ and not $(b \succcurlyeq a)$.

2.2.2 Order relations

The relations $\succ$, $\approx$, $\succcurlyeq$ and NC defined above are binary relations in that they involve two alternatives in the choice set $\mathcal{A}$. Before going on to talk about the decision maker we shall give a few properties of binary relations which will be useful in what follows.

Definition 2.1

A binary relation R on the set X is a partition of the set of pairs XxX. This partition creates two sub-sets of XxX, the first of which, denoted $\mathscr{R}$, is the sub-set of the related pairs and the second of which, denoted $\overline{\mathscr{R}} = \{XxX - \mathscr{R}\}$ is the sub-set of the unrelated pairs. (Given two sets X and A, we denote by $X - A$ the set of elements of X which do not belong to A).

For any pair $(a,b) \in \mathscr{R}$ we write $a R b$, *i.e. a* and *b* are related in *R*. The pairs of $\overline{\mathscr{R}}$ are those which are unrelated. The set $\mathscr{R}$ defines the relation, and all the pairs in the relation can thus be enumerated.

Examples. If we take $X = \{1, 2, 3\}$ and let $\mathscr{R}_1 = \{(1,2), (1,3), (2,3)\}$, then this relation R_1 is none other than $<$ since $(a,b) \in \mathscr{R}_1$ if and only if $a < b$. The relation R_2, defined by $\mathscr{R}_2 = \{(1,1), (1,2), (1,3), (2,1), (2,3), (3,3)\}$ is richer than R_1 since $\mathscr{R}_1 \subset \mathscr{R}_2$, this being the relation $\leq$.

Some definitions on binary relations:

Definitions 2.2

Reflexivity

A binary relation *R* is said to be reflexive if for all $a \in X$ we have $(a,a) \in \mathscr{R}$, *i.e. a R a.*

Irreflexivity

A binary relation *R* is said to be irreflexive if for all *a* $a \in X$ we have $(a,a) \notin \mathscr{R}$, *i.e.* not(*a R a*).

Symmetry

A binary relation *R* is said to be symmetric if $(a,b) \in \mathscr{R}$ implies $(b,a) \in \mathscr{R}$; *i.e. a R b* ==> *b R a.*

Asymmetry

A binary relation *R* is said to be asymmetric if for any pair $(a,b) \in \mathscr{R}$ implies $(b,a) \notin \mathscr{R}$, *i.e. a R b* ==> not(*b R a*).

Transitivity

A binary relation *R* is said to be transitive if $(a,b) \in \mathscr{R}$ and $(b,c) \in \mathscr{R}$ implies $(a,c) \in \mathscr{R}$, *i.e. a R b* and *b R c* ==> *a R c.*

Remark 2.3

Asymmetry is not the same property as non-symmetry. Thus, for example, the relation $\leq$ on the set of real numbers is not symmetric (since we do not have $a \leq b$ ==> $b \leq a$), but neither is it asymmetric (since $a \leq b$ does not imply $b > a$).

Proposition 2.4

All asymmetric relations are irreflexive.

Proof

If we make $a = b$ in the definition of asymmetry, we obtain $a \, R \, a \Longrightarrow$ not $(a \, R \, a)$, which is absurd and thus implies that we do not have $a \, R \, a$ (irreflexivity). Q.E.D.

Definition 2.5

Disjunction

Two binary relations R_1 and R_2 are said to be disjoint if $\mathcal{R}_1 \cap \mathcal{R}_2$ is empty; R_1 and R_2 are also said to be disjoint.

Definition 2.6

Comparison between binary relations: If there are two binary relations R_1 and R_2 defined on X, R_1 is said to be finer than R_2 if $\mathcal{R}_1$ contains $\mathcal{R}_2$. In other words $a \, R_2 \, b$ implies $a \, R_1 \, b$; within $\mathcal{R}_1$ there are the pairs of $\mathcal{R}_2$ and perhaps others besides.

The reader may immediately verify that the relation defined on the set of real numbers by $\leq$ is reflexive and transitive, that the relation defined by $<$ is transitive and irreflexive, and finally that equality is reflexive, symmetric and transitive; this leads us to introduce the following three definitions.

Definitions 2.7

A binary relation R is said to be a *preorder* if it is reflexive and transitive. A binary relation R is said to be an *order* if it is irreflexive and transitive. A binary relation R is said to be an *equivalence relation* if it is reflexive, symmetric and transitive. When R is an equivalence relation, for any given $x \in X$, the subset of X formed from the elements y such that $y \, R \, x$ is called an *equivalence class*. The equivalence classes perform a partition of X, *i.e.* they are disjoint and their union is equal to X.

Proposition 2.8

An order relation is necessarily asymmetric.

Proof

If the relation is not asymmetric, then there exists a pair (x, y) such that $x \, R \, y$ and $y \, R \, x$, and we then obtain $x \, R \, x$ by transitivity, which contradicts the non-reflexivity. Q.E.D.

2.2.3 The rationality of preferences

We now return to the decision maker and his choice set $\mathcal{A}$. We have seen that we can reasonably ask him to adopt alternatives between which he is indifferent; let this relation be $\approx$ with $\mathcal{I}$ as the corresponding sub-set of $\mathcal{A} \times \mathcal{A}$. In other words $(a,b) \in \mathcal{I}$ if and only if $a \approx b$ (a is indifferent to b). This same decision maker is aware of the sub-set $\mathcal{P}$ of couples of alternatives $\mathcal{A} \times \mathcal{A}$ such that $a \succ b$ (a is strictly preferred to b). We have also called $\succeq$ the union of the relations $\approx$ and $\succ$ ($a \succeq b$ if and only if a is preferred or indifferent to b).

Assuming that what we have said here make sense, the reader, with a little thought, should be able to conclude that:
- the indifference relation $\approx$ is reflexive (the decision maker is indifferent between x and x) and symmetric (if the decision maker is indifferent between x and y, he is also indifferent between y and x);
- the strict preference relation $\succ$ is asymmetric (if the decision maker strictly prefers x to y he does not strictly prefer y to x);
- the relations $\approx$ and $\succ$ are disjoint (if the decision maker is indifferent between x and y, he does not strictly prefer x to y and vice-versa).

The above conditions form a set of minimum requirements for the decision maker to be considered as rational. We now add to these unequivocal conditions another one which we shall discuss:
- the relation $\succeq$ is transitive (if the decision maker says that x is preferred or indifferent to y and if y is preferred or indifferent to z then x is preferred or indifferent to z).

Definition 2.9

From now on we shall give the name *'strong assumptions on the rationality of the decision maker'* to the following assumptions:
$\approx$ and $\succ$ are disjoint;
$\approx$ is reflexive and symmetric;
$\succ$ is asymmetric;
$\succeq$ is transitive.

Proposition 2.10

Under strong assumptions on the rationality of the decision maker, the relation $\succ$ is an order, $\approx$ is an equivalence relation and $\succeq$ is a preorder.

Proof

i) $\succ$ is transitive, since $x \succ y$ and $y \succ z$ imply $x \succeq y$ and $y \succeq z$. From the transitivity of $\succeq$ we deduce $x \succeq z$. Suppose that $x \approx z$; this implies that $z \approx x$ by symmetry and $z \succeq x$, which, with $x \succeq y$ gives $z \succeq y$ by transitivity. The latter assertion is absurd since we have $y \succ z$ which is incompatible with both $z \succ y$ (through the asymmetry of $\succ$) and with $z \approx y$ or $y \approx z$ (because $\approx$ and $\succ$ are disjoint). What we have is $x \succeq z$ and not ($x \approx z$) which implies $x \succ z$. Thus $\succ$ is transitive and asymmetric and hence irreflexive (Proposition 2.4) and is indeed an order.

ii) $\approx$ is transitive, since $x \approx y$ and $y \approx z$ imply $x \succeq y$ and $y \succeq z$, and by transitivity $x \succeq z$. Through symmetry we also have $y \succeq x$ and $z \succeq y$ and through transitivity $z \succeq x$. Suppose

$x \succ z$; this would contradict both $x \approx z$ (disjoint relations) and $z \succ x$ (asymmetry), and is consequently contradictory to $z \succcurlyeq x$. We therefore have $x \succcurlyeq z$ and not $(x \succ z)$. *i.e.* we do have $x \approx z$. Since $\approx$ is reflexive, symmetric and transitive it is indeed an equivalence relation.

iii) The relation $\succcurlyeq$ is reflexive since it contains $\approx$ which is reflexive. $\succcurlyeq$ is reflexive and transitive and is thus a preorder. Q.E.D.

We have thus just shown that with strong assumptions on the rationality of the decision maker, the preference model is a preorder on $\mathcal{A}$. *In most cases we shall therefore consider a set A and a preorder on this set $\mathcal{A}$ as a model of the decision maker's preferences.* We can even give the proof of a sort of inverse of proposition 2.10.

Proposition 2.11

Consider a set A with a preorder relation R; we can break down R into a symmetric part $\approx$ and an asymmetric part $\succ$ such that $\approx$ is an equivalence relation and $\succ$ an order.

Proof

By definition we write $x \succ y$ if $\{x R y$ and not $(y R x)\}$ and $x \approx y$ if $\{x R y$ and $y R x\}$. By definition, the relations $\succ$ and $\approx$ are disjoint. It is obvious that $\approx$ is symmetric and $\succ$ asymmetric. The relation $\approx$ is reflexive since R, being a preorder, is also reflexive. The relation R is indeed the union of $\succ$ and $\approx$ because, by definition, $x R y$ implies $x \succ y$ or $x \approx y$. The relation R is transitive by hypothesis. The relation R therefore satisfies the strong assumptions on the rationality of the decision maker and consequently (proposition 2.10), $\approx$ is an equivalence relation and $\succ$ an order. Q.E.D.

We have thus seen that a preorder on a set can be interpreted as the expression of the preferences of a (strongly) rational decision maker who is armed with an indifference relation and a strict preference relation. This leads us to state:

Definition 2.12

When the decomposition defined in proposition 2.11 is performed on a preorder R, the relation $\approx$ defined by $x \approx y$ if $\{ x R y$ and $y R x \}$ is called the symmetric part of the preorder, and the relation $\succ$ defined by $x \succ y$ if $\{ x R y$ and not $(y R x) \}$ is called the asymmetric part.

We now have to say exactly why stated that rationality assumptions are strong. It is generally agreed that the preference relation is strictly transitive; anybody who prefers a to b and b to c must be sufficiently certain of their preferences to conclude that they strictly prefer a to c. nevertheless, there does remain the possibility – pointed out by Roy (1985) – of a NC c (NC indicating non comparison). For example, the decision maker may strictly prefer a prize of a unique month in Paradise island at a given Christmas to a fortnight at a Colorado ski station every February for twenty years, and strictly prefer the skiing holiday deal to a week in

Florida in winter for twenty years; but it may well be impossible for the decision maker to compare the week in Florida and Paradise island holidays or, still worse, he may actually prefer the week in Florida for twenty years to a single stay in Paradise island. The exceptional and subjective over-valuing of the Paradise island holiday (through a desire to be identified with the jet set for example) upsets the preferences, just as does the difficulty of comparing the dissimilar 'objects' of a fortnight's skiing and a dream holiday. When two closely similar alternatives such as, here, two seaside holidays are compared, the comparison becomes easier and the financial aspect becomes primordial once again.

Note, however, that to preserve a semblance of reality we have had to assume that the choice of Paradise island is the result of a whim and therefore not very rational. In any case we only mention this rather difficult case 'by the way' and shall always assume that the strict preferences of the decision maker are in fact transitive.

The transitivity of indifference is quite another and more difficult problem. Consider first the apologue of the cup of tea (Luce, 1956). Suppose the decision maker prefers his tea with one lump (roughly 6 grams) rather than with no sugar. Clearly this decision maker will be express indifference towards a cup with 6g and one with 5.995g; similarly towards a cup with 5.995g and one with 5.990g and so on down to 0.005g. If the indifference relation is transitive we come closer and closer to deducing that the decision maker does not mind whether he has his tea with or without sugar. This modern fable can be applied to many domains where there is continuous evaluation, such as price: you are indifferent towards a price difference of a quarter on the price of a car but you strictly prefer a free new car to one costing twenty five thousand dollars.

There is however a limit; take for example high school marks. A pupil may remain indifferent towards an average of 15 and one of 14.95, and similarly between 14.95 and 14.9 etc.; however they will not express such indifference between 14 and 13.95 since these marks represent the difference between an ordinary pass and distinction. Here then we have transitivity of indifference within but not beyond certain limits. This brings us on to preference models with indifference thresholds, about which we shall have more to say in section 2.5. This transition from a 'physical' continuum to the discrete states of perception was earlier considered by Poincaré (Poincaré's paradox, 1902).

Doubtless the easiest way of overcoming the syndrome of the cup of tea is to say that *indifference is not transitive*. We can then state weak assumptions on the rationality of the decision maker.

Definition 2.13

Weak assumptions on the rationality of the decision maker.
The preferences $\succeq$ of the decision maker are said to satisfy the weak assumptions on rationality if:
strict preference $\succ$ is asymmetric and transitive;
indifference $\approx$ is reflexive and symmetric;
both the relations $\succ$ and $\approx$ are disjoint.

A preference structure satisfying the above assumptions is called a *quasi-order* as we shall see in section 2.5. First we shall see that with strong assumptions on rationality, a preorder relation can be manipulated through a functional representation.

2.3 Preorders and utility functions

The notions of order and preorder that we have just seen do not lend themselves easily to calculation. There does however exist an ordered set which is amenable to calculations: the set |R of real numbers. |R possesses the well-known preorder relations $\geq$ and $\leq$, of which the part corresponding to equality is the symmetric part and of which the asymmetric parts are respectively > and <. It is therefore fairly tempting to establish a relationship between a preorder on any set and |R, and this is what we shall do.

Note first of all that the preorders referred to above have the property that if we take any two numbers x and y in |R, we have either $x \leq y$ or $y \leq x$; this leads to the following definition.

Definition 2.14

A binary relation R on X is said to be *complete* if for any pair (x,y) of XxX we have either $x R y$ or $y R x$. A *complete preorder* is generally called a *weak order*.

Conversely, a *relation which does not possess this property is said to be partial*. This means that there exist at least two elements on which the relation R has no effect. If this is a preference relation, that indicates that x cannot be compared with y; we cannot say that x is preferred to y or vice-versa.

Proposition 2.15

If $\succeq$ is a complete preorder on $\mathcal{A}$ then $x \succeq y$ is equivalent to not($y \succ x$).

Proof

Let $\approx$ and $\succ$ be the symmetric and asymmetric parts of $\succeq$. It is always true that $x \succeq y$ is equivalent to $(x \approx y)$ or $(x \succ y)$ implies not($y \succ x$) since $\approx$ and $\succ$ are disjoint (in the case where $x \approx y$ or $y \approx x$) and $\succ$ is asymmetric (in the case where $x \succ y$). Suppose conversely that not($y \succ x$); as the preorder is complete, we either have $x \succeq y$ or $y \succeq x$. The only case contradicting our proposition would be where $y \succ x$, which is excluded. Q.E.D.

Definition 2.16

If $\succeq$ is a complete preorder on $\mathcal{A}$, a function U of $\mathcal{A}$ in |R is said to be a utility function representing the preorder $\succeq$ if and only if:

$$x \succeq y \iff U(x) \geq U(y) \tag{1}$$

Proposition 2.17

Let U be a utility function on $\mathcal{A}$ representing the complete preorder $\succcurlyeq$; we then have the following equivalences:

$x \succ y \iff U(x) > U(y)$
and
$x \approx y \iff U(x) = U(y)$

Proof

In terms of utility, $x \succ y$ is, by the definition of P, written $U(x) \geq U(y)$ and not($U(y) \geq U(x)$) or $U(x) > U(y)$. We proceed in the same way for the second equivalence, referring to the definition. Q.E.D.

Proposition 2.18

In the definition of the utility function we can replace equivalence (1) by:

$x \succ y \iff U(x) > U(y)$

Proof

If U is a utility function the above equivalence is indeed true. (Proposition 2.17). Suppose that $x \succ y$ is equivalent to $U(x) > U(y)$. Assume that we have $x \succcurlyeq y$. We know from proposition 2.15 that $x \succcurlyeq y$ is equivalent to not($y \succ x)$ or not($U(y) > U(x)$), *i.e.* $U(x) \geq U(y)$. Q.E.D.

We now have to consider the question of the existence of a utility function. Let us begin with an example. Let $\mathcal{A} = $ {Ford-Escort, Plymouth-Neon, Buick, Dodge-Stratus, Lincold, Cadillac} and assume that we have a very thrifty decision maker who chooses solely on price. His complete preorder will be Ford-Escort $\succ$ Plymouth-Neon $\succ$ Dodge-Stratus $\succ$ Buick $\succ$ Lincold $\succ$ Cadillac, the approximate prices being in the inverse order Pr(Ford-Escort) $< Pr$(Plymouth-Neon) $< Pr$(Dodge-Stratus), etc. The reader will have realized that in this case the function Pr is a utility function on $\mathcal{A}$ representing the decision maker's preorder. To be exact it is $-Pr$ which is a utility function since prices are being minimized. When a criterion is quantitative, the scale of values generally expresses the preorder or the inverse. Take the example of age criterion in a recruitment choice. Here the favorite is the youngest and so on in inverse order of age. It may happen however that people between the ages of thirty and forty are preferred to the others; in that case the preorder will follow neither the numerical scale nor its inverse.

To come back to automobiles, price is not the only criterion: taste also comes into play and for the writer of this paragraph the complete preorder on $\mathcal{A}$ is: Cadillac $\succ$ Lincoln $\succ$ Buick $\succ$ Dodge-Stratus $\succ$ Plymouth-Neon $\approx$ Ford-Escort. It is very simple to construct a utility function by ascribing a taste grade to each car.

Here we take U(Ford-Escort) = U(Plymouth-Neon) = 4, U(Dodge-Stratus) = 5, U(Buick) = 7, U(Lincold) = 8 and U (Cadillac) = 10.

A certain arbitrariness can be seen through this example since we could also have set U(Buick) = 6, U(Lincold) = 7 and U(Cadillac) = 9 and obtained a utility function representing the same order. *The utility function is not unique.*

We shall show that in a completely preordered choice set there always exists a kind of canonical utility function. To do this we need first of all a basic theorem.

Theorem 2.19

Let A be a finite set completely preordered by $\succcurlyeq$; the preorder is entirely represented by a path classifying all the elements, of the form:

$$a_1 \succcurlyeq a_2 \succcurlyeq a_3 \succcurlyeq \ldots\ldots a_{m-1} \succcurlyeq a_m$$

Proof

We prove this by recursion on the number of elements of A. If A has two elements the property is true since we have $a_1 \succcurlyeq a_2$ or $a_2 \succcurlyeq a_1$.

Suppose the property to be true for $m - 1$ elements. Consider a set A of m elements. The set $A - \{a_m\}$ only has $m - 1$ elements; it can thus be written, according to the recursion hypothesis and after any change of index of the elements:

$$a_1 \succcurlyeq a_2 \succcurlyeq a_3 \succcurlyeq \ldots\ldots a_{m-1}$$

Consider the set IN of indices IN = $\{i \,/\, 1 \le i \le m - 1$ and $a_i \succcurlyeq a_m\}$; this set is null or finite. If IN is null this means that whatever i we have $a_m \succcurlyeq a_i$. The set A is then written:

$$a_m \succcurlyeq a_1 \succcurlyeq a_2 \succcurlyeq a_3 \succcurlyeq \ldots\ldots a_{m-1}$$

If IN is not null, then we denote the highest index of IN by i_0. For $i > i_0$ we have $a_m \succcurlyeq a_i$ and hence the partial path:

$$a_m \succcurlyeq a_{i0+1} \succcurlyeq a_{i0+2} \succcurlyeq \ldots\ldots a_{m-1}$$

We shall have established our result if we prove that for all $i \le i_0$ we have $a_i \succcurlyeq a_m$. Let $i \le i_0$; we have $a_i \succcurlyeq a_{i0}$, and since $a_{i0} \succcurlyeq a_m$ this implies by transitivity $a_i \succcurlyeq a_m$, which concludes the proof. Q.E.D.

In fact the above result can be refined by introducing strict preferences and indifference. The following lemma will be required.

Lemma 2.20

Given a set A preordered by $\succcurlyeq$, $x \succ y$ and $y \approx z$ implies $x \succ z$.

Proof

We have $x \succeq y$ and $y \succcurlyeq z$ and we therefore obtain by transitivity $x \succcurlyeq z$. If we can show that $x \approx z$ is absurd this will lead to the required result. If we have $x \approx z$, then since we also have $y \approx z$, we get $x \approx y$ by transitivity, which contradicts $x \succ y$. Q.E.D.

We can now state the following corollary without proof.

Corollary 1 of theorem 2.19

A complete preorder $\succcurlyeq$ on a finite set $\mathcal{A}$ is entirely described by a path of the following form, where the $\succ$ and the $\approx$ occupy positions dependent on the preorder:

$$a_1 \succ a_2 \approx a_3 \approx a_4 \succ a_5 \succ ...a_{m-2} \approx a_{m-1} \succ a_m$$

Corollary 2 of theorem 2.19

There always exists a utility function representing a complete preorder on a finite set $\mathcal{A}$.

Proof

For a preorder described (as an example) by the path:

$$a_1 \succ a_2 \approx a_3 \approx a_4 \succ a_5 \succ ...a_{m-2} \approx a_{m-1} \succ a_m$$

we shall consider the rank of each element, taking tied rankings into account. Thus, in the above example, $U'(a_1) = 1$, $U'(a_2) = U'(a_3) = U'(a_4) = 2$, $U'(a_5) = 5$, etc. up to $U'(a_m) = m$. For any constant k, the function $U = k - U'$ is a utility function representing the preorder. Q.E.D.

Remark 2.21

The above corollary is very important since it not only proves the existence of the utility function, but also shows how to obtain it. Suppose the preference path to contain no indifference. Consider for example $a_1 \succ a_2 \succ a_3 \succ a_4 \succ a_5 \succ ... a_{m-2} \succ a_{m-1} \succ a_m$. Alternative a_1 is the first; in other words it is of rank 1 or $r(a_1) = 1$; a_2 is of rank 2 or $r(a_2) = 2$, and so on up to $r(a_{m-1}) = m - 1$ and $r(a_m) = m$. In this case the function $U(a_i) = m - r(a_i)$ is a utility function.

The existence of indifference complicates the situation. Take for example $a_1 \succ a_2 \approx a_3 \approx a_4 \succ a_5$. There are, among other, two simple solutions. The first solution is to do as we did in the proof of corollary 2, and set $r(a_1) = 1$, $r(a_2) = r(a_3) = r(a_4) = 2$ and $r(a_5) = 5$. A second solution, proposed by Kendall (1970) possesses certain advantages; it is called *rank averages*. When there are tied rankings, they are given a rank equal to the average of the ranks they would have had if there had been no tie. In our example this gives $r(a_1) = 1$, $r(a_2) = r(a_3) = r(a_4) = (2 + 3 + 4)/3 = 3$ and $r(a_5) = 5$. The rank thus obtained can be fractional; thus $a_1 \approx a_2 \succ a_3$ gives $r(a_1) = r(a_2) = (1 + 2)/2 = 1.5$ and $r(a_3) = 3$. One of the first advantages is that, in this case, the sum of the ranks is identical to that obtained without ties, *i.e.* $1 + 2 + 3 + ... + m = m(m - 1)/2$.

Definition 2.22

When the method of rank average is adopted, the utility function $U(a_i) = m + 1 - r(a_i)$ is called *canonical utility*.

Remark 2.23

Corollary 2 of theorem 2.19 becomes false if the choice set is infinite. For those amateur mathematicians who like counter-examples, we may say that this is the case for the following preorder of $|R^2$. Let there be two pairs (x_1, x_2) and (y_1, y_2) in $|R^2$; the preorder defined by Q is complete and is called the *lexicographic preorder*, since it corresponds to a dictionary search order:

$$(x_1, x_2) \succcurlyeq (y_1, y_2) \text{ if and only if } \{(x_1 > y_1) \text{ or } (x_1 = y_1 \text{ and } x_2 \geq y_2)\}$$

It is by no means easy to show that there exists no utility function in this case (see Fishburn, 1970); the reader may simply verify that this is indeed a complete preorder on $|R^2$.

From the example of automobile choice that the utility function is not unique; the various utility functions can, however, be related to each other.

Proposition 2.24

If U is a utility function representing the complete preorder $\succcurlyeq$ in a set A, then a necessary and sufficient condition for a function V of A in $|R$ also to be a utility function representing the same preorder is that there exist a strictly increasing application t of $U(A)$ in $|R$ such that $V = t \circ U$.

Proof

i) If U is a utility function, then it is such that:

$$x \succ y \iff U(x) > U(y) \text{ (proposition 2.18)}$$

but if t is strictly increasing, $U(x) > U(y)$ is equivalent to $t \circ U(x) > t \circ U(y)$ and so V is indeed a utility function.

ii) If V is a utility function, for all $z \in U(A)$ (there exists at least one element a of A such that $z = U(a)$) let $t(z) = V(a)$. Similarly, consider $(z_1, z_2) \in [U(A)]^2$ such that $z_1 > z_2$; we have $\{(z_1, z_2) \in [U(A)]^2 \text{ and } z_1 > z_2 \} \iff \{U(a_1) > U(a_2)\} \iff \{a_1 \succ a_2\}$ since U is a utility function, which for the same reason is equivalent on V to $\{V(a_1) > V(a_2)\}$ and hence equivalent to $\{t(z_1) > t(z_2)\}$ by definition of t. Q.E.D.

The above result can also be stated as *utility functions are uniquely defined up to a strictly increasing transformation*.

2.4 Ordinal and cardinal utility functions and evaluation of alternatives.

2.4.1 Ordinal utility function

We have seen in the paragraph above that when a utility function exists, which is always the case in discrete situations, then there exist an infinite number of them, defined to within a strictly increasing transformation.

Take a simple example which will be useful later. To evaluate the pupils in a class we have three marks in each of the subjects English (E), Math (M) and French F). We can identify these three disciplines as three different criteria. The marks of pupils *a*, *b* and *c* can be set out in the following decision matrix.

	E	M	F
a	06	07	05
b	01	01	18
c	10	06	09

In this case each criterion C_j has been expressed through a utility function U_j; thus we write $U_1(a) = 06$, $U_1(b) = 01$, $U_1(c) = 10$; similarly for the other criteria $U_2(a) = 07$, $U_2(b) = 01$, $U_2(c) = 06$ and $U_3(a) = 05$, $U_3(b) = 18$, $U_3(c) = 09$. If these marks represent the preorder of the decision maker for each of the criteria, this means that they only indicate the order $c \succ a \succ b$ in French, $a \succ c \succ b$ in Math and $b \succ c \succ a$ in English. *A utility function which only indicates the order and nothing else is said to be ordinal.* If this is indeed so in the case above, the following table will convey exactly the same information as the one above.

	E	M	F
a	02	14	15
b	01	10	18
c	04	12	16

And the utilities $U'_1(a) = 02$, $U'_1(b) = 01$, $U'(c) = 04$, $U'_2(a) = 14$, $U_2(b) = 10$, $U'_2(c) = 12$ and $U'_3(a) = 15$, $U'_3(b) = 18$, $U'(b) = 16$ do represent the same preorders as those in the first table. This observation may appear surprising to anyone used to giving or receiving marks; this is because generally the purpose of marks is not only to classify the candidates but also to indicate their 'absolute value'; in other words, it is not a matter of indifference to a to get 05 or 15 in French, even if his position is unchanged. We shall shortly be coming back to this aspect of the question. To conclude the subject of ordinal utility functions, assuming that our three pupils have been marked out of 20, consider the strictly increasing application of [0.20] on [0.20] defined by $t(z) = (1/20)*z^2$ and consider $V = t$ o U; we obtain a new table, still representing the same preorders.

	E	M	F
a	1.8	2.45	1.25
b	0.05	0.05	1.25
c	05	1.8	4.05

This table is just as good from an ordinal point of view as the previous two. This can all be basically summed up by only taking rows into consideration. By attributing three points to the first, two to the second and one to the last, as we did in the definition of a canonical utility function (definition 2.22), we obtain the following table containing all the ordinal information.

	E	M	F
a	2	3	1
b	1	1	3
c	3	2	2

2.4.2 Cardinal utility functions

While not attaching too much importance to the absolute value of marks, we may think that our French teacher figures that a and c are both content with equally mediocre performances, while *b* really is a good pupil. This is due to the large difference between the marks of *b* and those of *a* and c. The difference between *a* and *c*'s marks is only 4, whereas between *a* and *b* it is 13, more than 3 times as great. The teacher considers that this factor of three is significant, which is why we now state the following definition.

Definition 2.25

A decision maker is said to express interval cardinal preference or simply 'cardinals', if he is able to compare preference (if he states $x \succ y$ and $a \succ b$, he can also say that the difference between *x* and *y* is equal to, greater than or smaller than the difference between *a* and *b*).

Remark 2.26

The above definition is only of use if the choice set possesses at least three elements which are not mutually indifferent, and we shall assume this in the following section.

A cardinal utility function will thus be a function which respects differences as well as order. When a utility function represents cardinal preferences, the differences $U(x) - U(y)$ and $U(a) - U(b)$ are such that $U(x) - U(y) > U(a) - U(b)$ if and only if, for the decision maker, *x* is 'more preferred' to *y* than *a* is to *b*. This definition is fairly difficult to apply, and we now propose the following one which is essentially similar and has the merit of simplicity.

Definition 2.27

A utility function U representing the complete preorder of a decision maker is cardinal if, for any two pairs $(x \succeq y)$ and $(a \succeq b)$ in the complete preorder relation, the ratio $[U(x) - U(y)] / [U(a) - U(b)]$ only depends on the decision maker and not of *U* (*i.e.* it must be conserved in any utility function changes so long as the cardinal preferences of the decision maker do not change).

For school grades, and in a number of other domains, decision makers often feel they are able to give cardinal utilities. This assumption does not always stand scrutiny and in any case it requires much more information than ordinal utilities, since all differences have to be compared. We shall be coming back to this point in the next section.

Proposition 2.28

Suppose the decision maker has expressed a complete and cardinal preorder in the form of a cardinal utility function *U*; all the other cardinal utility functions representing the same preorder will be of the form $k_1*U + k_2$ where $k_1 > 0$ and k_2 is any number.

Proof

i) Let $V = k_1 U + k_2$ where $k_1 > 0$ and let there be two pairs such that $(x \succcurlyeq y)$ and $(a \succcurlyeq b)$ and $[U(x) - U(y)] / [U(a) - U(b)] = k$. $V(x) - V(y) = k_1[U(x) - U(y)]$ and similarly $V(a) - V(b) = k_1[U(a) - U(b)]$, and hence $[(U(x) - U(y)] / [U(a) - U(b)] = [V(x) - V(y)] / [V(a) - V(b)] = k$, proving that V and U represent the same cardinal preorder.

ii) Let U and V be two utility functions representing the same preorder. We set any two different elements a and b in the choice set. For any element x we have, by definition, $[U(x) - U(a)] / [U(b) - U(a)] = [V(x) - V(a)] / [V(b) - V(a)]$. Consequently $U(x) - U(a) = [[V(x) - V(a)] / [V(b) - V(a)]] [U(b) - U(a)]$, from which $U(x) = k_1 V(x) + k_2$ where $k_1 = [U(b) - U(a)] / [V(b) - V(a)]$ and $k_2 = U(a) - V(a) [[U(b) - U(a)] / [V(b) - V(a)]]$. Note finally that k_1 is positive since a and b are not indifferent (remark 2.26) and that the numerator and denominator are of the same sign. Q.E.D.

2.4.3 Quotient cardinal utility functions

The concept of cardinal utility set out above is sometimes called 'difference or interval cardinal utility function' to distinguish it from the quotient cardinal utility function which we now introduce. The quotient cardinal utility function assumes, in addition to the conditions of the cardinal utility function, the existence of a 'real zero'. When an alternative a_i is evaluated for the criterion j and receives a value $a_{ij} =$ 'real zero', this means *the absence of a property* which has an obvious physical character. This characteristic is called the 'acid test' of utility quotient (Bevan, 1980).

Consider the Celsius temperature scale; this is a cardinal utility scale, since we are able to state that the temperature difference between 40° and 20° is the same as that between 20° and 0°, *i.e.* 20° in both cases. Another temperature scale will also respect this equality of interval. Thus the temperature in degrees Fahrenheit $t_f = (9/5)t_c + 32$. Here a cardinal transformation can indeed be recognized.

However, since 0° does not mean an absence of temperature, this is not a 'real zero'. We cannot say that it is twice as hot at 40° than at 20°. And furthermore the zero Celsius is 32°F.

If on the other hand we consider the height of a shrub in inches, the value zero here indicates a 'real zero', *i.e.* the absence of the property of height. Thus we can say not only that the interval between 60 inches and 30 inches is twice that between 30 inches and 15 inches, but also that the 60 inch shrub is four times as tall as the 15 inch shrub. Any other scale, in centimeters for example, will have the same zero. These examples lead us to the following definition.

Definition 2.29

A decision maker is said to express quotient cardinal preferences if he can:
i) compare preference differences; *i.e.*. if he states $x \succ y$ and $a \succ b$, he can also state that the difference between x and y is equal to, greater than or smaller than that between a and b;
ii) there exists an element $a \in \mathcal{A}$ which can be used as an absolute reference and which does not possess the property on which preference in the other elements of $\mathcal{A}$ is based, and such that for any utility, $U(a) = 0$ ('real zero' in the 'acid test').

Proposition 2.30

If the decision maker expresses quotient cardinal preferences then any utility function representing the preferences satisfies:

i) the ratio $[U(x) - U(y)] / [U(a) - U(b)]$ is an invariant of the decision maker *i.e.* does not depends on U;

ii) the ratio $U(x)/U(y)$ is also an invariant of the decision maker.

Proof

Point i) results from condition i) of definition 2.29 and from definition 2.27. In fact, a quotient cardinal utility is first of all a cardinal utility.
To obtain ii), consider two utilities U and U' satisfying i); then for all x, y, a and b we have $[U(x) - U(a)] / [U(y) - U(b)] = [U'(x) - U'(a)] / [U'(y) - U'(b)]$. Let us call a the 'real zero': $U(a) = U'(a) = 0$. Let $a = b$; from the previous equation we then have $U(x) / U(y) = U'(x) / U'(y)$. Q.E.D.

Proposition 2.31

Two quotient cardinal utility functions U and U' representing the same preorder satisfy $U = kU'$ $(k > 0)$.

Proof

As we have cardinal utilities we have $U = k_1U' + k_2$ (proposition 2.28). The existence of the 'real zero' requires $U(a) = U'(a) = 0$, and hence $k_2 = 0$. Q.E.D.

Quotient cardinal utilities are, then, simply a subset of cardinal utilities. In what follows, unless we specify 'quotient cardinal', the notion of cardinal utility will always refer to interval cardinality.

2.4.4 Evaluating alternatives and constructing cardinal utilities

We have just seen that the notion of cardinal utility is a demanding one. How are such functions determined? First of all note that this question hardly arises when determining ordinal utilities; all the decision maker has to do is to rank his alternatives to obtain, *ipso facto*, (definition 2.22) a utility function. The worst that can happen is that this ranking by the decision maker does not respect rationality assumptions, and this can be checked by the analyst.

The situation is much more complex for cardinal utility functions. The first distinction to make is according to whether we are dealing with a qualitative or a quantitative criterion. In the case of a quantitative criterion such as price, the surface area of an apartment etc., the decision maker will already have a unit of measurement at his disposal. *This unit of measurement must not be confused with utility*. Thus, suppose one of the choice criteria for an apartment, for a childless couple, to be the surface area in square feet. Available apartments go from 200 to 2500 square feet. For our decision maker, relative to the criterion 'surface area', the utility of a 2500 square foot apartment is clearly not equal to twice that of a 1250 square foot apartment. If his optimum area is 1250 square feet, any larger apartment will appear too big. We can decide to attribute a utility of 1 to the area of 1250 square feet; to simplify matters, we will say that the worst case is 0 square feet, to which we give a utility of 0, while 200 square feet is very close to the worst, say 0.05. If we plot decision maker utility against surface area we obtain a curve like that in figure 2.2

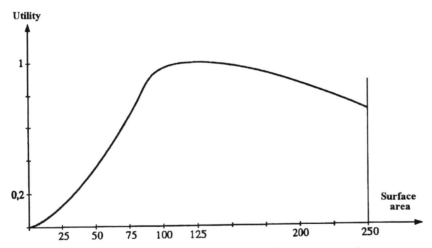

Figure 2.2 Decision maker's utility in terms of apartment surface area.

From the above example we can appreciate that the utility is very different from the units in which the measurements are actually expressed; utility expresses the psychological preferences of the decision maker. Cardinal utility, however, goes beyond this 'measurement of the strength of psychological preference', since it also measures preference differences, *i.e.* $U(a) - U(b)$. This is why certain authors distinguish between the utility function and the *value function* which actually measures differences in utility. Here we will not be making that distinction, which, within the present terms of reference, does not appear to us to be very useful, and we shall always refer to cardinal utility to mean that we are considering a utility which also serves to measure differences. this utility should be invariant through positive affine transformations (proposition 2.28). To come back to measurement, the reader should be aware that, even for monetary values, utility and measurement units are not one and the same thing. Take the case of the criterion 'salary' when considering multi-criterion choice of career. Suppose the possible monthly salaries vary, according to the position, from \$ 900 to \$ 5 500. For many individuals the utility curve will be shaped as in figure 2.3

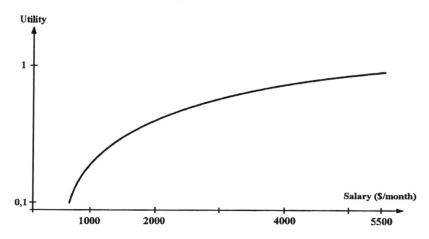

Figure 2.3 Utility in terms of salary.

The shape of the curve shows that the decision maker is very sensitive to a salary raise in the low salary range, and that this sensitivity lessens towards the top.

Thus, whether the criteria are quantitative or qualitative, the problem arises of constructing a cardinal utility function. The operation can be divided into several steps. We first of all have to assume that the decision maker is perfectly aware of the choice set; the next operation is to define the scale of measurement of the utility. Utility is often measured on a scale from 0 to 1, but this is not obligatory, and other scale may be adopted for practical or interpretative reasons. Once the scale has been defined, the decision maker has to calibrate it by describing those alternatives lying at noteworthy points on the scale; the first such alternatives to consider are generally the worst and the best, giving the extreme points on the scale. The other alternatives are then placed in accordance with preference differences (cardinality).

Various ways of setting about the construction of cardinal utility – once the scale has been defined – have been proposed; the simplest method, *direct evaluation*, consists of asking the decision maker direct questions of the type, 'Is the preference of *a* over *b* twice as strong as that of *c* over *a*?'. An affirmative answer would give $U(b) - U(a) = 2(U(c) - U(a))$. With a sufficiently large number of similar questions a reasonable approximation to the utility curve can be obtained (see figures 2.2 and 2.3). A second method consists of working directly on the curve and, through questions, having the decision maker approve its shape. Another way is to start with an alternative whose utility is taken as a unit and to have the decision maker express his satisfaction relative to this unit. For example, if the Ford-Escort is taken as the standard car for comfort, the decision maker must then say whether he would be 1.5 times as satisfied with a Pontiac. This is called the *ratio method*. One last simple method, the category *method*, consists of having the decision maker divide the scale into intervals of equal value; the decision maker then places alternatives within categories.

Finally, we shall consider four more advanced methods. In the *bisection method*, the utilities of the extreme alternatives are first determined; The decision maker is then asked to name the alternative with a satisfaction level exactly halfway between the best and the worst. Thus if the best alternative is given a utility of 100 units and the worst 0 units, the decision maker has to find an alternative worth 50 units, then one worth 25 units, 75 units and so on, each time dividing the intervals into two. The next method is called the *method of equivalent differences*; here the utility scale is divide into intervals of equal satisfaction. Going back to the example of the area of an apartment, by interviewing the decision maker we might deduce that an area increase of [700,800] – *i.e.*. from 700 to 800 square feet – results in a satisfaction increase equal to that accompanying increases of [800,950] and [1900,2500]. If these equalities are reflected in the intervals on the vertical axis, the utility curve can easily be drawn, and this is what has been done in figure 2.2. The third method is the *parallel world method* introduced by Camacho (1982, 1983), and given a formal axiomatic definition by Vansnick (1984). Let there be four alternatives a, b, c and d such that $a \succ b$ and $c \succ d$; to know that the ratio $[U(c) - U(d)] / [U(a) - U(b)]$ is equal to r/s, it is 'sufficient' that the decision maker be able to imagine that he has replicas of the alternatives considered available in parallel worlds. We define a^r as an alternative formed from r identical (or parallel) to a, *i.e.*. $a^r = (a, a,..., a)$ and is a vector with r coordinates. The above ratio r/s will then correspond to the indifference between the pairs (a^r,d^s) and (b^r,c^s) where r gains of a over b are balanced by s losses of d against c. What we have to do is ask the decision maker questions that will enable us to find the ratio r/s such that the previous indifference is observed. The method is formally attractive, but difficult to apply. Finally, in a more computer-oriented approach, Bana e Costa and Vansnick (1997, 1998) proposed a method for Measuring Attractiveness by a Categorical Based Evaluation TecHnique (MACBETH). In MACBETH, the decision maker gives the differences in value for each pair of alternatives according to a semantic scale (*e.g.* very weak, weak, moderate, strong, very strong, extreme). This leads to a 'difference matrix' measuring the difference in evaluation between the alternatives. By solving a linear program, MACBETH, in one sense, builds the simplest cardinal utility function that is consistent with the differences expressed by the decision maker. In fact, this is not always possible, and the differences given by the decision maker may be inconsistent, entailing the impossibility of obtaining a utility function satisfying the semantic differences. In the case of consistency, starting from the ordinal utility function, MACBETH offers a dialogue facility to compare the differences in utility and modifying the previous utility function to get an interval cardinal utility function. MACBETH is a very friendly package and offers many features for building utility functions with fixed values, for testing whether a numerical scale is consistent with given differences that cause the inconsistencies. MACBETH can also be used to determine weights (see section 4.7). The method, like the AHP method (see section 4.6), starts from a difference matrix, but instead of using the matrix to search for eigenvalues, it uses linear programming to build a utility function that is compatible with the differences that have been expressed.

Considerable work has been done on the determination and properties of cardinal utility functions (see for example Fishburn (1967), Allais (1979), McCord and de Neufville (1983 *a* and *b*), Vansnick ((1984), von Winterfeldt and Edwards (1986), Wakker (1986) and Bouyssou and Vansnick (1990)); however, the strength and quality of the results obtained are rather compromised by contradictory interpretations, a good account of which is to be found in von Winterfeldt and Edwards (1986) and Bouyssou and Vansnick (1990). Experiences prove that, according to the evaluation method, the utility values change entailing rather different choices by the subsequent aggregation procedures (Olson *et al.*, 1995).

However, a subject which has received still more attention than cardinal utility functions is that of the construction of linear utility functions relating to risk probabilities. From the work of von Neumann and Morgenstern (1944) came the axioms that decision maker preferences must satisfy for a so-called *von Neumann-Morgenstern utility* to exist. Numerous empirical experiments on the determination of von Neumann-Morgenstern utility in risk have been carried out, and these have always demonstrated that decision makers have not conformed to the model (see for example Allais (1953) *a* and b; Hershey *et al.* (1982); Kahneman and Tervsky (1979); Farquhar (1984); Cohen *et al.* (1987)). There is general agreement on the existence of several types of distortion due to the fact that humans do not evaluate risk by exactly following probabilities and associated utility expectancy; it is very likely that more complicated models involving criteria other than utility expectancy must be considered (see Cohen (1992)). There are also the inherent difficulties associated with decision maker thinking skills in the determination of utilities. In any case, what we do know of utility in the domain of risk is cause for skepticism about cardinal utility in the domain of the certain, and there are good reasons for supposing that the constructions described above are not always reliable, depending amongst other things on the method used. In addition, some of the experiments carried out in the field of the evaluation of additive utilities show that the decision maker is capable of substantially modifying his cardinal utility (*i.e.* he does not respect affinity) when it is suggested to him that he use one scale or another (*e.g.* percentages instead of values, or using a scale from 0 to 100 instead of one from 0 to 10). We shall encounter this problem of scale again on chapter 4.

It is also well known that presentation and context have a very strong effect on evaluation (Tversky and Kahneman (1974, 1981, 1988)). A review of the perturbing effect of cognitive bias on the evaluation of utility can be found in von Winterfeldt and Edwards (1986, Chapter 10). The difficulties increase further when we study the coherence between cardinal utility in the domain of the certain and the restriction to the domain of the certain of a von Neumann-Morgenstern utility that has been constructed in the domain of risk through lotteries; here the reader should consult the very complete work of Bouyssou and Vansnick (1990), together with von Winterfeldt and Edwards. In a word, and contrary to what may be expected, these two utilities do not in general stand any chance of coinciding. For this to happen it is necessary but not sufficient (Bouyssou and Vansnick (1990)) that an alternative x equidistant in utility between alternatives a and b (*i.e.* $U(x) - U(a) = U(b) - U(x)$) also correspond to a 50-50 lottery between a and b. Some workers have also tried to connect the two utilities, von Neumann-Morgenstern and cardinal, in the domain of the certain by constructing them simultaneously or by seeking a link between them; see Sarin *et al.* (1980), Krzysztofowicz and Koch (1989) and Bouyssou and Vansnick (1990).

This problem of cardinal utility is closely related to preference additivities, as we shall see later in chapters 4 and 6 where we deal with the construction of utilities relative to several criteria interconnected within an additive aggregation function.

2.4.5 Cardinal and ordinal utilities

A fundamental distinction that we have introduced is that between ordinal utilities which are uniquely defined up to a monotonic increasing transformation, and cardinal utilities which are uniquely defined up to a strictly positive affine transformation. In a definition equivalent to that in 2.27 it could be stated that the utility function of a decision maker is cardinal when it is uniquely defined up to a strictly positive affine transformation. We stress that this is a strong assumption since we are going from a function which only indicates an order and in which values play no role other than to indicate position – as we saw in the example of pupils – to one whose form is virtually fixed, since any modification by the decision maker which does not correspond to an affine transformation is equivalent to a change of preferences. Any change of scale which is not a transformation, a dilatation or a combination of the two will introduce a change in preferences which will result in changes in the mulitcriterion analysis outcome.

For this reason, the analyst must be constantly aware that decision makers often think they are expressing cardinality whereas they are actually only certain about one thing: the order. In any case, it is always useful to know which multicriterion methods are invariant through change of cardinal utility and which ones incorporate more or less cardinal utility. As far as the latter are concerned, the analyst must be much stricter on the quality of the information provided by the decision maker, who in turn must be perfectly sure about the scale he is using and the values he is providing. These are stiff requirements; the analyst may often express doubts on the statements of decision makers in this domain. Merely changing the scale – *e.g.* going from an evaluation of 0 to 20 to a judgement on [0,100] can lead to a change of the decision maker's cardinal utility while preferences (ordinal) remain unchanged.

Vigilance is all the more important when one considers that in practice *it is never hard to obtain numerical evaluations from a decision maker.* Man would seem to be a natural and spontaneous evaluator! This apparent ease must not blind us to the facts: it is much more difficult to evaluations robust enough to withstand changes of scale and context. This is already difficult enough when we are dealing with cardinal utilities; something close to faith is needed if we hope to extract cardinal utilities. In any case the analyst must show a critical spirit in this domain and above all analyze the robustness of the methods given the type of utility used.

2.5 Semi-criteria and pseudo-criteria

We saw in section 2.2 that a decision maker's preferences are, under strong rationality assumptions, characterized by a preorder. This is obviously a simplification as we have also seen (in the paradox of the cup of tea, section 2.2.3) that the assumption of transitivity of indifference is, in some cases, too strong. This leads us to introduce weak assumptions on the rationality of the decision maker. The difference between the weak and strong assumptions lies in the elimination of this transitivity of indifference.

The transitivity of indifference will be accounted for through the notion of threshold. Suppose that there exists an application U of $\mathcal{A}$ in $\mathbb{R}$ representing preferences. Let s be a positive real number (s being the threshold); we say that:

$$a \, I \, b \text{ or } a \approx b \text{ if and only if } |U(a) - U(b)| \leq s \tag{2}$$

In other words, as long as the distance between $U(a)$ and $U(b)$ does not exceed the *indifference threshold s*, the alternatives a and b are considered as indifferent. This idea works very well in the case of the cup of tea, the threshold s here representing the minimum concentration of sugar that the decision maker's taste buds are capable of detecting. As a logical outcome of the definition of threshold we have:

$$a \, P \, b \text{ or } a \succ b \text{ if and only if } U(a) > U(b) + s \tag{3}$$

The relation $(\approx, \succ)$ defined by (2) and (3) is a complete one. Furthermore, if $a \succ b$ and $b \approx c$ we have $a \succ c$. Similarly $a \succ b$ and $c \succ a$ implies $c \succ b$. *The relation $\succ$ is transitive and asymmetric.* The relation $\approx$ is reflexive, symmetric and non-transitive (figure 2.4). The relations $\succ$ and $\approx$ are disjoint, hence the weak assumptions on the rationality of the decision maker are verified.

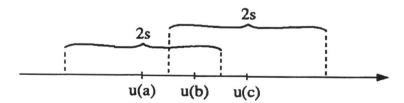

Figure 2.4 $\approx$ is not transitive (c $\approx$ b and a $\approx$ b and c $\succ$ a)

In addition to the weak assumptions on rationality, the reader will also see that for all $a, b, c, d, a \succ b$ and $b \approx c$ and $c \succ d$ implies $a \succ d$. To simplify the writing we shall denote this proposition by $PIP \subset P$, this notation recalling the presence of the binary relation in set theory (see section 2.2.2). It can also be verified easily that $a \succ b, b \succ c$ and $a \approx d$ implies $d \succ c$, i.e. $PPI \subset P$ (figure 2.5).

Figure 2.5 Properties of *P* and *I*

Given the weak assumptions on the rationality of the decision maker and the threshold preference properties that we have just been looking at, we can characterize this type of preference with the aid of a threshold and a utility function.

Proposition 2.32

If $\mathcal{A}$ is finite, any complete preference relation $\succeq$ satisfying both weak assumptions on rationality and:

i) $PIP \subset P$
ii) $PPI \subset P$ or
iii) $P^2 \cap I^2 = \emptyset$

is capable of being represented by a threshold preorder, *i.e.*. there exist *U* and *s* ≥ 0 such that (2) and (3) are satisfied.

Proof

See Vincke (1980).

A complete preference relation satisfying the conditions of proposition 2.32 is called *a semiorder*. (In French it is called a *quasi-order*: see Roy (1985) who uses properties i) and ii), whereas Vincke (1980 and 1989) uses properties i) and iii); the reader should verify that conditions ii) and iii) are equivalent). In addition it should be noted that the assumption of transitivity in the weak assumptions on decision maker rationality is not necessary here because $PIP \subset P$ implies transitivity for $b = c$.

Remark 2.33

Proposition 2.32 extends to certain conditions of infinite space: see Vincke (1977).

Still assuming that $\succ$ and $\approx$ express weakly rational preferences, we now go on to consider the relation $\succeq = (\succ \cup \approx)$. On examining all possible cases, this can be seen to satisfy the following two relations for all *a*, *b*, *c* and *d*:

Definition 2.34

$\succeq$ is said to satisfy the *Ferrers relation* if $a \succeq b$ and $c \succeq d$ implies $a \succeq d$ or $c \succeq b$.

Definition 2.35

$\succeq$ is said to be semitransitive if $a \succeq b$ and $b \succeq c$ lead to $a \succeq d$ or $d \succeq c$.

We can illustrate definitions 2.34 and 2.35 with the aid of a graphic representation of preference using arrows (figure 2.6). We assume that $\succeq$ is complete, which implies $a \succeq b$ is contradictory to $b \succ a$ (proposition 2.15).

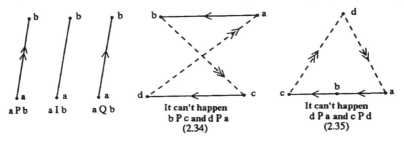

Figure 2.6 Graphic illustration of the properties of $\succeq$.

We can see in figure 2.6 that crossing $a \succeq b$ and $c \succeq d$ with $b \succ c$ and $d \succ a$ would, from the point of view of preferences, be incoherent (the Ferrers relation). Similarly, the existence of a d such that $d \succ a$ and $c \succ d$ is unacceptable, from a rationality point of view, while $a \succeq b$ and $b \succeq c$ (semitransitivity). The above rationality properties are sufficient to characterize a *semiorder*.

Proposition 2.36 (Scott and Suppes, 1958)

If the choice set $\mathcal{A}$ is assumed to be finite, then the complete relation $\succeq$ is a semiorder if and only if it satisfies 2.34 and 2.35.

Proof

See Fishburn (1970).

In many cases, a constant indifference threshold can be a drawback. One's indifference threshold to money, for example, may well depend on one's total wealth. (Though the anecdote of the banker Lafitte picking up a pin on his driveway could suggest that education has a bigger influence than wealth!) However, we shall consider the following conditions in which s is a numerical function with positive values:

$a\ I\ b$ or $a \approx b$ if and only if $-s(U(a)) \leq U(a) - U(b) \leq s(U(b))$ (4)
$a\ P\ b$ or $a \succ b$ if and only if $U(b) + s(U(b)) < U(a)$ (5)

The relation $\succ$ so defined is transitive and still satisfies $PIP \subset P$.

Definition 2.37

A complete preference relation satisfying weak assumptions on rationality and $PIP \subset P$ is an *interval order*.

Proposition 2.38

On a finite choice set $\mathcal{A}$, a binary relation is an interval order if and only if there exists a utility U and a positive value numerical function s such that (4) and (5) are verified.

Proof

See Vincke (1980).

Remark 2.39.

If the condition for coherence is imposed:

$U(a) > U(b)$ implies $U(a) + s(U(a)) > U(b) + s(U(b))$

then it can be shown that, with a change in utility function, we can return to a constant threshold or semiorder (Vincke,(1980 and 1989)). This is the case in particular if $s(U(x)) = k\ U(x)$ where $k > 0$.

Figure 2.7 shows a case where coherence condition 2.39 is violated. From the figure we can read that $d \succ a$ even though $a \approx b$ and $d \approx b$; similarly, $a \succ c, d \succ a$ and $b \approx c, b \approx a, b \approx d$; this is not a very intuitive situation.

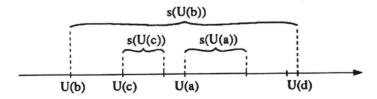

Figure 2.7 Condition 2.39 is not satisfied.

If now we consider the relation $\succeq = (\succ \; \cup \approx)$ associated with weakly rational preferences, we can characterize the interval order.

Proposition 2.40 (Fishburn, 1970)

Suppose the choice set $\mathcal{A}$ is finite; the complete relation $\succeq$ is an interval order if and only if it satisfies Ferrers' relation.

Proof

See Fishburn (1970).

If we wish to get even closer to reality, we might object to the previous models on the grounds that, though the problem of the cup of tea has disappeared, there still remains a sharp jump on passing from indifference to strict preference. To attenuate this transition, we can introduce a new *relation* $\succ_f$ of *weak preference* (Roy, 1985).

We then obtain the model:

$$a\, I\, b \text{ or } a \approx b \text{ if and only if } -s_1(U(a)) \leq U(a) - U(b) \leq s_1(U(b)) \tag{6}$$
$$a\, P_f\, b \text{ or } a \succ_f b \text{ if and only if } U(b) + s_1(U(b)) < U(a) \leq U(b) + s_2(U(b)) \tag{7}$$
$$a\, P\, b \text{ or } a \succ b \text{ if and only if } U(b) + s_2(U(b)) < U(a) \tag{8}$$

where s_1 and s_2 are positive value numerical functions. The function s_1 is the *indifference threshold* while s_2 defines a *preference threshold*. (Roy, 1985). A graphic representation of these various concepts is shown in figure 7.1.

Definition 2.41 (Roy, 1985)

The preferences defined by the above model are said to form a *pseudo-order* if and only if:

i) the function s_1 is constant,
ii) for all a $s_2(U(a)) \geq s_1$,
iii) $U(b) > U(a)$ implies $U(b) + s_2(U(b)) \geq U(a) + s_2(U(a))$.

Remark 2.42

It is possible to consider a variable indifference threshold s_1, but to obtain a pseudo-order, coherence condition 2.39 must be added: $U(a) > U(b)$ implies $U(a) + s_1(U(a)) \geq U(b) + s_1(U(b))$.

As before, the pseudo-order structure can be characterized by $\succ$, $\approx$ and $\succ_f$.

Proposition 2.43 (Roy, 1985)

A complete preference structure $(\approx, \succ, \succ_f)$ on a finite set $\mathcal{A}$ (*i.e.*. any pair (a,b) in $\mathcal{A} \times \mathcal{A}$ satisfies $a \approx b$ or $a \succ b$ or $a \succ_f b$) is a pseudo-order if and only if it satisfies the following conditions:

i) $\approx$ is symmetric, $\succ$ and $\succ_f$ are asymmetric

ii) $(\approx, (\succ \cup \succ_f))$ is a semiorder,

iii) $(\succ, \approx_p)$ is a semiorder ($a \approx_p b$ if and only if not($a \succ b$) and not ($b \succ a$)),

iv) $PIP_f \subset P$, $P_f IP \subset P$, $PP_f I \subset P$, $IP_f P \subset P$.

Proof

Vincke (1980).

The reader will have realized from the above result that twin threshold models are by no means a trivial affair. It is easy to see why results are most often presented on the basis of strong rational preferences with the associated preorder structure. For certain applications and software programs we shall nevertheless be using the models described in this section.

Note too that for the sake of simplicity *we have only been considering complete binary relations in the above section*. The semiorder and pseudo-order structures defined for partial relations have still more complicated and difficult properties than those defined here because we lose the functional representation U together with the real number image which naturally induces completeness. The reader may gain an idea of the difficulties involved from Roy and Vincke (1982), Roubens and Vincke (1984, 1985) and Roy (1985).

Finally, note that the preference structures studied up to this point consider that two pairs $a \succ b$ and $c \succ d$ express incomparable preferences, *i.e.* we cannot say that 'a is more preferred to b than c is to d'. Nevertheless, in general $U(a) - U(b)$ is different from $U(c) - U(d)$. If we wish to consider this difference in algebraic terms, this will mean that we have to add to the preference structure defined on $\mathcal{A}$ a preference structure defined on $\mathcal{A} \times \mathcal{A}$. Algebraically speaking this is a difficult task which has been the subject of research, an idea of which can be had by consulting Roy (1985, p. 162 et seq.), Vansnick (1984 and 1987) and Roy and Bouyssou (1987). Modeling becomes even more complex if we assume that the relation defined on $\mathcal{A} \times \mathcal{A}$ is a fuzzy relation (Perny (1992), Perny and Roy (1992), Perny (1998); see also the references in the survey papers by Perny and Pomerol (1999) and Greco *et al.* (1999)). Another proposition, intermediate between a preorder on the pairs in $\mathcal{A} \times \mathcal{A}$ and a preorder in $\mathcal{A}$, consists of introducing, preferences on what one does not like alongside the usual preferences. We then obtain what Fishburn (1992a) calls 'multiattribute signed orders'. Finally, a simpler (non-algebraic) way of tackling the same question is to assume first of all the existence of a utility function and to examine the difference $(U(a) - U(b))$. This then leads to the notion of cardinality (see section 2.4.4) and beyond this to additivity (section 6.4).

2.6 Models and aims of multicriterion decision making

2.6.1 Criteria

As we said in the first section, we always start off from a set of alternatives and attributes.

We shall say that *a criterion is constituted when the decision maker or the analyst has been able to assign the preorder to the decision maker's preferences, relative to the attribute in question*. When this preorder is complete, which is not always necessarily the case, then as we have seen, in the discrete case in which we are interested, this is equivalent to saying that we know a utility function of the decision maker relative to the attribute in question. Choosing the 'right' utility function is never easy, especially for qualitative criteria.

To summarize, *we use 'criterion' to mean a preorder expressing the decision maker's preferences on the choice set. This preorder only takes into account a single dimension of the problem (an attribute or evaluation axis)*. In simple cases the preorder is complete and the criterion will then correspond to the utility function. Each element a_{ij} in the decision matrix will then have the form $a_{ij} = U_j(a_i)$. When finer preference modeling with a semiorder or pseudo-order is possible, we shall refer to semicriteria or pseudo-criteria, and we shall make use of the functional representations in the above section.

From a formal point of view, it is not necessarily an easy task to construct a criterion or its associated utility function (see section 2.4.4); neither is this always easy from a practical point of view, and Bouyssou (1989) demonstrates just how difficult it can be to construct a 'noise nuisance' criterion in the problem of choosing a location for an airport. We shall come back to this question in Chapter 11.

2.6.2 The various problem formulations

Our starting ingredients are now all specified: a choice set and criteria which will give a decision matrix. In the commonest case the preorders will be represented by utility functions and the matrix will have the numerical coefficients $a_{ij} = U_j(a_i)$.

What can we look for out of this? Several answers are possible; let us first examine the technical side.

We can look for the 'best choice' within the choice set $\mathcal{A}$. Failing a best choice we can look for the smallest possible subset of 'excellent' alternatives, each representing a possible decision. This is the *selection problem* corresponding to the P_α *problem* of Roy (1975, 1985). As the notion of best choice has no canonical meaning in multicriterion analysis (cf. chapter 3), this problem lies more within Simon's limited rationality (1983), and what the decision maker looks for are those 'satisficing' alternatives likely to be selected in the final decision. In the selection problem, then, we look for as small a subset $\mathcal{A}$ as possible of satisfying alternatives. What the analyst has to provide is either the subset of satisfying alternatives or a selection procedure enabling the subset to be determined; in the latter case we come back to the voting problem (see 2.1.5).

The second problem formulation is more ambitious, and has as its objective a ranking of the whole set of alternatives in $\mathcal{A}$. In other words, we wish to obtain a complete preorder on $\mathcal{A}$. This is the problem of social choice (see 2.1.5). *The ordering problem is contained within Roy's ranking problem (1985), or P_γ.* The result we want is a preorder on the whole of the alternatives in $\mathcal{A}$, and so the multicriterion method can simply be used to provide this classification; but what is more generally required is *a procedure enabling us to go from the preferences expressed by each criterion to a single global preference of the decision maker (a unique complete preorder on $\mathcal{A}$). Such a procedure is called a preference aggregation procedure.* When each criterion is expressed in the form of a utility function, what we then need is a function which will transform the n utility functions associated with the n criteria into a single utility function representing the decision maker's preorder. This function is called an *aggregation function.*

A third problem formulation is to assign the alternatives to classes which have been defined a priori. This *classifying or sorting problem was named P_β* by Roy (1985). For examination candidates, for example, there might be three classes: those who pass, those who fail, and those who pass the oral examination. We then require either the list of alternatives in each category or the assigning procedure. When the classes have not been defined *a priori*, and we require similarities, we are entering another domain closer to data analysis than to multicriterion analysis.

Note that these three problem formulations are not independent from each other and that in particular the ranking of alternatives (P_γ) can serve as a basis for solving the P_α problem. To do this it is only necessary to define the thresholds. Similarly, the classes in P_β are often defined by limiting values on a numerical scale. We can thus use the ranking for classifying. In Roy (1985) and Roy and Bouyssou (1993a), analyses in depth of these problems are given in a treatment which goes well beyond the basic ideas we have introduced here. The very fact that the first step in all these procedures is itself an important piece of information on the nature of the possible alternatives and also constitutes prior thinking about the decision, may in itself be an interesting result. Research here concerns problem formulation P_δ (Roy and Bouyssou, 1993a).

In practice, many multicriterion methods, especially those of the seventies and early eighties, relied on the problem formulation based on ranking. These methods were influenced by the model of social choice and consisted essentially of aggregation procedures.

This direction of research was modified with the appearance of personal computers which enabled certain hitherto theoretical methods actually to be applied, and this led in turn to the discovery of new problems and new ideas.

2.6.3 Progressive information multicriterion methods

The role of the decision maker can be seen in a new light with the use of computers in multicriterion methods.

In the classical paradigm, the decision maker provides his choice set, his preferences for each attribute (the criteria) and his problem formulation. He then waits in silence for the result of multicriterion analysis. All the information within the mind of the decision maker is given once and for all at the start of the process. This is also the case in voting and in social choice.

Thus, in the classical approach, the decision maker supplies the basic information, then disappears from the scene. The analyst entertains the illusion that he will find the best choice by relying on the rationality of the model. In other words, the choice of the decision maker will be dictated by the information he has provided at the start, together with the logic of the model; If the decision maker is not convinced about the result – and often he is not – too bad, it is he who is wrong. The analyst is excessively in control.

The personal computer has now afforded the decision maker the possibility of becoming involved in multicriterion analysis, and this has led to a modification to the basic paradigm of the decision maker's role. In the new paradigm, the decision maker can provide information bit by bit as the multicriterion analysis unfolds. This information is of two types: modifications to or refinements of the initial information (criteria and even the choice set itself); and – more recently – information on the overall preorder of the decision maker.

This new paradigm lies firmly within the domain of decision supports, and we therefore assume that the choice refers, not to any intrinsic rationality in the model, but to the (limited) rationality of the decision maker. *The decision maker, without being aware of it, brings along a preorder of preferences on $\mathcal{A}$. This preorder, which we shall call the decision maker's preorder, is obviously unknown both to decision maker and analyst*, and the role of multicriterion analysis will be to make it progressively appear. In other words, the process of multicriterion analysis will lead to the construction of a ranking (the P_γ problem) or to a selection (the P_α problem). This end result depends not only on the logic of the model, but also on that of the decision maker as revealed when he provides information at the various stages in the process. Against this new background the situation can be summed up by saying that the multicriterion method *allows the decision maker to reveal his own preferences progressively* (decision maker's preorder). This process, however, takes place within the logical framework set by the model, which is why we can say that there is 'aid to decision'. We shall call these multicriterion methods *progressive information methods*. The decision maker is involved all the way through the multicriterion analysis in giving information about his real preferences. We could say that the multicriterion method helps the decision maker to construct his overall preferences just as the knowledge engineer helps the expert to model his expertise.

The distinction between progressive information methods and those without progressive information appears fundamental to us, revealing as it does the paradigm shift referred to above; a shift from methods where the decision is contained in the model to methods for aid to decision where a conclusion is impossible without the aid of the decision maker.

We shall have a good deal to say about these progressive information methods which use the possibilities – especially the graphic and visual possibilities – of computers (see the methods of Angehrn and Korhonen, chapter 9, for example). These interactive methods are a sub-group within the family of interactive decision support systems (DSS). For an general account of everything concerning DSS, see Lévine-Pomerol (1989) and Vanderpooten (1990b) for an analysis of interactivity in multicriterion methods; these ideas will be taken up in chapters 9 and 11.

2.6.4 The negotiation problem

The various problems examined in section 2.6.2 are based on technical distinctions on the aim of the analysis. In the field of multicriterion analysis, the process itself, and not just the result, should be of fundamental interest. In other words it is perfectly legitimate, and even recommended, to view multicriterion analysis as a negotiation process inside or outside an organization. This point of view deserves some explanation. In any decision process involving several people, the latter, even if they do not have conflicting interests, will at least have different sensitivities which will result in different evaluations or choice criteria. *The best way of coming to a decision or an agreement is not to concentrate on the object of the decision, but rather on the criteria and the procedure.* For example, it is useless to take part in arguments on the possible sites for a new airport if one has no intention of discussing criteria or of seeking some balance between these criteria. This balance may itself be provided by a multicriterion method on which everybody has agreed.

Other examples we could take include salary negotiations and modernization negotiations. In no case can we escape from analysis of the criteria – in this case productivity and number of jobs – of the parties involved. Clearly once the criteria have been constructed and accepted by the parties, the discussion will center on the multicriterion procedure enabling actions acceptable to all to be arrived at. Multicriterion modeling is particularly suitable for this type of process and negotiation, whereas monocriterion 'optimization' modeling is both unsuitable and 'technocratic' in the sense that it is he who designs the model – generally a representative of the state or management – and who, *de facto*, takes the decision.

The above considerations spill over into negotiation research, particularly the *fundamental result that a successful negotiation requires a shift of the discussion from issues towards procedures* (evaluation of the arguments of the parties, utility comparisons etc.) This is what Fisher and Ury (1982) refer to as separating the negotiators from the problem. Our opinion is that this is a fundamental point to which multicriterion analysis brings very useful methodology and tools, and which deserves to be used much more often.

2.7 Evaluation of alternatives and normalization

We have defined a criterion as a preorder (or a semiorder or pseudo-order) in the choice set $\mathcal{A}$. Most multicriterion methods, however, require a numerical decision matrix. In other words, given that a numerical scale is always completely ordered, implicitly, *this transformation of the decision maker's preorder into a numerical scale always comes down more or less to constructing a utility function.* Unfortunately, this utility function is not unique, as we have seen (section 2.3).

It is not an easy matter to evaluate each alternative a_i relative to a given criterion j to obtain the coefficients a_{ij} of the decision matrix. For the moment we shall leave aside the problem of how to construct the criteria proper (according to Bouyssou's expression (1989)), *i.e.*. of the consensus which must emerge between parties on what should be put in the criteria and how to evaluate the 'noise nuisance of the airport' or 'the company image' or 'the social risk of the project'. We shall develop this aspect of the problem in chapter 11.

The problem posed here assumes that the alternatives are known with certainty, and that the decision maker agrees to evaluating the alternatives on a numerical scale; we thus obtain *ipso facto* a decision maker's utility function relative to the attribute under consideration.

To throw more light on the problem, take the example of the criterion 'price'; we can evaluate an alternative in dollars, thousands of dollars or millions of dollars, but in contrast to what the reader may think, the choice does not come back to a simple question of common sense. A simple example will prove this: I have the choice of three portfolios, *a*, *b* and *c*. I have two criteria, the amount of the investment, which I shall want to minimize, and the yield, which I shall want to maximize.

In dollars, the decision matrix will be as follows:

	INVEST.	YIELD
a	100 000	10 000
b	500 000	70 000
c	1 000 000	150 000

To deal with minimization of the investment, we change the sign and maximize, obtaining:

	INVEST.	YIELD	SUM
a	−100 000	10 000	−90 000
b	−500 000	70 000	−430 000
c	−1 000 000	150 000	−850 000

If the decision maker is happy taking the sum (or the average) of the two criteria, the result is that shown in the third column, and the choice of the decision maker will be $a \succ b \succ c$. If on the other hand the decision maker elects to express the investments in thousands of dollars and the yields in dollars, the following new table is obtained:

	INVEST.	YIELD	SUM
a	−100	10 000	9 900
b	−500	70 000	69 500
c	−1 000	150 000	149 000

The decision maker's choice then becomes $c \succ b \succ a$ – the exact opposite of the previous result!

This trivial example proves two things; first, the choice of scale is *not* trivial, and second, there is in certain multicriterion methods such as the sum adopted here, *compensation* between the various values obtained by a single alternative according to various criteria. This compensation assumes that comparable scales are used, which is why values are generally chosen to lie between 0 and 1. This operation is called normalization.

Definition 2.44

We shall qualify as ordinal any multicriterion decision method in which the results are not affected by any change in ordinal utility function on the criteria. In other words, if one or more of the utility functions associated with the criteria are modified by a strictly increasing mapping, the method will still give the same results.

Definition 2.45

By analogy, we shall call, *cardinal a multicriterion method in which the results are not affected by any change in cardinal utility function on the criteria.* In other words, if one or more of the utility functions associated with the criteria are modified by a strictly positive affine mapping, the method will still give the same results.

Clearly the *question of choice does not arise in an ordinal multicriterion decision method*. It does however arise in all other cases, and in particular those involving the concept of compensation which we shall be discussing in the ensuing chapters. Here it is desirable for the criteria to be evaluated on scales comparable in type, range, measuring unit, any zero position, mode and dispersion. We shall not go into the generalities of this normalization problem here; the reader could consult Roberts (1979). We shall simply enumerate below the classical normalization methods.

The problem of evaluation is an irritating one, since there generally exists no canonical scale for qualitative criteria alone. Take, for example, the evaluation of the images of various companies ; would you give marks between 0 and 5 or 0 and 20? Similarly, for a portfolio yield, would you base the evaluation on annual income or percentages? For the ensuing multicriterion decision method these questions are by no means anodyne.

With this in mind, we can now begin to deal with normalization procedures. For a given criterion j, evaluations of the m alternatives are given by a_{ij}, $i = 1,...,m$. A normalization procedure will transform the vector $(a_{1j}, a_{2j},...a_{mj})$ into a normalized vector $(v_{1j}, v_{2j},...v_{mj})$.

For the sake of simplicity we shall assume that we MAXIMIZE all criteria. A criterion can always be maximized by changing the sign of the evaluation since $\text{Min}_i(a_{ij}) = -\text{Max}_i(-a_{ij})$. In certain cases, and in particular if the values of a_{ij} are strictly positive, we can convert from minimization to maximization by using the transformation $v_{ij} = 1/a_{ij}$, enabling us to go from $\text{Min}_i(a_{ij})$ to $\text{Max}_i(v_{ij})$. This transformation has the advantage of conserving quotient cardinality (see section 2.43.).

We shall also assume that for all i and for all j we have $a_{ij} \geq 0$. This does not make the problem any less general, since if fit is not the case, it is sufficient to consider $\text{min}_{ij}(a_{ij}) = k < 0$ and to add $-k$ to all the terms of the matrix. This transformation is obviously neutral as far as both cardinal and ordinal utilities are concerned (but not quotient cardinals).

In figure 2.8 are gathered together the four main vector normalization procedures. Starting with vector $a = (a_1, a_2,...,a_m)$ where $a_i \geq 0$ we obtain the normalized vector $v = (v_1, v_2,...v_m)$.

	Procedure 1	Procedure 2	Procedure 3	Procedure 4
Definitions	$v_i = \dfrac{a_i}{\max a_i}$	$v_i = \dfrac{a_i - \min a_i}{\max a_i - \min a_i}$	$v_i = \dfrac{a_i}{\sum_i a_i}$	$v_i = \dfrac{a_i}{\left(\sum_i a_i^2\right)^{1/2}}$
Normalized vector [1]	$0 < v_i \leq 1$	$0 \leq v_i \leq 1$	$0 < v_i < 1$	$0 < v_i < 1$
Modulus of v	variable	variable	variable	1
Proportio-nality conserved	yes	no	yes	yes
Interpretation	% of maximum a_i	% of range $(\max a_i - \min a_i)$	% of total $\sum_i a_i$	ith component of unit vector

[1] Except for some pathological values of a

Figure 2.8 *The principal methods of vector normalization*

In multicriterion decision analysis, each column in the decision matrix formed from the vector $a_{.j} = (a_{1j}, a_{2j},...,a_{mj})$ is normalized. Procedure 1 is the most widely used; it is simple to interpret and it respects proportionality. The second is an improved version of the first designed to ensure that the evaluations cover the interval [0,1], *i.e.* that, for each criterion, the worst value is 0 and the best 1. It respects cardinality but, unlike the first, not proportionality, *i.e.*. $a_{ij}/a_{i'j}$ is not necessarily equal to $v_{ij}/v_{i'j}$.

Procedure 3 is also frequently used, especially in the hierarchic method of Saaty (1980), and also, as we shall see later, for normalization of weightings, it offers the same advantages as procedure 1 while giving smaller, more concentrated values. Procedure 4 is less intuitive for a non-mathematical decision maker, but has the advantage of enabling comparisons of vectors of norm 1, but it produces the same concentrations as procedure 3. In practice it is seldom used; the program IDEAS of Vetschera (1988) is the only application known to the authors.

All these normalization procedures may be considered as different variations of a single unifying method: the choice of a measuring scale on which to plot distances to two 'origin' points in two spaces:

$$r_{ij} = d(a_{ij}, O_1)/d(a._j, O_m) \tag{9}$$

where $a_{ij} \in |R, O_1 \in |R, a._j \in |R^m$ and $O_m \in |R^m$, being O_1 and O_m the origins.

Minkowski's metric m_p between the points $x = (x_1, x_2, ..., x_n)$ and $y = (y_1, y_2, ... y_n)$ in $|R^n$ is defined by:

$$m_p = [\sum_i |x_i - y_i|^p]^{1/p}$$

The most commonly used values taken for p are $p = 1$, $p = 2$ and $p = \infty$. If we take $p = 1$ (block distance), formula (9) becomes:

$$r_{ij} = |a_{ij} - 0| / \sum_i |a_{ij} - 0| = a_{ij} / \sum_i a_{ij}$$

which corresponds to the third procedure in figure 2.8.

For $p = 2$ (euclidean distance), we obtain:

$$r_{ij} = a_{ij} / [\sum_i a_{ij}^2]^{1/2}$$

which gives procedure 4.

Finally for $p = \infty$, since the limit of $[\sum_i |x_i - y_i|^p]^{1/p}$ is equal to $\text{Max}_i |x_i - y_i|$ (the Tchebycheff distance), substitution into (9) gives:

$$r_{ij} = a_{ij} / \text{Max}_i \, a_{ij}$$

which corresponds to the first procedure in figure 2.8.

One last question is worth considering: unimodal utility functions. With a criterion such as price, a decision maker's preferences generally extend from the lowest to the highest point in the scale. One end of the scale indicates the worst choice, the other the best. But this is not necessarily the case for all criteria. Suppose we want to launch a sweet drink onto the market; we consider that one of the choice criteria of the customer will be the amount of sugar per litre. Clearly there exists an optimum quantity of sugar, say 50g, which will satisfy the taste of the majority of consumers. If the possible quantities of sugar extend from 5g to 150g, the optimum is seen not to be at the end of the scale; the satisfaction of the average consumer increases from 5g to 50g, then decreases from 50g to 150g. The value of 50g is the mode of the utility function, hence the expression 'unimodal'. If we represent the situation on a graph (figure 2.9), we see that the utility function for sweetness is not monotone, unlike that for price, (and even for price the function is not necessarily linear – see section 2.4.4). For unimodal preferences, the utility increases to a maximum, then decreases.

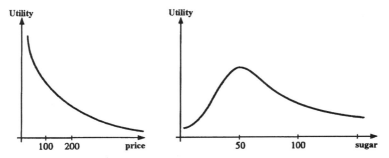

Figure 2.9 Monotone and unimodal preferences.

Unimodal preferences can – rather conveniently for the subsequent treatment – be assimilated to monotone utility. Let a_j^* be the mode of the function, *i.e.*. the optimum satisfaction for the criterion j; to find the monotone utility we simply let:

$$v_{ij} = |a_j^* - a_{ij}| \quad \text{and minimize } v_{ij}.$$

We have now introduced all the tools necessary to tackle the problem of multicriterion choice. In the next chapter we shall explain the essential results that can be obtained from the basic model without introducing any aggregation of the criteria.

3 ANALYSIS OF DOMINANCE AND SATISFACTION

One of the things we shall do in this chapter is to introduce a fundamental concept in multicriterion analysis: the Pareto optimum. Suppose a decision maker has to choose from a choice set $\mathcal{A}$, a car for example. To do this he will evaluate the problem with the help of several criteria such as price, quality, comfort etc. For each of these attributes he is able to classify the alternatives in a (strongly) rational manner, *i.e.* he has available n preorders which can be totally represented by utility functions U_j. As we have already seen, of he chooses the cheapest object (price criterion), he has little chance of ending up with the most comfortable car (comfort criterion). If on the other hand he chooses the most comfortable car, there is little likelihood of this being the cheapest.

It will not take him long to realize that gaining on one criterion involves losing on another (roundabouts and swings); criteria are said to be in conflict or contradictory (more or less). Finding an expensive uncomfortable car may give rise to the hope of finding another both cheaper and more comfortable, which the rational decision maker will choose without hesitation. However in many cases it is not possible to get a lower price for the same degree of comfort, and then the car being offered can in a certain sense be considered as optimal; it is said to be optimal in Pareto terms (see 1.4). This notion of Pareto optimum is the subject of this chapter, and we shall begin with the essential notion of product preorders.

3.1 Product preorders and dominance.

3.1.1 Product preorders: definition and symbols

Let $\mathcal{A}$ be a choice set on which one or more decision makers possess several preorders (criteria). We denote these preorders by $\succcurlyeq_j$, $j = 1, ..., n$ and by $\approx_j$ and $\succ_j$ the associated symmetrical and asymmetrical relations.

Definition 3.1

The product preorder of the n preorders $\succcurlyeq_j$ is the preorder defined by:

$a \succcurlyeq b$ if and only if for any $j = 1, ..., n$ *one has* $a \succcurlyeq_j b$.

It can easily be verified that the above relation is a preorder, *i.e.* that it is reflexive and transitive.

Example 3.2

Let $A = |R^2$ and $x = (x_1, x_2)$, $y = (y_1, y_2)$; we define the preorders:
$x \succcurlyeq_1 y$ if and only if $x_1 \geq y_1$
and $x \succcurlyeq_2 y$ if and only if $x_2 \geq y_2$
The product preorder is defined by:
$x \succcurlyeq y$ if and only if $x_1 \geq y_1$ and $x_2 \geq y_2$

What should be noticed immediately in this example is that, although $\succcurlyeq_1$ and $\succcurlyeq_2$ are complete preorders, the product preorder is partial. In other words, there are for the product preorder pairs which are incomparable; *e.g.* we do not have $x \succcurlyeq y$ or $y \succcurlyeq x$ for the pairs $x = (7, 2)$ and $y = (5, 8)$.

In practical terms, if each preorder corresponds to one criterion, the product preorder is that for which all the criteria are in agreement.

The set $|R^2$ bearing the product preorder defined above is the prototype of the product preorders and we shall be constructing most of our simple examples and counter-examples in $|R^2$ (see next section).

Notation 3.3

If the preorders $\succcurlyeq_j$ are complete and represented by utility functions U_j, we let $U(x) = (U_1(x), U_2(x), ..., U_n(x))$; this is a vector in $|R^n$.

Proposition 3.4

The canonical product preorder of $|R^n$ being defined by $x = (x_1, x_2, ..., x_n) \geq y = (y_1, y_2, ..., y_n)$ if and only if for all $j = 1, ..., n$ one has $x_j \geq y_j$, and the utilities U_j representing the complete preorders $\succcurlyeq_j$; we therefore have:
$x \succcurlyeq y$ if and only if $U(x) \geq U(y)$ where $\geq$ is the product preorder of $|R^n$.

Proof

For all j, we have $x \succcurlyeq_j y \Longleftrightarrow U_j(x) \geq U_j(y)$, leading to the required result. Q.E.D.

This proposition proves that when the preorders are represented by utilities, all the results can be obtained in $U(A)$ which is a subset of $|R^n$ with its canonical preorder.

Convention and notation 3.5

In $|R^n$ the canonical product preorder will always be represented by $\geq$, and the reader should take care not to confuse this with the preorder of $|R$, according to whether $\geq$ applies to a real number or a vector. The *asymmetric part* of the product preorder $|R^n$ is denoted $>$ and is defined by:

$x > y$ if and only if for all $j = 1,..., n$ $x_j \geq y_j$ and there exists j_0 such that $x_{j0} > y_{j0}$.

On the other hand, *the product of the asymmetric parts* of the n preorders of $|R^n$ is denoted $>>$:
$x >> y$ if and only if for all $j = 1,..., n$ one has $x_j > y_j$

3.1.2 Properties of product preorders

The distinguishing feature and main drawback of product preorders is that they are partial. The following properties are true for all preorders but are only useful for partial preorders.

In what follows we assume that $\mathcal{A}$ has n preorders $\succcurlyeq_j$, which may be represented by utilities U_j. The product preorder is denoted $\succcurlyeq$.

Definition 3.6

a is said to dominate b in the wide sense if $a \succcurlyeq b$. In other words, as $\succcurlyeq$ is the product preorder, this means that a $\succcurlyeq_j b$ for all $j = 1, ..., n$. *a is said to dominate b in the strict sense* if $a \succ b$, where $\succ$ is the asymmetric part of $\succcurlyeq$.

Definition 3.7

An alternative a of $\mathcal{A}$ is said to be *efficient, or to be a Pareto optimum* for the preorder $\succcurlyeq$, if there exists no alternative $b \in \mathcal{A}$ such that $b \succ a$, *i.e.* if there exists no alternative in $\mathcal{A}$ which strictly dominates a.

If $\succcurlyeq$ is the product preorder of the $\succcurlyeq_j$, an element a is efficient point if in $\mathcal{A}$ there cannot be found an element b which is preferred for all the $\succcurlyeq_j$ and is strictly better for one of them. In terms of criteria, 'a is a Pareto optimum' means that in the choice set, no single criterion can be strictly improved without another being diminished. Remember (*cf.* section 1.2) that if each criterion represents the satisfaction of an agent, an efficient point is a point where the satisfaction of one agent cannot be improved without diminishing that of at least one other. This is the notion which was introduced by Pareto.

Definition 3.8

Given $\mathcal{A}$ and $\succcurlyeq$, the set of efficient points is called the *Pareto set*. (For a graphic representation, see the following section).

The essential point here is that, for the decision maker who has to make a selection (P_α), the choice set can, with no problems, be replaced by the Pareto set. In organizational problems, it is sometimes useful to keep those alternatives which are the dominated ones within a complete ordering. Thus if we consider the alternatives $\{a, b, c, d\}$, and if, say, d is dominated by a but not by b or c, the choice of a rational decision maker can be $a \succ d \succ b \succ$ c; this is the 'brilliant second' phenomenon (Schärlig, 1985). Though alternative d is dominated by a, it is more satisfying than b or c, which are not dominated.

Definition 3.9

A subset C of $\mathcal{A}$ *is complete for* $\succcurlyeq$, if, for all $x \in \mathcal{A}\text{-}C$, there exists $y \in C$ such that $y \succ x$.

In other words, if the decision maker is made to choose from C instead of from his initial choice set $\mathcal{A}$, he is at no disadvantage (at least, within the problem of selection); if he had chosen x, it would still be possible for him to take y which is strictly better.

Definition 3.10

A complete subset C is minimal if no strict subset of C is complete. A minimal complete set is the smallest set from which the decision maker can choose without being at a disadvantage.

Proposition 3.11

i) All complete sets contain the set of efficient points.
ii) If the set of efficient points is complete, then it is complete and minimal.

Proof

i) If C is complete, then for all $y \notin \mathcal{A}$, there exists $x \in C$ such that $x \succ y$. Consequently, if $y \notin C$, y does not belong to the set of efficient points.
ii) Let C be complete, strictly contained in the set of efficient points; from i) C will itself contain the efficient points, which is absurd. Q.E.D.

Proposition 3.12

i) A minimal complete set cannot contain two elements a and a' such that $a' \succ a$.
ii) The only possible complete minimal set is the set of efficient points.

Proof

i) Let a and a' belong to a complete minimal set C such that $a' \succ a$. It can easily be verified that the set $C - \{a\}$ is still complete, which contradicts the minimal condition.

ii) Let A be minimal and complete; from proposition 3.11 it will contain the set of efficient points. Suppose this is not the case; there will then exist $b \in C$, which is not an efficient point. By definition there exists $c \in A$ such that $c \succ b$. From i) $c \notin C$ which in turn implies – since C is complete – the existence of $c' \in C$ such that $c' \succ c$. By transitivity $c' \succ b$, which from i) is absurd. Q.E.D.

As a consequence of the above propositions, it is essential to note that in selection problems, the set of efficient points is the right choice set for the decision maker since, on the one hand the latter has nothing to lose if he exchanges his choice set A for the set of efficient points (completeness), and on the other hand this is the smallest possible set having this property (minimality).

We end this section by showing that, *in the discrete case, the set of efficient points or Pareto set is always complete and thus complete and minimal*, from proposition 3.11.

Proposition 3.13

If A is finite then the set of efficient points is complete.

Proof

Let O be the set of efficient points. If $A = O$ then O is complete; otherwise let $x \in A - O$. There exists $y_1 \in A$ such that $y_1 \succ x$. If $y_1 \notin O$, there exists $y_2 \in A$ such that $y_2 \succ y_1$ and so on until all the elements in A are used up (finite number) or until we have $y_p \in O$. By transitivity $y_p \succ x$, proving that O is complete. Q.E.D.

Remarks

3.14 In the finite case in which we are interested, the rational decision maker's choice will consist of selecting an element from the Pareto set. The non-efficient point (or dominated) elements are of no interest except in the case of the 'brilliant second' referred to above.

3.15 If the preorder $\succeq$ is complete, the Pareto set is the set of maximal elements, *i.e.* the elements at the head of the chain (see theorem 2.19). In monocriterion optimization, it is the set of elements which maximizes the unique criterion.

3.16 In a partial product preorder, the Pareto set will contain elements which cannot be compared with each other (or which are mutually indifferent, unless there is minimality). There are generally many of these. It is vital to understand that if the decision maker's preorders (criteria) and choice set A are made up of reliable information, *the rationality of the model will be entirely expressed by the Pareto set*

O. Without supplementary information, all the efficient points (those which cannot be compared with each other) represent a right choice which is rational for the decision maker. Then it is only considerations outside the model which will allow a choice to be made between the efficient points. These considerations may take the form of additional information provided by the decision maker (progressive information methods, *cf.* 2.6.3), or *a priori* information on the relative importance of criteria.

3.2 Cones and preorders

The notions introduced in the above section have highly visual properties which can be appreciated by any reader who is willing to make a small detour into the world of cones. This is not indispensable to the rest of the work; on the other hand readers whose appetite is whetted by what follows or is interested in research into the mathematical properties of Pareto sets can consult Sawaragi *et al.* (1985) or Steuer (1986) and, for more recent works, Skulimowski (1987, 1996) and Jahn (1999).

We have seen how the space $\mathbb{R}^n$ plays a central role in the study of preorders when these are represented by utility functions (this being the simplest and commonest case). $\mathbb{R}^n$ has the structure of a vector space, and for the sake of simplicity we shall only be considering vector spaces in the following part of this section.

Definition 3.17

Let E be a vector space; a preorder $\succcurlyeq$ is said to be compatible with vector space structure if:
$x \succcurlyeq y$ implies $x + z \succcurlyeq y + z$ for all $z \in E$
and $x \succcurlyeq y$ implies $\alpha x \succcurlyeq \alpha y$ for all positive or null scalars α.

Proposition 3.18

A necessary and sufficient condition for the existence of a preorder $\succcurlyeq$ compatible with vector space structure is that there exist a part F of E such that $0 \in F$, $F + F = F$ and $\alpha F = F$ (for all strictly positive scalars).

Proof

i) The condition is necessary. $\succcurlyeq$ is a compatible preorder; let $F = \{x \mid x \succcurlyeq 0\}$; 0 here is a neutral element of the vector space. If x and y belong to F, we have $x \succcurlyeq 0$ and $y \succcurlyeq 0$, from the compatibility $x + y \succcurlyeq y$ and by transitivity $x + y \succcurlyeq 0$, whence $x + y \in F$. $F + F$ is therefore contained in F and, since $0 \in F$ because of the reflexivity $(0 \succcurlyeq 0)$, F is contained in $F + F$, leading to the equality. If $x \in F$, $x \succcurlyeq 0$, from the compatibility $\alpha x \succcurlyeq 0$ and $\alpha x \in F$. Hence we have αF contained in F. If $x \in F$, then from the compatibility this leads to $x/\alpha \in F$ whence $x \in \alpha F$; we then have finally $\alpha F = F$.

ii) The condition is sufficient. Let F be a part satisfying the properties of the proposition statement. Let: $x \succcurlyeq y$ if and only if $x - y \in F$.

Since $0 \in F$, we can deduce that $x - x \in F$ and $\succcurlyeq$ is reflexive. We now show that $\succcurlyeq$ is transitive; Let $x \succcurlyeq y$ and $y \succcurlyeq z$; this is equivalent to saying that $x - y \in F$ and $y - z \in F$. Taking the sum we have $(x - y) + (y - z) = x - z \in F$, implying $x \succcurlyeq z$. Therefore $\succcurlyeq$ is a preorder and we see immediately that $\succcurlyeq$ is compatible. Q.E.D.

Remark 3.19

Any preorder on $|R^n$ compatible with vector space structure is defined by the part $F = \{x \mid x \succcurlyeq 0\}$, *i.e.* the set of positive or null elements. Thus for the canonical preorder of $|R^2$ (*cf.* example 3.2), the set of positive elements is $H = \{x = (x_1, x_2) \mid x_1 \geq 0 \text{ and } x_2 \geq 0\}$; this set is called the *positive orthant*. Similarly, the positive orthant of $|R^n$, $\{x = (x_1, x_2, ..., x_n) \mid \text{for all } j \text{ one has } x_j \geq 0\}$ defines the canonical preorder of $|R^n$.

The next definition is required because in the book we shall be needing certain notions relating to convexity.

Definition 3.20

A subset C of a vector space is convex if whatever $x \in C$ and $y \in C$ implies, for all $\alpha \in [0,1]$, that $\alpha x + (1 - \alpha)y \in C$. In other words the segment $[x,y]$ is contained in C as long as x and y belong to C.

Definition 3.21

A set F such as $0 \in F$ and $\alpha F = F$ for all strictly positive α is called a *pointed cone*. We shall simply call it a cone.

Figures 3.1 displays several types of cones.

Proposition 3.22

For a cone F the following two properties are equivalent:
    i) F satisfies $F + F = F$;
    ii) F is convex.

Proof

i) $\Longrightarrow$ ii) Let $x \in F$ and $y \in F$ for all $\alpha \in]0,1[$, αx and $(1 - \alpha)y$ belongs to F since F is a cone. From i), $\alpha x + (1 - \alpha)y \in F$. The cases $\alpha = 0$ and $\alpha = 1$ are obvious.

ii) $\Longrightarrow$ i). Let $x \in F$ and $y \in F$; since F is a cone, $2x$ and $2y$ belong to F. Since F is convex, for all $\alpha \in [0,1]$, $\alpha(2x) + (1 - \alpha)(2y)$ belongs to F. Let $\alpha = \frac{1}{2}$; we then have $x + y \in F$. Q.E.D.

We can now state that a preorder compatible with vector space structure is always associated with a convex cone.

Proposition 3.23

The preorder $\succeq$ associated with the cone F is complete if and only if $F \cup -F = |\mathbf{R}^n$, where $-F = \{x \mid -x \in F\}$.

Proof

The condition is sufficient. Let x and $y \in |\mathbf{R}^n$; we have $x - y \in F$ or $x - y \in -F$, *i.e.* $y - x \in F$. This proves that the preorder associated with F is complete.

To show that this is necessary, consider any x of $|\mathbf{R}^n$. If the preorder $\succeq$ is complete we have, either $x \succeq 0$ or $0 \succeq x$, hence x belongs to F or $-F$. Q.E.D.

The property of proposition 3.23 is not satisfied by the positive orthant H which is known to represent the partial ,preorder of $|\mathbf{R}^2$.

The most useful result in visualizing multicriterion properties is that relating cones to the set of efficient points.

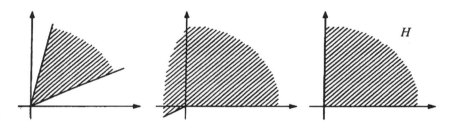

Figure 3.1 A convex cone, a non-convex cone and the positive orthant H in $|\mathbf{R}^2$

Proposition 3.24

If the preorder $\succeq$ is defined by a convex cone F such that $F \cap -F = \{0\}$, then a is efficient point if and only if $\mathcal{A} \cap (a + F) = \{a\}$.

Proof

Saying that $x \in a + F$ is equivalent to saying that $x - a \in F$, or $x \succeq a$. The condition is sufficient since $\mathcal{A} \cap (a + F) = \{a\}$ implies that there does not exist any element x in $\mathcal{A}$ such that $x \succeq a$ and hence, *a fortiori*, $x \succ a$. The element a is therefore an efficient point.

The condition is necessary. If a is efficient point, there exists no element x in $\mathcal{A}$ such that $x \succ a$. But $x \succ a$ is equivalent to $x \succeq a$ and not($a \succeq x$) (definition 2.12). This is expressed as $x - a \in F$ and $(x - a) \notin -F$. If $F \cap -F = \{0\}$, then $x \succ a$ is equivalent to $x - a \in F$, since if it belongs to F it cannot belong to $-F$ unless $x = a$, which is excluded. Therefore, a is efficient point implies that there does not exist $x \in \mathcal{A} \cap (a + F)$, except a itself. Q.E.D.

This simple result can be refined in numerous ways (see Skulimowski, 1987), but is sufficient here to illustrate the main results: positive orthants and classical cones satisfy $F \cap -F = \{0\}$ and it is easy to visualize the efficient points of a set $\mathcal{A}$ for the canonical preorder of $|R^n$. See Figure 3.2 below, in $|R^2$.

In the first diagram of figure 3.2, if $\mathcal{A} = \{a, b, c\}$, we see that a is not a Pareto optimum since $b \in a + H$; on the other hand, c is a Pareto point since $c + H$ does not contain any point in $\mathcal{A}$. In the second diagram the situation is identical since $(a + H) \cap \mathcal{A}$ is not void whereas $(c + H) \cap \mathcal{A}$ is. The reader will see that *the set of efficient point or Pareto points forms the 'north-east' boundary* of the choice set $\mathcal{A}$. When preorders are represented by utility functions, this result is important because from proposition 3.4, the image of the Pareto points in $\mathcal{A}$ is made up of the north-east boundary of $U(\mathcal{A})$.

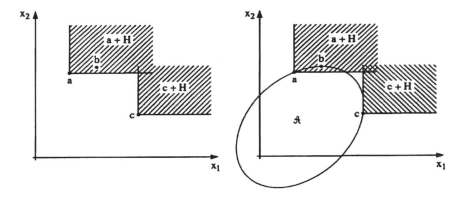

Figure 3.2 Graphic representation of Pareto points

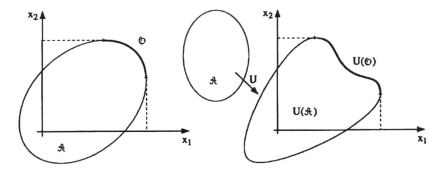

Figure 3.3 Graphic representation of the Pareto set

In the first diagram of figure 3.3, we have drawn the Pareto set O asociated with a subset $\mathcal{A}$ of $|R^2$. We can see that O does indeed represent the north-east boundary of $\mathcal{A}$. In the second diagram, $\mathcal{A}$ is any set whatever and so the mapping U (notation 3.3) sends $\mathcal{A}$ into $|R^n$, which here is $|R^2$. The image of the Pareto points forms the north-east boundary of $U(\mathcal{A})$.

The above result can also be used to show complete non-minimal efficient sets in $|R^2$. So, in the left-hand diagram of figure 3.4 the Pareto set is empty and the set C is non-minimal complete, any section above α creating a smaller complete set. Note that $\mathcal{A}$ is neither finite (3.13) nor compact. When $\mathcal{A}$ is compact, it can be shown, under certain conditions, that the Pareto set is minimal complete (see Ekeland, 1979). In the right-hand diagram of figure 3.4, the Pareto set O is not empty, but it is not complete either: the point a is not dominated by an element of O.

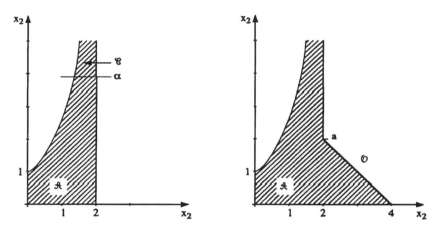

Figure 3.4 Non-minimal complete sets.

Another useful consequence of proposition 3.24 concerns the comparison of preorders. If F_1 is contained in F_2, then F_1 will have fewer positive elements and this corresponds to a *preorder which is coarser or less precise than F_2*.

Proposition 3.25

If F_1 is contained in F_2 and if $F_i \cap -F_i = \{0\}$ for $i = 1,2$, then the set of efficient points O_2 is contained in O_1.

Proof

If there exists no $x \in \mathcal{A}$ such that $x \succ_2 a$, this is equivalent (proposition 3.24) to $x - a \notin F_2$, and therefore $x - a \notin F_1$ and $a \in O_1$. Q.E.D.

What should be remembered is that *the more incomplete (or the coarser) the preorder, the smaller the positive cone and the larger the set of efficient points*. In the limit, for a complete preorder, the set of efficient points will be small and reduced to an optimum (or to mutually indifferent elements ensuring an optimum – see remark 3.15). When two criteria differ very greatly, this will mean that, for the product preorder there will be many alternatives which cannot be compared; an alternative which is right for one criterion will be wrong for the other and cannot be compared with an alternative which is bad for the first criterion and good for the second. The product preorder is coarse and the number of efficient points will be high. On the other hand, if the criteria do not differ very much, the number of efficient points will be small.

With the set $\mathcal{A}$ in Figure 3.5, the Pareto set is reduced to a point $a = (1,1)$ for the canonical preorder defined by F_1. O_2, on the other hand, is equal to the segment $[a,b]$ with $b =(1,0)$ as can be verified from proposition 3.24.

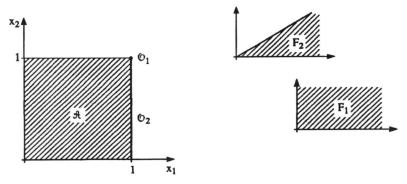

Figure 3.5 Comparison of Pareto sets.

3.3 Pre-analysis of dominance

To simplify the problem of decision, we consider the preorders associated with the criteria and the product preorder induced (section 3.1). Suppose we wish to maximize all the criteria. An element a of $\mathcal{A}$ will be dominated in the strict sense by b if $b \succ a$ for the product preorder. *In terms of criteria this means that b is better than a for all criteria and strictly better for at least one of them.* Looking at the numerical decision matrix, this means that $U_j(b) \geq U_j(a)$ for all $j = 1, ..., n$ and $U_{j_0}(b) > U_{j_0}(a)$ for at least one criterion j_0.

An efficient point or optimum alternative, in the Pareto sense, is an alternative which is not strictly dominated; in other words, no alternative can be found which is better, in the wide sense, for all the criteria, and strictly better for one of them. Another way of expressing this is to say that if I wish to improve a single criterion in relation to an efficient point alternative, then I shall have to lose out on at least one other. We have seen that the set of efficient points is complete and that a rational decision maker's choice can therefore be restricted to the subset of efficient points (*cf.* 3.1).

We now illustrate these ideas with an example which we shall be referring to several times in the rest of the book.

3.26 The example of personnel selection

A company has short-listed nine applicants for a position. These alternatives A_i, $i = 1, 2,..., 9$ are called: Albert, Blanche, Charles, Donald, Emily, Frank, Georgia, Helen and Irving. The personnel recruitment department decide to consider five criteria; the first three are quantitative.

– Criterion # 1: Time spent in higher studies, expressed in years.

– Criterion #2: Professional experience within the qualification in question, expressed in years.

– Criterion #3: Age, expressed in years.

For the other two criteria, which are qualitative, the recruiting manager has decided to express a complete preorder by marks from 0 through 10.

– Criterion #4: Evaluation from the interview expressed as a mark out of 10.

– Criterion #5: result of psychometric tests expressed as a mark out of 10.

All criteria are to be maximized except the third: age. Figure 3.6 shows the decision matrix with the evaluations for each applicant relative to each criterion. For ease of use we have kept the age criterion in its natural form, to be minimized, though theoretically it ought to be present as a utility function to be maximized; this would have meant using negative numbers for age, Albert for example having a utility of -28 for criterion #3.

Alternatives	Criteria					Pre-analysis of dominance
	C_1 MAX	C_2 MAX	C_3 MIN	C_4 MAX	C_5 MAX	
1 Albert	6	5	28	5	5	
2 Blanche	4	2	25	10	9	
3 Charles	7	7	38	5	10	
4 Donald	5	7	35	9	6	
5 Emily	6	1	27	6	7	
6 Frank	5	7	31	7	8	Dominated
7 Georgia	6	8	30	7	9	
8 Helen	5	6	26	4	8	
9 Irving	3	8	34	8	7	

Figure 3.6 The decision matrix and pre-analysis of dominance

We recommend in practice, to give the ages in natural form, even though this requires some mental gymnastics, the preferred applicants being those with the lower ages and the dominated ones those with higher ages – the reverse reasoning compared to that for the criteria to be maximized.

First of all we can see that there exists no alternative which is dominant in the wide sense, *i.e.* an alternative which dominates all the other alternatives in the wide sense. Such an alternative would be the best choice in the wide sense for all the criteria and would therefore be the best choice.

Secondly, there is one dominated alternative: Frank is worse than or equal to alternative #7 (Georgia); she is younger than him and equal or better on all other criteria. Georgia is said to dominate Frank. No other alternative is dominated: they are all Pareto optima or efficient alternatives. Apart from Frank, the choice of whom would be irrational with the information given, all other choices are reasonable, according to one's point of view.

It is a simple matter to devise a screening process which will eliminate all the strictly dominated alternatives from any decision matrix. In the worst case, using the simplest algorithm, for m alternatives there will be $m(m-1)/2$ comparisons of values multiplied by n criteria. For a very large number of alternatives the calculation could become rather long; for dimensions encountered in practice, around a hundred alternatives and fewer than twenty criteria, this poses no real problems even on microcomputers. In a classical study, Kung, Lucio and Preparata (1975) gave the exact complexity bounds using the best algorithms known. Using our notation (m alternatives, n criteria) we consider a function $C(m,n)$ which gives the optimum number of value comparisons required to find all the Pareto optima with the best algorithm. Kung *et al.* found that:

$C(m,n) \geq E[\log_2(m!)]$ where E denotes the integral part,

and $C(m,n) \leq O[m(\log_2(m))^{n-2}]$.

Several values of these values for bounds are given in figure 3.7.

m	Lower bound	Upper bound				
		$n = 6$	$n = 10$	$n = 20$	$n = 50$	$n = 100$
10	22	1 217	148 293	$2.42 \ 10^{10}$	$1.06 \ 10^{26}$	$1.24 \ 10^{52}$
20	62	6 978	2 434 727	•	•	•
50	215	50 731	51 472 767	•	•	•
100	525	194 841	379 631 102	•	•	•
1 000	8 530	9 863 837	$9.7 \ 10^{10}$	•	•	$\sim 10^{100}$

Figure 3.7 Number of comparisons required to find the efficient points

The other interesting question we can ask is, how many optimum Pareto alternatives remain when the dominated alternatives have been eliminated from a given decision matrix? We have already seen (*cf.* end of section 3.2) that this will depend both on how much the criteria converge or diverge, and on the number of criteria. If we assume that the criteria are completely independent (this is rare in practice), then Rosinger (1991) showed that the number of pairs of alternatives which are comparable in the product preorder, *i.e.* such that $a \succcurlyeq b$ or $b \succcurlyeq a$, is of the order of $O(1/2^n)$. This is a number which decreases very rapidly, in other words the product preorder is highly partial, leading to the existence of a large number of Pareto points. For decision matrices *mxn* filled randomly (*e.g.* a random draw) the number of Pareto optima can be calculated by recurrence. Calpine and Golding (1976) worked on this problem. The main result is that for around ten criteria most of the alternatives are Pareto optimum, even for a large number of alternatives (*m* =1000) – see figure 3.8.

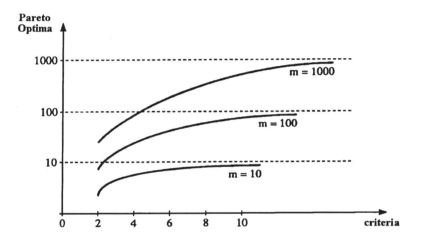

Figure 3.8 The number of Pareto optima (after Calpine and Golding)

As we have mentioned several times, a decision maker dealing with a selection problem (P_α) will not choose a dominated Pareto alternative. This is why, *in most of the programs that we shall be examining, the first job is to look for dominated alternatives*. These will then be eliminated from the decision matrix, provided the decision maker agrees. We shall call this search for dominated alternatives the dominance pre-analysis: it is usually carried out before any multicriterion analysis proper. We must also repeat the extremely important fact that, in the absence of any other information from the decision maker, a simple statement of ordinal utilities characterizing the criteria does not allow us to carry out any operation beyond that of eliminating the dominated alternatives. In other words, a rational decision maker will be able to find a good reason, relative to a particular criterion, for choosing any one of the alternatives among the Pareto optima. Certain Pareto optima will, nevertheless, provide a better compromise than others (see chapter 5).

3.4 Pre-analysis of satisfaction

In cognitive psychology, a notion that has been well known for quite a time is that of *'aspiration level'*; according to Siegel, this was first introduced by Dembo in 1931. The idea is that an individual's aspiration level will control his behavior. Siegel (1957) moreover showed that the aspiration level lies on the cardinal utility curve at the point where the slope is greatest, marking a change in preference strengths.

We shall introduce a related notion in the domain which concerns us. This idea is once again linked to the work of Simon (1955) on limited rationality (*cf.* section 2.6.2), and his concept of a 'satisficing' solution. Simon refers to the search for a solution which ends once the decision maker is satisfied; this notion is clearly opposed to that of optimization since the decision maker stops looking when he feels satisfied without knowing whether he might find a better choice.

Instead of the notion of aspiration level we shall use that of *satisfaction level*, expressing a utility level such that a Simon-decision maker stops searching when he thinks that what he has obtained is *satisficing*. We might think, like Siegel, that the decision maker stops when his aspiration level is reached or exceeded; the two notions – aspiration level and satisfaction level – would then appear more or less the same. But as Simon (1955) points out, satisfaction level actually appears to be highly mobile. It may vary according to various information, especially that concerning results already obtained and the ease with which they have been obtained.

We shall use the notion of satisfaction level in the form of a *satisfaction threshold* or level for each criterion (to be maximized or, with the notion suitably adapted, to be minimized). We shall assume that, for each criterion given in the form of a complete preorder (hence of a path, see theorem 2.19), the decision maker knows of an alternative, or a utility level not necessarily represented by an alternative, which he would be happy with. This will obviously not be the best choice for the given criterion, but he would feel happy provided he obtained the satisfaction level in question. If the preorder is represented by a utility U_j, then the decision maker can define his satisfaction threshold by a number $\hat{u}_j$ such that if an alternative satisfies $U_j(a) \geq \hat{u}_j$ it is satisfying, whereas if $U_j(a) < \hat{u}_j$, the alternative is to be rejected.

With one satisfaction level defined for each criterion we obtain an overall level $\hat{u} = (\hat{u}_1, \hat{u}_2, ..., \hat{u}_n)$. An intelligent pre-analysis will consist of eliminating all those alternatives which are dominated by $\hat{u}$; the choice set is thus reduced. A more drastic choice is to keep in the choice set only those alternatives which dominate the vector $\hat{u}$, *i.e.* those for which the decision maker obtains at least his satisfaction level. This idea can be refined eliminating only those alternatives which are below the satisfaction level for certain decisive criteria; this echoes an idea of Kepner and Tregoe (1981) with their 'must', 'want' and 'ignore' criteria; once the non-satisfying criteria have been eliminated, multicriterion analysis can be carried out on the remaining ones.

Let us see how this works using the personnel recruitment example introduced above. We consider that the applicant must have done at least four years of higher studies and not be more than 35 years of age. For criterion #1 (number of years of higher studies) our satisfaction threshold is accordingly 4 and for criterion #3 the threshold is 35 not to be exceeded. It is unnecessary for us to set a threshold for the other criteria; this will, by default, be the worst case for each criterion considered. As we see in figure 3.9, there are two alternatives which do not dominate $\hat{u}$: Charles and Irving; for these two, the satisfaction level is not reached for at least one of criteria #1 and #3. These two alternatives could be eliminated from the final choice.

Alternatives	Criteria					Satisfaction pre-analysis
	C_1 MAX	C_2 MAX	C_3 MIN	C_4 MAX	C_5 MAX	
1 Albert	6	5	28	5	5	
2 Blanche	4	2	25	10	9	
3 Charles	7	7	38	5	10	Non Satisficing
4 Donald	5	7	35	9	6	
5 Emily	6	1	27	6	7	
6 Frank	5	7	31	7	8	Dominated
7 Georgia	6	8	30	7	9	
8 Helen	5	6	26	4	8	
9 Irving	3	8	34	8	7	Non Satisficing
Satisfaction level:	4 LOW	1 LOW	35 HIGH	4 LOW	5 HIGH	

Figure 3.9 Pre-analysis of satisfaction

We do not get the same result here as we did for pre-analysis of domination, which eliminated Frank. The two sorts of reasoning are independent and complementary.

Note that if the decision maker's requirements are too high, it is quite possible that no alternative will dominate $\hat{u}$; the decision maker will then have to reduce his requirements. In our example the reader can verify that with $\hat{u} = (5, 6, 35, 6, 7)$ only Frank and Georgia remain satisfying, and with $\hat{u} = (7, 6, 35, 6, 7)$ nobody dominates $\hat{u}$. Note too that, though dominated by Georgia, Frank is a good candidate (the phenomenon of the brilliant second introduced in 3.1.2).

Reasoning based on the interactive variation of satisfaction levels can be used in methods of multicriterion analysis where we have progressive information. Interactive and progressive modification of satisfaction levels aimed at reducing the choice set forms the basis of the PRIAM multicriterion choice method developed by Lévine and Pomerol (1986), and we shall be referring to this in section 9.2. An earlier method proposed by Tversky (1972a) is known as *Elimination by Aspects* (EBA). For each of the attributes the decision maker defines certain aspects which he wishes to feature in the final solution, and he eliminates those alternatives which do not have these aspects. Having done this for one attribute, he goes on to the next one and reiterates. Using our notation, this will consist of choosing an initial criterion, say criterion #1, and fixing $\hat{u}_1$. We eliminate the alternatives a such that $U_i(a) < \hat{u}_1$. We then go on to a second attribute and do the same thing, and so on until only one alternative remains, and this is chosen. The method combines lexicographical elimination (section 5.6) and progressive information.

3.5 Methods of discrete multicriterion decision

Before attacking this subject it will be helpful to say a few words about the state of the art. The first survey of the field was by MacCrimmon (1973) in a revised version of work carried out in 1968. We then find the systematic study of Hwang and Yoon (1981), together with the books by Rietveld (1980) and Goicoechea *et al.* (1982). The discrete field receives wide treatment in recent work by Changkong *et al.* (1985). Nijkamp and Voogd (1985) and in the book by Schärlig (1985). Vincke (1986) is mainly devoted to multicriterion decision in Western Europe and Ozernoy (1988) in the ex-USSR. Despontin *et al.* (1983) compiled a list – which at the time was nearly complete – of 96 multicriterion methods (both discrete and continuous, we hasten to add!). Vincke (1992) provides a reasonably accurate view of the main methods and the principles on which they are based. Roy and Bouyssou (1993a) have written a very comprehensive book on multicriterion methods which was partly issued in Roy and Bouyssou (1987). We can also add various references such as Wallenius (1975), Khairullah and Zionts (1981), Massam and Askew (1982), Gershon and Duckstein (1983a), Despontin *et al.* (1986) and Olson (1996). These authors give empirical comparisons of certain multicriterion methods; typically the articles concern half a dozen or so methods chosen by the analyst. Finally we can cite the systematic review of discrete multicriterion methods undertaken by one of us (Barba-Romero, 1987).

We have already mapped out the field of multicriterion decision: the use, or not, of progressive information methods, the use of ordinal or cardinal utilities. We shall be using these concepts to guide the reader. Other features are also of interest from a practical point of view: the number of alternatives and/or criteria that the method is capable of handling, the role of compensation and weighting, the aims of the method (see section 2.6). All these items have their importance, and they will crop up at appropriate points in the book; however, they are too disparate to form a single satisfactory framework for classification.

This is why we have opted for groupings based on theoretical internal structure and on the way the various most commonly used methods can be brought together.

We hope the reader will thus form a realistic and global idea of multicriterion decision, even though we do not enter into all the details. Readers who need more detailed information can use the numerous references we provide.

We do not pretend in this book to describe all the multicriterion methods without exception; this would be practically impossible since there are now so many, some of which are even confidential; but we do hope to present a reasonably complete sample. On a more modest scale, we shall try to give a complete overview of the different families of methods, ignoring only a few atypical obsolete methods. Methods with practical applications will have pride of place, especially if they have been tried and tested.

4 WEIGHTING METHODS AND ASSOCIATED PROBLEMS

4.1 Weights and weighted sums

4.1.1 Weights

Quite often in multicriterion analysis, a decision maker suddenly takes the view that one criterion is either more or less important than another; this may be for various reasons including personal preference (which may be reasonably objective or completely subjective). We shall call this measure of relative importance of criteria as seen by the decision maker, the *weight*. This chapter is devoted to examining the nature of weights, the methods which can be used to evaluate them and the consequences of how well or badly they are used.

Before we begin we must settle a question of terminology. We shall denote by w_j the weight attributed to a criterion j, whatever the nature of the latter (qualitative, quantitative, ordinal or cardinal). We shall also use the term *weight vector* $w = (w_1, ..., w_j, ..., w_n)$ when referring to all of the weights. The decision matrix (see section 2.1.4) together with the weight vector constitutes all the information required in principle to 'solve' the multicriterion choice problem using methods without progressive information (see 2.6.3).

Suppose for the moment that the weights have been determined and that we have to attribute a numerical value to each one. How can we go about this? We shall begin by describing the most widely known aggregation method; in fact many people believe there is no other way of making a decision in the presence of multiple criteria. This is the weighted (linear) sum, whose main advantage is that it is both intuitive and simple to apply, which are good enough reasons for beginning here our study of aggregation methods. When we go on to analyze the limitations of the method and its implicit assumptions we shall be able to introduce the difficulties accompanying multicriterion decision and to show how important is the evaluation of the weights.

4.1.2 Weighted (linear) sum

As we have said, the method is very simple, but we must nevertheless carefully specify the starting data and the transformations they undergo; we shall then describe the method itself.

A) Starting data

A1) We assume we have m alternatives a_1, a_2, a_3, ..., a_m and n criteria C_1, C_2, C_3, ..., C_n. Each criterion Cj is represented by a utility function U_j and we write $a_{ij} = U_j(a_i)$ where a_{ij} is the coefficient of line i column j of the decision matrix. The value of a_{ij} results either from the construction of an actual utility function, or from a natural evaluation in the case of a quantitative criterion such as price or age; for reasons which will become clear, we shall assume in every case that this utility is cardinal quotient (see 2.4.3).

A2) We also assume that each criterion C_j carries a positive or null weight w_j. Since a null weight for criterion C_j is equivalent to eliminating the criterion, hence in what follows, for all j we have $w_j > 0$. We shall also assume that the weights are of the 'cardinal quotient' type, *i.e.* that any other weight vector w', equivalent to w in the mind of the decision maker, satisfies $w' = kw$ (k strictly positive).

B) Data transformation

B1) First we normalize all the coefficients a_{ij} by any of the three methods described in section 2.7 which conserve proportionality. The new values will lie between 0 and 1, the best evaluation being that closest to 1.

B2) We normalize the weights w_j so that their sum is equal to 1, simply by dividing each w_j by $\sum_j w_j$.

In what follows we assume that the coefficients a_{ij} and weights w_j have been normalized as explained.

C) Applying the weighted sum method

For each alternative a_i we evaluate:
$$R(a_i) = \sum_j w_j a_{ij} \quad (i = 1, 2, ..., m) \tag{1}$$

The chosen alternative a_i will be the one which gets the highest value of $R(a_i)$. If there is a tie, we take any one of them. Since the values $R(a_i)$ are real numbers, they are naturally ranked and in any ranking problem, the alternatives can be classified according to the values of $R(a_i)$ obtained. The expression 'weighted sum' is fairly clear, since we are taking the sum of the evaluations obtained, for each alternative, over the various criteria. The expression *'linear sum'* is sometimes used to indicate that the alternatives a_{ij} are to the power 1; when we say 'weighted sum' this should be taken to include the word 'linear'.

Example 4.1 Personnel selection

We shall summon our six applicants who overcome the domination and satisfaction pre-analysis (see example 3.26). Figure 4.1 shows the importance the decision maker attaches to each criterion on a scale from 0 to 5 (for the moment we shall not go into how these weights have been obtained).

Alternatives	Criteria				
	1 MAX	2 MAX	3 MIN	4 MAX	5 MAX
Albert	6	5	28	5	5
Blanche	4	2	25	10	9
Donald	5	7	35	9	6
Emily	6	1	27	6	7
Georgia	6	8	30	7	9
Helen	5	6	26	4	8
Weights	5	5	2	4	4

Figure 4.1 Data for the selection example with weights

The information contained in figure 4.1 makes up the input data on which we shall carry out the normalizing transformations described in *B*). Note that criterion C_3 must be minimized, to do which we shall maximize the inverse $1/a_{i3}$. (We could have taken $-\text{Max}(-a_{i3})$ as in section 2.7, but we would then have had to add a constant k and to use $k - a_{i3}$ in order to obtain positive values; unfortunately this operation destroys proportionality, which is why we favor the inverse. We shall be coming back to this problem later on). Having replaced the ages by their reciprocals, we normalize all the criteria in the same way by dividing by the sum (procedure #3 in figure 2.8). We normalize the weights in the same way. The result of these calculations appears in the table in figure 4.2.

Alternatives	Criteria				
	C_1	C_2	C_3	C_4	C_5
Albert	0.188	0.172	0.168	0.122	0.114
Blanche	0.125	0.069	0.188	0.244	0.205
Donald	0.156	0.241	0.134	0.220	0.136
Emily	0.188	0.034	0.174	0.146	0.159
Georgia	0.188	0.276	0.156	0.171	0.205
Helen	0.156	0.207	0.180	0.098	0.182
Weights	0.25	0.25	0.10	0.20	0.20

Figure 4.2 Data preparation (normalization)

The third and final step of the weighted sum method is to calculate the evaluation for each candidate according to formula (1); we obtain:

R(Albert)   = 0.154 R(Emily)   = 0.134
R(Blanche) = 0.157 R(Georgia) = 0.207
R(Donald) = 0.184 R(Helen)   = 0.165

The result is that Georgia will be chosen. The method gives the overall ranking of the applicants, which is as follows:

1st	Georgia	4th	Blanche
2nd	Donald	5th	Albert
3rd	Helen	6th	Emily

The weighted sum method is based on certain theoretical assumptions and the user should be aware of these to avoid inappropriate application; the chief assumption is that there exists a decision maker's cardinal utility function which is additive over the criteria; this is quite a strong assumption, as we shall see in chapter 6, as it pre-supposes the independence of the criteria as well as the inter-criteria comparison of values achieved by the alternatives. Here the weights express the substitution rate between criteria.

Definition 4.2

If the decision maker's overall utility is given by the weighted sum $R(a_i) = \sum_j w_j a_{ij}$ (formula 1), then the ratio w_k/w_j is called the rate of substitution between the two criteria k and j, since, other things being equal, if the utility $U_k(a_i)$ decreases by δ_k, $U_j(a_i)$ will have to be increased by $\delta_k(w_k/w_j)$ for the decision maker's overall utility not to change.

Using the notation $R(a_i) = \sum_j w_j a_{ij}$ we obtain $w_k/w_j = \dfrac{\partial R}{\partial a_{ik}} \bigg/ \dfrac{\partial R}{\partial a_{ij}}$ which well expresses the idea of marginal substitution rates

Without going into details (we shall be doing this in chapter 6), anybody thinking of using the weighted sum method should be aware of the conditions attached: cardinality of the data (utilities and weights), prior normalization and assumptions justifying the additivity of the decision maker's preferences.

4.1.3 The effect of normalization

Prior normalization of the data is not a neutral operation, and the final result of aggregation may well depend on the normalization method used. We shall give an example, but note first of all that the normalization of weights plays no part in ranking the alternatives. For example, if we replace values w_i by αw_i ($\alpha > 0$), the sum $R(a_i)$ is replaced by $\alpha R(a_i)$, which does not change the ranking. Now we return to the normalizing of the alternatives.

Example 4.3 Influence of the method of normalizing alternatives.

Consider the multicriterion problem with the following data:

		Criteria	
		Max C_1	Max C_2
Alternatives	a_1	9	3
	a_2	1	8
Weights		0.4	0.6

The weights have already been normalized to a sum of 1. If we apply normalization procedure #1 of figure 2.8 (division by Max) to the utilities, we obtain the results below for the weighted sum.

		C_1	C_2	R
Alternatives	a_1	1	0.375	$0.4*1 + 0.6*0.375 = 0.524$
	a_2	0.111	1	$0.4*0.111 + 0.6*1 = 0.644$
Weights		0.4	0.6	

If on the other hand we use procedure #3 (division by the sum), we obtain the following figures.

		C_1	C_2	R
Alternatives	a_1	0.9	0.273	$0.4*0.9 + 0.6*0.273 = 0.524$
	a_2	0.1	0.727	$0.4*0.1 + 0.6*0.727 = 0.476$
Weights		0.4	0.6	

In the first case, alternative a_2 comes out on top, whereas in the second it is alternative a_1.

Using the terminology of section 2.4, we say that *the result of aggregation by the weighted sum method depends on the utility function chosen for each criterion, or else that the method is not ordinal*, since the ranking of the alternatives has not changed whereas the final result has (definition 2.44). This is a very preoccupying matter when it comes to choosing units, as we have already seen in the simple example at the beginning of section 2.7. Similarly, the result of example 4.1 are influenced by our opting to maximize $1/a_{i3}$ rather than $-\text{Max}(-a_{i3})$ for criterion #3. The reader can verify that if we had maximized $k - a_{i3}$, then normalized by procedure #1, we would have obtained slightly different results (Donald and Helen would have changed places).

Why then is a method which is so demanding in theoretical assumptions, and so prone to influence by arbitrary choices at the time of application, so widely (and sometimes badly) used? Our opinion is that this is because, in addition to being easy to calculate, it is immediately comprehensible to any decision maker, even the least mathematical. We shall see in chapter 11 that simple methods secure the confidence of the decision maker.

The weighted sum has been abundantly used in numerous contexts. Thus the so-called Fishbein-Rosenberg variant is used in market surveys; after normalization, a simple weighting of the qualitative criteria is applied and the value obtained is divided by the price of the product to give a sort of quality/price ratio which is used to rank the products (see Despontin *et al.*, 1986, for an interesting application to consumer associations). Furthermore, the weighting method often lies behind numerous other much more complex aggregation methods such as the permutation method (see section 8.4) or Saaty's hierarchy analysis method (section 4.6). We shall also be referring to various multicriterion decision software packages which use linear weighting as the sole method of aggregation. Chapter 10 is devoted to software.

4.1.4 the weighted product

One way of partially avoiding the influence of the normalization procedure on the final result is to use the weighted product method, whose principle is close to that of the weighted sum. Values are multiplied instead of added, and each alternative being evaluated as follows:

$$P(a_i) = \left(a_{i1}^{w_1} \right) \times \left(a_{i2}^{w_2} \right) \times \ldots \times \left(a_{in}^{w_n} \right) = \pi_j \left(a_{ij}^{w_j} \right) \tag{2}$$

We shall show that aggregation by weighted product is much less prone to influence from normalization procedures than that by weighted sum.

We shall say that a normalization procedure is *'cardinal quotient', or for simplicity we shall refer to linear normalization* if, with the notation of section 2.7 with v_{ij} the normalized value of a_{ij}, v_{ij}/a_{ij} does not depend on i, *i.e.* that, for each fixed criterion j, this ratio is constant. This is the case for procedures #1 (division by $\text{Max}_i(a_{ij})$), #3 (division by $\sum_i (a_{ij})$) and #4 (division by $(\sum_i (a_{ij})^2)^{1/2}$).

Proposition 4.4

The aggregation obtained by weighted product is linear normalization procedure invariant.

Proof

Let $v_{ij}/a_{ij} = 1/k_j$. Applying formula (2) we have:

$$P(a_i) = \pi_j \left(v_{ij}^{w_j} \right) = \pi_j \left(\left(a_{ij}/k_j \right)^{w_j} \right) = \pi_j \left(a_{ij}^{w_j} \right) \times \left(1/\pi_j \left(k_j^{w_j} \right) \right)$$

Since $\left(1/\pi_j\left(k_j^{w_j}\right)\right)$ does not depend on i, it will not affect the relative ranking of $P(a_i)$, which will as before only depend on $\pi_j\left(a_{ij}^{w_j}\right)$, proving that the final ranking is not affected by normalization. Q.E.D.

The above proposition cannot only be applied to the weighted sum described in the previous sections: the order of the sums $R(a_i) = \sum_j w_j v_{ij} = \sum_j w_j (a_{ij}/k_j)$ may be modified since the values of k_j cannot be removed from the sum. If we concern ourselves with substitution rates, we have $\dfrac{\partial P}{\partial a_{ik}} = P\,w_k / a_{ik}$ and hence

$$\frac{w_k}{w_j} = \frac{\partial P}{\partial a_{ik}} \cdot \frac{1}{a_{ij}} \bigg/ \frac{\partial P}{\partial a_{ij}} \cdot \frac{1}{a_{ik}} \quad \text{or} \quad \frac{w_k}{a_{ik}} \frac{\partial P}{\partial a_{ij}} = \frac{w_j}{a_{ij}} \frac{\partial P}{\partial a_{ik}}$$ which means that if a_{ik} decreases by

δ_k, a_{ij} must increase by $\dfrac{a_{ij}}{a_{ik}} \cdot \dfrac{w_k}{w_j} \cdot \delta_k$. In other words, the substitution rate is equal, not to w_k/w_j, but to $(w_k/a_{ik})/(w_j/a_{ij})$; it is dependent on the level of utility. This argument, which psychologically makes sense, is advanced by Lootsma (1994) to defend the method of multiplicative aggregation in REMBRANDT (*cf.* section 4.6 below).

A practical consequence of proposition 4.4 is that, when the weighted product is used, each criterion can be expressed in any desired unit (provided any change of unit is linear), without any need to normalize back to [0,1].

On the other hand, a drawback of the method (as in the case described by Schärlig and Pasche (1980) in a problem of choice of site for a production unit), is that it considerably favors/disfavors the final evaluation of any alternative which, relative to one criterion, is far from the average; see also Schuijt (1994). In other words, the weighted product over-values extremes, leading to undesirable results. This is all because a weighted product is in fact nothing more than a weighted sum when logarithms have been taken; we have:

$$\log(P(a_i)) = \sum_j w_j \log(a_{ij}).$$

The neutrality of the linear normalization procedures in the sum above is paid for by the logarithmic aspect, which is not linear and hence modifies the evaluation intervals.

Finally, note that, as for the sum, going from w_j to αw_j ($\alpha > 0$) involves going from $P(a_i)$ to $[P(a_i)]^\alpha$ which does not change the order provided P is positive.

4.2 Geometrical interpretation

This section is not strictly necessary, but for any reader who has studied vector spaces it will help to clarify the logic behind simple (linear) weighting and a certain number of the associated problems.

Using the notation of section 3.3, for an alternative x in $\mathcal{A}$ we let $U(x) = (U_1(x),$ $...,U_j(x), ..., U_n(x)) \in U(\mathcal{A}) \subset |R^n$. If we assume that, for any alternative x, the decision maker utility is given by the sum $V(x) = \sum_j w_j\, U_j(x)$, then $V(x)$ is actually equal to the scalar product $V(x) = w.U(x)$. The linearity appears more clearly in $|R^n$ if we let $U_j(x) = X_j$, and $X = (X_1, ...X_j, ...X_n)$, since we then obtain $V(x) = \sum_j w_j\, X_j = wX$.

The preferred alternative of the decision maker will be one of the solutions to the problem:

maximize $V(x)$ over the set of alternatives $x \in \mathcal{A}$.   (3)

In other words, in $|R^n$, any solution a of (3) is such that $U(a)$ maximizes the linear form $w.X$. Figure 4.3 represents the corresponding situation in two dimensions. The tangent to $U(A)$ has the equation $w.X = w.U(a)$ and the vector w is perpendicular to this tangent.

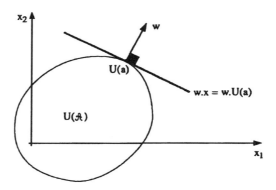

Figure 4.3 Search for the optimum alternative through the weighted sum

Remember (with the notation of 3.5) that we let $w \gg 0$ if, for all coordinates j we have $w_j > 0$. In figure 4.3 we can see that the point $U(a)$ is an efficient point of $U(\mathcal{A})$. This property is general, as the following proposition shows.

Proposition 4.5

If there exists $w \gg 0$ such that a maximizes problem (3), then a is an efficient point.

Proof

Suppose that a is not an efficient point; there will then exist b such that $b \succ a$. This requires (see proposition 3.4) that for all j we have $U_j(b) \geq U_j(a)$ and there exists j_0 such that

$U_{j0}(b) > U_{j0}(a)$. Multiplying by w and summing we obtain $\sum_j w_j\, U_j(b) > \sum_j w_j\, U_j(a)$ which contradicts the fact that *a* maximizes problem (3). Q.E.D.

The inverse proposition is false for two fundamental reasons; the first is easy to understand through the following example.

Example 4.6

The following decision matrix has been normalized and we have $\mathcal{A} = \{A, B, C\}$.

		Criteria	
		Max C_1	Max C_2
	A	1	0
Alternatives	B	0.4	0.4
	C	0	1

From proposition 3.22, the three alternatives are efficient points in $\mathbb{R}^2$. The situation is represented in figure 4.4 i).

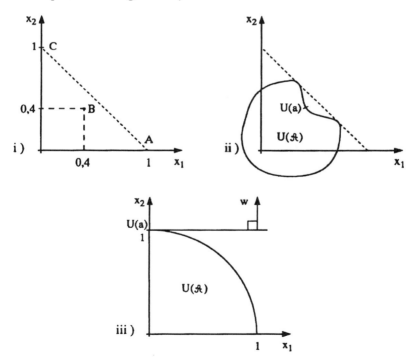

Figure 4.4 Positions of various efficient points in $\mathbb{R}^2$.

A rational decision maker may very well decide that he strictly prefers B to A and C and that A and C are mutually indifferent. It is obvious in figure 4.4 however that B cannot maximize a linear form on $\mathcal{A}$ (shown here by a straight line). We can show this by calculation. Let w_1 and w_2 the weights; we have $V(A) = w_1$, $V(B) = 0.4(w_1 + w_2)$ and $V(C) = w_2$. Can we have $0.4(w_1 + w_2) \geq w_1$ and $0.4(w_1 + w_2) \geq w_2$? This requires $0.4w_2 \geq 0.6w_1$ and $0.4w_1 \geq 0.6w_2$, *i.e.* $2/3w_2 \geq 3/2w_2$ which is absurd for positive weights. Alternative B cannot therefore be the maximum of a weighted sum.

The impossibility that we have revealed clearly has some connection with convexity (definition 3.20), and this can be seen in figure 4.4ii) where the efficient point $U(a)$ will never be reached by a linear form (here, a straight line) to be maximized since it lies in a hollow of the non-convex set $U(\mathcal{A})$. We shall be meeting this property again in chapter 8, example 8.2 and proposition 8.3.

The inverse of proposition 4.5 is false for a second reason; this is that, even for convex sets, we do not always have $w \gg 0$. Thus if $U(\mathcal{A})$ is a disc (figure 4.4iii), the point $U(a)$ is an efficient point and it maximizes $V(x) = 0 * U_1(x) + 1 * U_2(x)$, *i.e.* $w_1 = 0$ and $w_2 = 1$. This difficulty of null coordinates for w disappears in our case when $\mathcal{A}$ is finite; we can therefore state an inverse to proposition 4.5

Definition 4.7

Let $A = \{a_1, ..., a_i, ..., a_m\}$ be a finite set; A^c is the convex *envelope* of A, and is the set of all x such that $x = \sum_{i \in I} \lambda_i a_i$ where I is any part of $\{1, 2 ..., m\}$ and where, for all i we have $\lambda_i \in [0,1]$ and $\sum_{i \in I} \lambda_i a_i = 1$.

It can easily be verified that the convex envelope of A is the smallest convex figure which contains A and that it is a (convex) polyhedron. We show this situation in figure 4.5.

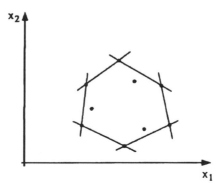

Figure 4.5 The convex envelope of the finite set in $\mathbb{R}^2$

We can now state without proof the proposition which contains both proposition 4.5 and its inverse, The proof involves non-standard separation properties in vector space (see Rockafellar, 1970, and Jaffray and Pomerol, 1989).

Proposition 4.8

Let A^c be the convex envelope of the finite set A in $|R^n$. A point $a = (a_1, ..., a_j, ..., a_n)$ in A^c is efficient, for the canonical preorder of $|R^n$, if and only if there exists $w \gg 0$ such that a maximizes $w.x$ on the set of $x \in A^c$.

The important thing to bear in mind from all this is that, although we might assume that the weighted sum should give the decision maker's preorder, in actual fact we eliminate certain efficient points (example 4.6). The only points in $U(\mathcal{A})$ that can be obtained with the weighted sum are those belonging to the Pareto boundary of the convex envelope of $U(\mathcal{A})$. Thus in example 4.6, B does not belong to the Pareto boundary (segment $[C,A]$) of the convex envelope of $\mathcal{A}$ (triangle ABC).

4.3 Determining weights

It is clear, without the need for further examples, that the values attributed to the weights can have a determining influence on the result of aggregation in the two weighting methods examined above. Indeed, this is what happens in most aggregation methods that we shall be looking at, with rare exceptions such as lexicographical methods where only the order of weights is relevant (chapter 5). This importance of the values of the weights is a natural result of their role in multicriterion decision of indicating the importance the decision maker attributes to each of the criteria. It is therefore of the utmost importance to try to evaluate the weights so that they reflect the decision maker's preferences as accurately as possible.

In the field of economics, publications have given wide treatment to the modeling of consumer or decision maker preferences. In multicriterion analysis, on the other hand, estimation of the relative importance of criteria is a subject that has much more recently been studied. Before actually describing the methods of evaluation which have been proposed, we shall introduce a few general ideas on the subject.

First of all we need to know what is the exact nature of the weights w_j: cardinal or ordinal.

Definition 4.9

The weights are said to be *ordinal* if only their ranking counts (the largest, the second largest etc.). They are said to be *cardinal* if their exact numerical value w_j plays a role. Some authors use the expression *'coefficient of importance'* to denote ordinal weights either in order to distinguish them from cardinal weights or to indicate that they are not being considered as substitution rates (definition 4.2).

According to whether they are ordinal or cardinal, weights will not play the same role, particularly in the matter of tradeoff. In this domain we can be rigorous, like Jacquet-Lagrèze (1983) or Bouyssou (1986) and state that an aggregation method is compensatory when the increase in value of one alternative, relative to one criterion, is able to compensate the decrease relative to another criterion. We shall be expressing this idea in the next chapter in definition 5.21. However, we can intuitively imagine that there are differences in degree of compensation and that the criteria are more or less intimately 'mixed' during aggregation. The strongest compensation is always associated fairly closely with cardinality.

As we have mentioned, there are some rare methods where it matters little whether the weights are ordinal or cardinal. This is the case for the lexicographical method, this type of method being in general non-compensatory (proposition 5.48). In other methods such as ELECTRE the weights must represent a cardinal scale of intervals since they will be used, as in the case of the weighted sum, to add the importance of the criteria.

In totally and indisputable compensatory methods, such as the weighted sum, the weights w_j must be quotient cardinal as they will be used as substitution rates between criteria (definition 4.2). In this type of method there is the problem of the link between the weights and the units used to measure a_{ij} values. In the simple example in section 2.7 we saw that when we do not apply the prior normalization B) of section 4.1.2, the values of the weights are necessarily linked to the chosen units. This problem of the links between weights and units of measurement has been the subject of vigorous debate in the context of certain methods of weight evaluation such as Saaty's AHP method (see section 4.6). The question deserves to be examined carefully since *most decision makers are wrongly and ingenuously inclined to assign weights without reference to the units chosen for fixing the values of a_{ij}*, as though they could be independent (see Roy, 1987b, 1990; Roy and Figueira, 1998).

Various methods of evaluating weights have been proposed and there are even some comparative studies. It was thought for some time (Eckenrode, 1965 and Nutt, 1980) that the actual method used was not unduly important, and that provided one remained within the classical range of methods the actual choice was merely a matter of expediency or personal taste. Later studies, however, (Schoemaker and Waid, 1982) have clearly proved that the method of evaluating weights has an effect on the final outcome. We shall come back to this in section 4.8 and chapter 11, but first, in the next sections, we shall examine the various methods of weight evaluation. We shall start with the entropy method, whose main advantage lies in its 'objectivity' relative to the decision maker, with the data of the problem (a_{ij}) determining the relative importance of the criteria.

The so-called direct evaluation methods, from simple ranking to the more recent and more complicated methods of successive comparisons due to Churchman and Ackoff (1954), will be examined in section 4.5. Indirect methods will be studied in the next two sections, the first of which (4.6) is on a type of methods which we shall call 'eigenvalues'; among these is the very well-known hierarchical analysis of Saaty, based on pairwise comparisons of criteria. We shall then (4.7) describe methods of approximation relying on pairwise comparison of alternatives. Finally, in the last section we shall comment on some general aspects of the methods we .

have described and we shall invoke the difficulties and open problems related to the weight evaluation problem.

4.4 The entropy method

In 1982, a well-known worker in multicriterion analysis proposed an 'objective' method of determining weights (Zeleny, 1982 chapter 7). In this method the values of the weights are determined, without the direct involvement of the decision maker, in terms of the values a_{ij} in the decision matrix. Although this may appear to be in complete contradiction to our previous statement that weights should represent the relative importance the decision maker attaches to the criteria, we shall see that there are good reasons for giving some consideration to this method.

The essential idea is that the importance relative to a criterion j, measured by the weight w_j, is a direct function of the information conveyed by the criterion relative to the whole set of alternatives. In concrete terms, this means that the greater the dispersion in the evaluations of the alternatives a_j for j, the more important the criterion j. In other words, the most important criteria are those which have the greatest discriminating power between alternatives.

The question then becomes one of finding a suitable measure of dispersion. Shannon's theory of information (1949) provides a concept which is right in line with our requirements: the entropy of an information channel. The entropy method works as follows:

a) We start off with the original evaluations a_{ij} for each criterion j, and these are then normalized by procedure #3 (division by $\sum_i a_{ij}$).

b) We calculate the entropy E_j of each criterion:
$$E_j = -k \sum_i a_{ij} \log(a_{ij})$$

Where k is a constant which we adjust so that for all j we have $0 \le E_j \le 1$; $k = 1/\log(m)$, for example, would be quite suitable.

c) The closer together the values of a_{ij} the higher the entropy E_j of a criterion. Now, this is exactly the opposite of what we require to measure discriminating power, and we therefore take the opposite, which we shall call 'measure of dispersion'
$$D_j = 1 - E_j$$

d) Finally we normalize the sum of the weights and we write:
$$w_j = D_j / \sum_j D_j$$

Let us now see how all this works in our personnel selection example.

Example 4.10 Personnel selection, evaluation of the weights by the entropy method.

The evaluations a_{ij}, normalized by dividing by the sum, are shown in table 4.6 which simply repeats the data of figure 4.2.

Alternatives	Criteria				
	C_1	C_2	C_3	C_4	C_5
Albert	0.188	0.172	0.168	0.122	0.114
Blanche	0.125	0.069	0.188	0.244	0.205
Donald	0.156	0.241	0.134	0.220	0.136
Emily	0.188	0.034	0.174	0.146	0.159
Georgia	0.188	0.276	0.156	0.171	0.205
Helen	0.156	0.207	0.180	0.098	0.182

Figure 4.6 Table of normalized values of alternatives

From these data we can calculate the entropies E_j, the dispersion D_j and the weights w_j for each criterion. The results are shown in figure 4.7.

Criteria	$E_j = -(1/\ln 6)\Sigma_i \, a_{ij} \ln a_{ij}$	$D_i = 1 - E_i$	$w_j = D_j / \Sigma_j D_j$
C_1	0.995	0.005	0.04
C_2	0.908	0.092	0.66
C_3	0.997	0.003	0.02
C_4	0.973	0.027	0.19
C_5	0.988	0.012	0.09

Figure 4.7 Weights calculated by the entropy method.

As we have already said, there is at least one reason to justify use of this method; first of all, the mechanical calculation of entropy does indisputably exclude any subjectivity on the part of the decision maker in determining the weights. This idea is interesting in cases of conflict where the parties involved are arguing over the values of the weights; the method may there prevail on account of its perfect neutrality. Should we consider that the total absence of subjectivity in evaluating weights is reason enough for the method to be used? Whatever the case, if the above evaluation appears too disembodied, we can always invite the decision maker to multiply the values of weights w_j obtained by the entropy method, by a factor x_j representing the preferences of the decision maker. The final result $y_j = w_j x_j$, once normalized, will be the weights used. In this way 'objectivity' and decision maker preference are wed.

Another 'objective' method of determining weights has been proposed by Diakoulaki *et al.* (1992); this is based on how great the correlation is between the columns of the decision matrix. If r_{jk} is the coefficient of correlation between column j and column k, the weight of criterion j is then defined as:

$w_j = \sigma_j \sum_k (1 - r_{jk})$, where σ_j is the standard deviation of column j.

Thus, the more the information provided by a criterion j differs from that provided by the other criteria, the greater the weight of criterion j, which will have a high variance. It was by combining these weights with a method of distance from the ideal that Diakoulaki *et al.* devised the multicriterion decision method CRITIC.

Finally, another 'objective' way to assess the importance of the criteria consists in endowing the set of criteria of a fuzzy measure and then to measure the importance of each criterion as a member of a coalition by means of the Shapley value (Grabish and Roubens, 1999).

4.5 Direct evaluation methods

A large range of methods can be placed under this heading. The term indicates that the decision maker assigns weight values directly, which would seem to require no further explanation were it not for a number of more or less subtle complications. These are the oldest methods, and this is why they have been so widely studied in various contexts including measurement, psychometrics, decision theory, operational research and preference theory.

4.5.1 Simple ranking

We shall start with the simplest method: the ranking pure and simple of criteria. The only information asked of the decision maker is his order of preference for ranking the criteria. We give the value 1 to the least important criterion, 2 to the next most important and so on up to the n most important which is given the value n (the number of criteria). Ties are allowed, and the tied criteria are given the average of the values they would have obtained if there had been no tie, following the method of Kendall (1970) (see remark 2.21). The values obtained are then normalized.
We can now apply this to out personnel selection example and see what we get.

Example 4.11 Weight evaluation by the method of direct evaluation, in the personnel selection example.

We rank the criteria from the best to the worst, including ties. Suppose the decision maker's choice is as follows:

1st	Professional experience (EXP)
2nd	Higher studies (HST)
3rd/4th	Interview (INT)
3rd/4th	Psychometric tests (PSY)
5th	Age (AGE)

The criteria INT and PSY thus have equal importance. From this we can assign the following values to the criteria:

HST	EXP	AGE	INT	PSY
4	5	1	2.5	2.5

where 2.5 is the average of 3 + 2. Finally, dividing by the sum 15 we obtain the weights:

HST   EXP   AGE   INT   PSY
0.26   0.33   0.07   0.17   0.17

This method has the advantage of simplicity, with the decision maker only having to provide ordinal information and with very few calculations necessary, but there is a serious disadvantage: it precludes the possibility of weights taking all possible values between 0 and 1. This can be realized if we call r_j the values assigned to each criterion j by Kendall's method; we have already observed (remark 2.21) that the sum of the values of r_j is equal to the sum of the first n integers, *i.e.* $n(n + 1)/2$. After normalization, the weights w_j are equal to $w_j = 2r_j / [n(n + 1)]$ and since $1 \leq r_j \leq n$, it follows that: $2/[n(n + 1)] \leq w_j \leq 2/(n+1)$. For $n = 5$, for example, no weight can be greater than 1/3 or smaller than 0.067. For this reason, the method does not appear very realistic.

One evaluation method which does not have the above drawback deserves mention; this is the probabilistic method of evaluation of Rietveld (1984); see also Rietveld and Ouwersloot (1992) for an up to date summary of this method and an application to the siting of nuclear reactors. It uses the same information as the simple ranking, *i.e.* decision maker's ranking of the criteria. The procedure assumes a uniform distribution of the probability of the vector w lying anywhere within the simplex $S = \{w / \ w_j \geq 0, \ \sum_j w_j = 1\}$, with the order of criteria $(w_1 > ... w_j > ... w_n)$ respected and with an upper bound assigned to each w_j, *i.e.* we write $w_{n-j+1} < w_{n-j} < (1 - \sum_{1 \leq k \leq j} w_{n-k+1})/(n - j)$. The values of these upper bounds favor the most important criteria because Rietveld considers that the function which associates the value w_j to the rank of a criterion is concave, which roughly speaking means that the intervals between the values of w_j decrease as the criteria become less and less important. From this distribution Rietveld calculates the expectancies which give the weights and finds $w_{n-j} = (1/n) \sum_{0 \leq k \leq j} (1/(n - k))$. The program DEFINITE (see chapter 10) uses this method of evaluation.

Finally, Vansnick (1986) proposes another method in which the importance of the criteria, *i.e.* their rank, is transformed through a linear program into weights reflecting this preorder.

4.5.2 Simple cardinal evaluation

In the method of simple cardinal evaluation or rating, the decision maker evaluates each criterion according to a pre-defined scale of measurement (*e.g.* from 0 to 5, from 0 to 100 etc.). Thus in our personnel selection example, if we ask the decision maker to evaluate the importance of the criteria from 0 to 5, we may assume he may obtain:

HST	EXP	AGE	INT	PSY
5	5	2	4	4

which is then normalized by dividing by the sum, giving:

HST	EXP	AGE	INT	PSY
0.25	0.25	0.1	0.2	0.2

which are the weights we used for the weighted sum.) A variation, called the *'ratio method'* (von Winterfeldt and Edwards, 1986), consists of asking the decision maker to evaluate the relative importance of the criteria relative to the least important among them. For example, criterion j is 2.5 times as important as criterion k. Respecting the coefficients of proportionality and bringing the sum to one, we obtain the cardinal weights. The combination of discrete evaluation of cardinal utilities (section 2.4.4) and estimation of weights by the ratio method, together with aggregation by simple weighting, are among the simplest possible solutions to the problem of multicriterion choice. It is this combination which is used in Edwards' SMART system ; see Edwards (1977) and von Winterfeldt and Edwards (1986).

These methods of simple cardinal evaluation require more information from the decision maker than ranking, since they involve cardinality (see 4.8). On the other hand, they do not have the drawback of restricting the value of the interval, at least in theory; as we have seen, it is often the decision makers who, failing to take into account the effect of dividing by the sum, limit themselves by a tendency to 'give too high a value to criteria which they consider of little importance, and too low a value to the preferred criteria' (Nutt, 1980; Weber and Borcherding, 1993).

In this type of method, the psychological inertia of the decision maker – an entirely human characteristic – can bias the process dangerously and result in an incoherence which is hard to eliminate, even when ingenious processes, generally graphical, are used to minimize these drawbacks and, above all, to make the information asked of the decision maker more intuitive. For example, Simos (1990) suggested to use playing cards with the name of the criteria. Then, the decision maker ranks the cards in a deck by separating the criterion cards by as many white cards as necessary to indicate the distance between criterion cards. Roy and Figueira (1998) shown that this method, as well as simple ranking (see 4.5.1), reduces the range of the weights. This drawback is overcame in the algorithm of Roy and Figueira (1998). In the latter, the decision maker gives the gap between the criterion cards and the ratio between the worst and the best criteria. With these data, one can manage to use the whole range of values for the weights and remove the restricted range inconvenience of Simos' method. Anyway, measuring the difficulty of questions that decision makers are asked for determining the weights, Larichev and Nikiforov (1987) and Larichev and Moshkovitch (1997) consider that ordinal weight data are considered to be acceptable and familiar whereas cardinal weight data are considered complicated.

It should also be borne in mind that decision maker evaluations can vary considerably for reasons which have nothing to do with the problem, such as the order in which the criteria are presented to him, his pre-suppositions on the use to which his values will be put, any semantic connotations of the symbolic scales used (a lot, not much, average, good etc.), the precise time when he is questioned, etc.

Numerous studies in broad agreement with each other have been carried out in this area and widely published in decision psychology (see, *e.g.*, Poulton, 1977 and Fischoff, 1980). As far as weights are concerned, experiments have confirmed that evaluations are highly variable. Vieli (1984) showed that weights estimations varied when decision makers were questioned at different times. Varying the scale used to measure utilities caused modifications in weights which were not always in agreement with the hypothesis of additivity of preferences; see von Nitzsch and Weber (1993), Weber and Borcherding (1993) and Lootsma (1994) and the references they give. In fact these studies show that decision makers do not at all take into account scale range bias effects. Goldstein (1990) and Mousseau (1992b) showed that the weights indicated by a decision maker depend on the type of alternatives presented to him; weights of criteria changed according to whether the alternatives were chosen from among the best or the worst. This 'framing effect' has been well attested, and particularly the role of the 'reference point' (Tversky and Kahneman, 1991). This is related to the well known phenomenon of risk aversion for gains and risk seeking for losses (Tversky and Kahneman, 1982) and also leads to bias in weight determinations (Schapira, 1981).

4.5.3 The method of successive comparisons

Churchman and Ackoff (1954) were the first to propose this method. It has since become a classic of direct evaluation methods. However, it does require some effort on the part of the decision maker, but with the result that the inconsistencies of the simple cardinal method are eliminated.

We shall describe a variant of the method which we consider better than the original; this was proposed several years later by Knoll and Engelberg (1978). They called it RCAT (the 'Revised Churchmann Ackoff Technique'). Without going into all the details, the method is in six steps:

1) Criteria are ranked.
2) Criteria are evaluated on a cardinal scale.
3) The criteria are systematically compared on the basis of a union of successive criteria, starting with the first. The decision maker has to compare the first with the second, then with the second plus the third, and so on until a change in preference is arrived at; the same thing is then done with the second criterion, then the third and so on.
4) The consistency between all the previously given cardinal values and all the comparisons made by the decision maker is checked.
5) Any values not consistent with the comparisons are modified.
6) The values obtained are normalized.

Steps 1 and 3 involve the decision maker, and the others are carried out automatically by computer. Let us see how this works, once again with our personnel selection example.

Example 4.12. Evaluation of weights by the RCAT method in the example of personnel selection.

In the first step, we ask the decision maker to rank the criteria according to their importance. We assume we have the same ranking as in section 4.5.1.

1st	Professional experience (EXP)
2nd	Higher studies (HST)
3rd/4th	Interview (INT)
3rd/4th	Psychometric tests (PSY)
5th	Age (AGE)

where the criteria INT and PSY are tied.

The second step is to assign values to the criteria, respecting their order; for example, 5 to the first, 4 to the second etc, and sharing the points if there are ties (as in 4.5.1). We obtain:

HST	EXP	AGE	INT	PSY
4	5	1	2.5	2.5

The third step, and the heart of the method, is to ask the decision maker to make comparisons between single criteria and aggregates of criteria. We obtain the following table:

EXP: HST + INT + PSY + AGE   HST: INT + PSY + AGE   INT: PSY + AGE
EXP: HST + INT + PSY   HST: INT + PSY
EXP: HST + INT

Beginning below with the first column on the left, the decision maker has to go up the column until the criterion on the left (EXP) is considered to be less important than the conjunction of those on the right. This is repeated for the other two columns. In our example, suppose the decision maker's responses were as follows:

1) EXP > HST + INT but EXP < HST + INT + PSY
2) HTS < INT + PSY
3) INT = PSY (already known from the initial ranking).

If we now compare these results with the cardinal evaluations, we see that:

3) INT = PSY → 2.5 = 2.5 correct.
2) HST < INT + PSY → 4 < 2.5 + 2.5 correct.
1) EXP > HST + INT → 5 > 4 + 2.5 not correct.

We now have to modify the weights for the preference inequality to be satisfied. The program uses proportionality relationships to find suitable weights (details of how the algorithm works are given in the references quoted); we find:

	HST	EXP	AGE	INT	PSY
	2.58	5	0.65	1.61	1.61

All the inequalities are satisfied with this set of weights. Normalizing the sum to unity we have:

	HST	EXP	AGE	INT	PSY
	0.23	0.44	0.06	0.14	0.14

One of the drawbacks of this method is that it is relatively complicated, but this is largely overcome by the use of a computer. There remains, nevertheless, the fact that the information the decision maker has to give is not straightforward. But most important of all are the theoretical implications: the type of questions addressed to the decision maker, and the use of 'coalitions' of criteria and the associated sums of coefficients implies an assumption that the criteria are independent and additive. This means – to anticipate chapter 6 – that we pre-suppose the existence of an additive utility function on the part of the decision maker. As this assumption is also necessary to legitimize the aggregation by weighted sum, these two methods – RCAT evaluation and weighted sum – go well together. There are a good many examples of the use of this method in actual cases; including Stimson (1969) for a problem of public health investment selection, and Knoll and Engelberg (1978) for the choice of new police radar equipment.

In conclusion, we shall briefly comment on other recent work where we find some interesting generalizations which tend to draw the methods and real-life situations closer together and to liberate it from the theoretical assumptions we have just seen. Mond and Rosinger (1985), for example, describe an interactive method of directly evaluating weights; in this method, comparisons are made between any groups of criteria freely chosen by the decision maker. Solymosi and Dombi (1986) also proposed a method based on the interactive comparison of groups of criteria, which they treat as pairwise comparisons (see next section), subject to certain coherence conditions (semiorder).

4.6 Eigenvalue methods

Under this denomination are gathered a series of weight evaluation procedures based on calculation of the eigenvector of the largest modulus of a matrix of pairwise comparisons of the criteria. Though the method was first published in a previous article by Klee (1971) (the DARE method), the most well-known work is by Saaty, in the late seventies, as AHP (Analytic Hierarchy Process). The methods form an active field of investigation rich in controversy and in variants which are being increasingly used in various applications. We shall mention some of the variants at the end of this section, but first we describe the main points through the AHP method.

As we have explained above, AHP starts off by evaluating a vector of weights $w = (w_1, w_2, w_3, ..., w_n)$ attached to the criteria of any multicriterion decision problem. We begin by comparing each criterion i with each criterion j, in concrete form that we shall describe later. This comparison yields values a_{ij} which we place in a square matrix of dimension n called the pairwise comparison matrix $A = (a_{ij})$. The idea of introducing pairwise comparisons between criteria is based on the tactic 'divide to rule', since it is thought that the decision maker finds it easier to make comparisons than to take stock of the whole set of criteria, as is implicitly necessary in direct evaluation methods. It is in any case well known that the brain is unable to handle more than 7 ± 2 items in the short term memory (Miller, 1956), which throws seriously into doubt the possibility of interactively manipulating weight vectors if the number n of criteria exceeds 7.

In the book published in 1980, aimed at popularizing his method which had previously been the subject of scientific papers (Saaty, 1977), Saaty gives a practical reason for his choice of the following scale among twenty eight others that he tried:

Value of a_{ij}	when criterion i compared with j is:
1	equally important
3	slightly more important
5	strongly more important
7	demonstrably more important
9	absolutely more important

The intermediate values 2, 4, 6 and 8 can also be used if necessary. If criterion i is neither greater than nor equal to j, a_{ji} is first evaluated as previously and we then write $a_{ij} = 1/a_{ji}$.

The practical reasons for choosing this scale of values and this method of proceeding include:
- there is a wide range of possibilities which do not exceed the capacity of the short term memory;
- integral values are used and we pass from one level to another through increase of one or two units;
- equivalence between i and j is characterized by the value 1.

As, reasonably enough, the criterion i is as important as itself, the coefficients a_{ii} of the main diagonal take the value 1. As we have seen, we always have $a_{ij} = 1/a_{ji}$, and so we merely have to evaluate the upper triangular part of the matrix A, *i.e.* the $n(n-1)/2$ coefficients a_{ij} such that $j > i$.

Binary comparison matrices are so-called reciprocal matrices which have very useful properties on which the effectiveness of the AHP method partially depends. Before looking at these properties, we must introduce the concept of consistency.

We shall say that a pairwise comparison matrix is consistent when, for all i and j, $a_{ij} = w_i/w_j$. Roughly speaking, this means that a_{ij} (which reflects the relative importance of i compared to j) is exactly equal to the weight ratio (*i.e.* to the quotient of the absolute importance that we are trying to evaluate). It is highly desirable for the decision maker to possess this property of consistency when he is making his comparisons. The AHP method does not assume that the property is satisfied; it actually supposes rather the opposite, thereby taking real account of flesh and blood decision makers who are all more or less inconsistent!

It would also be desirable that the transitivity of the evaluations be verified, *i.e.* that, for all i, j, k, $a_{ij}*a_{jk} = a_{ik}$. In fact, it can be shown that this property of transitivity is equivalent to the above property of consistency; we shall therefore only refer to the latter.

Suppose we have the property of consistency; in this case let us denote by W the comparison matrix $W = (w_{ij}) = (w_i/w_j)$. It is then easy to verify that:

$$W.w = \begin{pmatrix} w_1/w_1 & w_1/w_2 & . & w_1/w_n \\ w_2/w_1 & w_2/w_2 & . & w_2/w_n \\ \hdotsfor{4} \\ w_n/w_1 & w_n/w_2 & . & w_n/w_n \end{pmatrix} \times \begin{pmatrix} w_1 \\ w_2 \\ ..... \\ w_n \end{pmatrix} = nw$$

This means that n is an eigenvalue of W and that w is an associated eigenvector. It can further be shown that, if W is a positive coefficient matrix, this eigenvalue n is positive and of greatest value (dominant), which itself implies that w is the dominant eigenvector. In what follows, because w is defined to within one multiplying coefficient, we shall normalize w, *i.e.* after division by the sum we shall have $\sum_j w_j = 1$.

When we do not have consistency, provided the values a_{ij} are not too perturbed, for reasons of continuity we have that:

$A w = \delta_{max}$ where δ_{max} is the dominant eigenvalue.

In other words, w still remains the eigenvector associated with the dominant eigenvalue. It can also be shown that $\delta_{max} \geq n$, with equality if and only if there is consistency. This enables us to define a *coefficient of inconsistency CI*:

$$CI = \frac{\delta_{max} - n}{n - 1}$$

which we shall compound with a *coefficient of random inconsistency* (*CRI*) which is obtained by calculating CI for randomly filled reciprocal matrices.

n	CRI
2	0
3	0.58
4	0.90
5	1.12
6	1.24
7	1.32
8	1.41
9	1.45

We can thus determine the inconsistency ratio *IR*: *IR* = *CI/CRI*. When *IR* < 10% this means that the inconsistency is acceptable.

To summarize, the AHP method takes place in three steps:
a) The decision maker is asked to fill out the matrix *A* of pairwise comparisons.
b) The dominant eigenvector *w* and the inconsistency ration *IR* are computed.
c) If *IR* < 10% we accept *w*, otherwise we ask the decision maker to review his comparisons.
Here now is the result of applying the method to the usual personnel selection example.

Example 4.13 Evaluation of weights in the personnel selection example using Saaty's AHP method.

We ask the decision maker to compare each criterion with those which follow, giving the elements a_{ij} in the decision matrix *A*. We use the previously defined scale from 1 to 9, and reciprocals when *j* is preferred to *i*. Suppose the decision maker's evaluations are as follows:

	HST	EXP	AGE	INT	PSY
HST		1/3	7	5	5
EXP			8	5	5
AGE				1/4	1/4
INT					1
PSY					

The coefficient of 7 in the table means that the criterion 'Higher studies' is 'much more important' than 'Age', and so *on* for the other criteria. The complete matrix *A* is:

$$A = \begin{pmatrix} 1 & 1/3 & 7 & 5 & 5 \\ 3 & 1 & 8 & 5 & 5 \\ 1/7 & 1/8 & 1 & 1/4 & 1/4 \\ 1/5 & 1/5 & 4 & 1 & 1 \\ 1/5 & 1/5 & 4 & 1 & 1 \end{pmatrix}$$

whose dominant eigenvector after normalization is:

	HST	EXP	AGE	INT	PSY
w = (	0.30	0.48	0.04	0.09	0.09)

The associated dominant eigenvalue is δ_{max} = 5.296, giving an inconsistency ratio of:

IR = *CI/CRI* = (1/1.12) ×[(5.296 − 5)/(5 − 1)] = 0.066, which is acceptable.

The AHP method enables one of the major difficulties of multicriterion analysis to be tackled: weight evaluation. The term 'Hierarchy' in the name of the method alludes among other things to the possibility of setting up a hierarchy of criteria (see section 4.7). Actually, the AHP method arrives at the criteria weights through a hierarchy so that what is finally produced is a multi-level weighted sum. The aggregation method is thus well defined and promoters of the method sell it as a complete multicriterion decision method; here we have only defined the principal part which consists of assigning weights through the dominant eigenvector of the pairwise comparison matrix. Saaty's complete method is sold in the form of a software package, EXPERT CHOICE, and we shall be discussing it in section 10.4.

Before finishing this section we should add a more general comment. Looked at in detail, the AHP method supports a strong theoretical interpretation based on the theory of graphs, to be found in Harker and Vargas (1987). However it does not satisfy certain theoretical conditions such as the axiom of irrelevant alternatives (*i.e.* that the final ranking can be perturbed by introducing new alternatives; see section 5.2). Echoes of the arguments surrounding the non-respect of this axiom can be found in Belton and Gear (1983), Schoner and Wedley (1989), Dyer (1990) and Holder and Saaty (1990). In fact Saaty's method shares this defect with most of the ordinal methods, since ranking of the criteria is largely ordinal in the AHP method with the absolute values of the coefficients a_{ij} having less influence than their relative values. However, two of the method's merits, which have been widely commented upon by Saaty and his collaborators, are, firstly, that within certain limits the method can detect and handle the inconsistency of human decision makers (Saaty and Vargas, 1984c, Saaty, 1987, Harker and Vargas, 1987), and secondly that it will accept a hierarchy of criteria, unlike those methods which require global comparisons of the alternatives (Vargas, 1990). Unfortunately weighting depends on the way in which the criteria have been grouped (Barzilai and Golany, 1990).

Other methods have also been proposed for obtaining a vector of weight w from a comparison matrix A, including minimizing a quadratic error ($\sum_i\sum_j (a_{ij} - (w_i/w_j)^2)$ (De Graan, 1980), but Saaty claims that the AHP method is the best, in that it is less affected than the others by decision maker incoherence during evaluation of a_{ij} values (Saaty and Vargas, 1984a and 1984b). These results could be challenged by the recent proposal of another metric (Cook and Kress, 1988). It has also been suggested to take the geometrical mean of the lines of the matrix A (Barzilai *et al.*, 1987); the weights thus obtained are then normalized so that their product is equal to 1.

The AHP method has been successfully used in a wide range of applications from corporate planning (strategy planing, choice of projects, choice of investments, choice of equipment, commercial prospecting, auditing etc.) to the resolution of international conflicts. Many references to these applications can be found in Zahedi (1986).

Finally there have been some more recent variants of the AHP method; these include the GEM (Graded Eigenvalue Method) of Takeda *et al.* (1987) and the 'Gradual' method of Harker (1987), which both offer the possibility of not insisting on all $n(n-1)/2$ criterion comparisons, since it is not very realistic to do so when n is very large or in repetitive situations.

On the other hand, the nine-level scale of AHP has been criticized; thus Lootsma (1987, 1992b) used psychometric reasons to propose measuring the decision maker preferences on a scale with divisions in geometrical progression. The a_{ij} values of the AHP, following the scale of geometrical ratio 2 suggested by Lootsma, takes the values −8, −4, −2, 0, 2, 4, 8 and their inverses. In the REMBRANDT software (Lootsma, 1992a), comparisons are made for a given criterion j between the alternatives on this geometrical scale. The score matrix, similar to the AHP A matrix, obtained from pairwise comparisons is transformed into utilities U_j through a logarithmic regression which consists of minimizing $\sum_{i<j} [\ln(a_{ik}) - \ln(w_i) + \ln(w_k)]^2$. For a given alternative a, the utilities $U_j(a)$ are then aggregated according to the formula:

$$U(a) = \pi_{j=1\ldots n} [U_j(a)]^{w_j}$$

where the weights w_j are calculated as in the AHP method. REMBRANDT can thus be assimilated to a multiplicative AHP method.

4.7 Methods of comparison of alternatives

We can attempt to obtain information on the weights from global comparisons between alternatives. The first method is that of *compensation* and is based on the notion of tradeoff rates (definition 4.2). The decision maker is given alternatives that only differ on two criteria, say criteria 1 and 2. Out of these two alternatives $a = (a_1, a_2, x)$ and $b = (b_1, b_2, x)$, the decision maker says which he prefers; say he prefers a and that we have $a_1 > b_1$. We then decrease a_2 until the decision maker becomes indifferent between a and b; the ratio $(a_1 - b_1)/(a_2 - b_2)$ is then equal to w_1/w_2.

Another fairly similar method, called *evaluation by price*, consists of asking the decision maker how much he would be willing to pay to switch from the best to the worst value of a given criterion. The weights are then proportional to the prices he indicates. More sophisticated is the *'swing method'* of von Winterfeldt and Edwards (1986). This consists of considering alternatives for which the evaluations of every criterion except one are at their worst level. The criterion which is not at its worst level is 'swung' to its highest level. Out of these n alternatives the decision maker chooses the one he likes best, and the corresponding 'swung' criterion will receive the highest weight. The operation is then repeated with the decision maker being asked to choose, out of the remaining alternatives, the one which gets him the 'second highest gain in preference', this indicating the second weight, and so on. We thus obtain ordinal weights. If in addition we ask the decision maker to evaluate the size of the gains in preference obtained through each 'swung' criterion, using the highest gain of all (the first choice) as the unit, this will immediately give a scale of cardinal weights. Following similar ideas, if we give the differences in evaluation between the swung alternatives, the MACBETH package can be used to determine the weights. In MACBETH a graphic interface facilitates comparisons between the swung alternatives.

Another method using swung alternatives has been proposed by Larichev and Moshkovich (1991, 1997) for the ZAPROS system (see Larichev *et al.*, 1993 and

Olson, 1996, chapter 9). This is based on comparisons of alternatives in which all the attributes but one are at their best level. The attribute that is not at its best level takes all possible values (the number of values is limited). By directly questioning the decision maker and forcing transitivity a complete preorder called the joint ordinal scale (JOS) is obtained for the set of alternatives for which all the attributes but one are at their best level. For example, on a scale from 1 (bad) to 3 (good), with three criteria we obtain a JOS ranking:

rank	criteria	
1	3 3 3	
2	3 2 3	
3	3 3 2	
4	3 3 1	alternatives
5	3 1 3	
6	2 3 3	
7	1 3 3	

The JOS ranking gives an indication of the weight of an attribute. Thus in our example attribute #1 has a higher weight than attributes #2 or #3 because the alternatives with value 3 (good) for attribute #1 are ranked in the first positions. For any alternative, say $a = (2,3,2)$, the values can be replaced by the ranking obtained by each criterion in the JOS ranking. Thus 2 is in the sixth rank, 3 is in the first rank and the other 2 is in the third rank. In terms of ranking, alternative a then becomes $(6,1,3)$, similarly alternative $b = (3,3,1)$, after passing through the ranks, becomes $(1,1,4)$. Larichev and Moshkovich propose that the criteria then be reordered with the best ranks in front, a thus becoming $(1,3,6)$ and b $(1,1,4)$. A partial preorder is then deduced from this new presentation of the alternatives, b then dominating a. This is actually a kind of Borda aggregation, except that the preorder is partial since, for example, $(1,3,6)$ and $(1,6,3)$ are not comparable.

Alongside the relatively direct methods that we have just seen, there also exist some interesting indirect methods, and in the field of marketing there has appeared a whole range of methods of evaluation or estimation of weights which are closely related to the ideas examined in this chapter and which deserve some consideration. These are the methods whose purpose is to evaluate the importance (the weights w_j) to the consumer of the various characteristics or features (the criteria) of a given commercial product in order to explain indirectly how the consumer chooses between different brands (the alternatives) of a product.

The first work in this field was on the classic system LINMAP of Srinivasan and Shocker (1973a and b) and the improved version of Pekelman and Sen (1974). In both these methods, the above problem is modeled and solved by minimizing the distance to the ideal product, leading eventually to an easily solvable linear program (see 8.5.3). Horsky and Rao (1984) performed a deeper theoretical analysis of this problem and showed that the estimators obtained by the linear program are consistent, *i.e.* that they converge in probability towards the weights w that we require to estimate.

To apply the above model the decision maker must declare his (ordinal) preference between the various pairs of brands. In this case, therefore, we may speak of global or holistic comparison, since the decision maker must have in his head all the attributes while comparing brands. This is in fact how we do choose one brand rather than another. This information can be difficult to give if there is a large number *m* of alternatives.

The reader may get the idea that the procedure we have just described is the very opposite of multicriterion decision, since we begin with choices between the alternatives to obtain the weights, while multicriterion decision involves finding the weights in order to determine the 'best' alternative. Nevertheless, though this may appear rather surprising, the idea forms the basis of some multicriterion decision methods, including the UTA method of Jacquet-Lagrèze and Siskos (1982) and the method proposed by Zionts (1981). In sections 6.5 and 8.2 respectively we shall see that in these methods it is not all the alternatives which are compared (which would presuppose that the multicriterion problem had already been solved!), but only a sub-set of alternatives which serves to 'reveal' the preferences of the decision maker.

4.8 Other problems

In this last section we shall mention various more specialized questions relating to weight evaluation. Some workers are currently investigating these questions, but it should not be assumed that they are in any way secondary or academic; on the contrary, they are vital of we are to improve the quality of decision making in real situations. A bibliographical survey of weighting by Mousseau (1992a) has 140 entries, which gives a good idea of how actively this fundamental aspect of multicriterion aggregation is being investigated.

First we must mention the structuring of criteria and its repercussions on weights. The views expressed in the well-known book of Keeney and Raiffa (chapter 2) on this topic, though presented against a more general background, remain very relevant. Keeney and Raiffa say that while constructing the set of criteria to be included in a given multicriterion decision problem, we should bear in mind that any structure must be in a form which can be used in later analysis. We may note in passing that Keeney and Raiffa suggest methods of construction of the set of relevant criteria: a meeting of experts or a survey of people involved in the decision. A natural way of structuring the criteria is that proposed by Manheim and Hall (1967) though the idea is a very general one: that is to use a top to bottom hierarchy which divides a global criterion into sub-criteria and so on until a level is arrived at where the decision maker considers that he can actually reason.

If we look at the five criteria of our personnel selection example in the light of this hierarchical structure, one possibility is that of figure 4.8. The figures in brackets in figure 4.8 are the weights associated with the criteria. We have given the same weights to the lowest level criteria as in section 4.1, though we could have used any other weight vector among those previously determined. From these weights and by means of an obvious aggregation ('obvious' pre-supposes the existence of a hypothetical underlying utility function, which is by no means trivial: see chapter 6), we obtain the higher level weights of 0.4 and 0.6 for the criteria RESumé and psychological PROfile. At the higher level of the criterion SUItability for the job, the weight is 1. It should be noted that once the weights for the lowest level criteria have been fixed (with the sum equal to 1), the various groupings possible at higher levels will not change the final evaluation of the alternatives. This merely represents a structuring of criteria to enhance the clarity of the model and possibly ease the task of the decision maker.

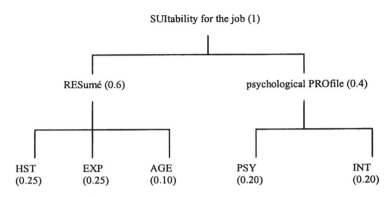

Figure 4.8 The hierarchy of criteria in the personnel selection example

With the assumptions we have made the relations between weights and criteria seem clear. Now we can ask the question: if we are to reflect decision maker preferences as accurately as possible, would it not be better to begin by assigning weights to the second level and then to share them out between the third and lowest level? This would require a definition of 'partial' weights relative to each sub-branch in the hierarchy – here at the third level relative to the second, and a final re-composition of the partial weights in level three and those in level two to finally arrive at the definitive weights. *This descending process is called a hierarchical attribution of weights which we shall distinguish from the non-hierarchical methods which we have been talking about up to here.*

These problems have been experimentally studied in various places (Stillwell *et al.*, 1987, and Borcherding *et al.*, 1991), and the conclusions are all in broad agreement: the hierarchical attribution method is richer than the non-hierarchical method in that it leads to a higher weight variance. Unfortunately, significant differences in the final values of the weights and more or less substantial inconsistencies are observed, according to the method of weight evaluation used and the model adopted for hierarchical organization of the criteria; this is hardly reassuring. Nevertheless the concept of a hierarchy of criteria appears to be quite relevant in reality and is more and more widely used. We shall be coming back to it in chapter 9 in connection with artificial intelligence. Saaty with his AHP method considers the hierarchical model as a central part of his methodology. Many discrete multicriterion decision software packages offer the possibility of hierarchical attribution of the criterion weights.

Lastly, note that when the hierarchical method is used, the degree of detail in the division of the criteria will also affect the weights given by the decision maker; it has been observed that a divided criterion always attracts a greater weight than one that has not been divided (Weber, Eisenführ and von Winterfeldt, 1988). Thus, in our example, if we had divided 'Professional EXPerience' into several criteria such as 'Technical experience' and 'Experience in responsibility', then asked our decision maker to directly attribute the weights of the new criteria, it is very likely that the total resultant weight for 'Professional EXPerience' would have been strictly greater than 0.25.

A second center of interest has been the weakening of the assumption of cardinality of weights, an assumption which has nearly always been underlying up to here. That assumption results from the desire to reduce the amount of information required from the decision maker. We are then led to consider more robust weight structures in which the weights are defined on an ordinal scale (only the ranking order of the criteria counts) or else form a semiorder (see section 2.5). Several methods of multicriterion decision which have appeared in the last few years, such as ORESTE and MELCHIOR (chapter 7) and QUALIFLEX (section 8.4) follow these ideas. It should however be appreciated that if ordinal weights are used, results as robust and conclusive as those from cardinal weights will not be achieved, the latter enabling the working model to be more finely attuned to the actual situation, albeit requiring a greater effort on the evaluation side.

A third area of research is on methods of achieving a good analysis of sensitivity to weights, this is highly desirable given their great influence on the final result. The problem dealt with here is how the stability of the final result depends on variations in the weights. This stability was recently analyzed by Mareschal (1988) for a large number of multicriterion decision methods, those which he calls *r-order* additive methods (which include the weighted sum and the PROMETHEE method among others). What this author does is to try to define theoretically the weight stability intervals, *i.e.* the interval within which a weight can vary (the others weights remaining constant) without the final result changing. Here we also find the work of Bana e Costa (1986, 1988) which contains a sensitivity analysis of weights for the special case of three criteria, with the regions of stability shown graphically. Bana e Costa (1986, 1988, 1991) even turn this around to propose defining an over-ranking relation on the alternatives (see chapter 7) in terms of the surface of the set of weights supporting the ranking $a \succ b$ compared with the surface of weights supporting the opposite ranking. Still in the context of three criteria, the same author also proposes a method for determining weights for decision makers who are in disagreement (the negotiation problem), using an integration over the set of conflicting weights. Virtually all multicriterion decision software which uses weights (see chapter 10) incorporates a weight sensitivity analysis procedure. Some applications, such as Belton and Vickers' VISA (1990), are specifically designed for this.

The fourth and last aspect that we want to point out is the static nature of weights as considered up to here; it is possible for them to vary during the decision process, especially when the latter is viewed in constructivist terms (see section 11.1). The entropy or LINMAP methods do in a way use this idea of the weights depending on the situation out in the field (here, the alternatives). But the question is taken much further than this by Mandic and Mamdani (1984) in their analysis of weights that vary with the level of values reached by the alternatives according to each criterion (the a_{ij} values). A single example among the thousands which could be given is our recruitment problem, where the criterion of age will count a lot more if two candidates have both followed similar courses of study and both had a satisfactory interview, than if one of them has studied to a much lower level than the other. The importance of a criterion thus depends on the alternative being considered. A relatively simple way of taking this very real phenomenon into account is to make use of production rules, as in expert systems; we shall be looking into this in chapter 9. Another approach is to adjust the weights with the aid of an expert system; this idea is to be found in Reyes and Barba-Romero (1986).

5 ORDINAL MULTICRITERION METHODS

5.1 Introduction

We saw in section 2.4 that the utility functions which represent the order are defined unique up to a strictly increasing transformation. *We shall therefore say that a discrete multicriterion decision system is ordinal if the result it provides is invariant by strictly increasing transformation of the utilities.* In other words, the result of such a method only depends on the decision maker's preorders relative to each criterion. No change in utility function or unit, and no normalization can affect the result so long as the preorders do not vary.

Why restrict ourselves to ordinal methods? There is an excellent reason: utility functions are 'fragile' whereas preorder is, in general, more robust. Once a decision maker has fixed his preorder, at least for the asymmetric part (strict preference: see the definition in 2.12), he will not easily change his mind. Things are naturally less certain as regards indifference – though in the discrete case that we are considering, the apologue of the cup of tea (*cf.* section 2.12) would be harder to apply. We saw (section 2.4.4) that the cardinal utilities which, in numerous cases are pre-supposed by the theory, are not easy to construct and that the decision maker is not necessarily consistent (*cf.* also section 4.8). What seems to be proved by experiment (*e.g.* Olson *et al.*, 1995) is that cardinal utilities are very fragile and that *the decision maker changes cardinal utility without realizing it* (*i.e.* when he is perturbed by a change of units or scale, he gives utilities which he thinks represent the same preorder but which are actually not invariant by strictly positive affine transformation). For example on going from a scale of 0 to 10 to one of 0 to 100, there is a change of cardinal utility. Variations are also observed if the same decision maker is questioned at different places and times. We shall only mention in passing the effects of context and cognitive bias which can perturb utilities (see section 2.4.4). These effects have been so widely reported and are so strong that in our opinion they are themselves capable of affecting the global preferences of the decision maker and not merely the cardinal utilities.

All these doubts may legitimately be harbored on the quality of cardinal utilities are grist to the mill of ordinal methods, which is why we are devoting a chapter to the latter. These methods are not entirely free of drawbacks, otherwise they would have inhibited the development of all other methods which use cardinality in one way or another. We shall see their limitations in the course of this chapter. The methods have great historical importance since they were the first to be studied long before any others, and we shall be using them to introduce some new concepts.

5.2 Borda's method

First we shall describe the simplest of the Borda aggregation methods, the one to which any reference to the Borda method generally refers. Then we shall look at several variants more or less related to this.

5.2.1 The classical Borda aggregation procedure.

This method of decision for a jury of several people was first proposed in a note to the French Académie des Sciences by Chevalier Jean-Charles de Borda in 1781 (see section 1.2). De Borda (1733-1799) was a scientist whose main interest was in practical applications; he was a man of many parts, as was the fashion in the 18th century (physicist, mathematician and sailor). Thus he is well-known for various discoveries in the fields of ballistics and fluid mechanics (Académie des Sciences, 1772).

The idea of the method is to add the rankings obtained by a given alternative in relation to each of criterion. For a given criterion, 1 point is given to the alternative which comes first, 2 points to the second, three to the third and so on. The social choice, or aggregate preorder, is obtained by summing all the points obtained for all the criteria for each of the alternatives and ranking first that which has the fewest points, second that which has the next highest number of points and so on.

For the sake of coherence, we shall slightly modify this basic idea so that the alternative coming first will be that with the most points, not the fewest, thereby obtaining utility functions rather than their opposites.

Suppose now we have to analyze m alternatives. We choose m integers $k_1 > k_2 > k_3 \ldots > k_m \geq 0$ that we shall call *Borda coefficients*. For each criterion j, the alternatives are ranked according to a complete preorder, and we call r_{ij} the rank of alternative i for the preorder associated with criterion j. In accordance with corollary 2 of theorem 2.19, the alternatives, relative to criterion j, form a preference chain of the type:

$$a_{i1} \succ a_{i2} \succ a_{i3} \approx a_{i4} \succ \ldots \succ a_{im-1} \succ a_{im}$$

where $\succ$ denotes strict preference and $\approx$ indifference. We shall denote by rk the function associating k_1 to a_{i1}, k_2 to a_{i2} and so on as long as there are only strict preferences. If there are ties (indifferences), each tied alternative is given the average of the coefficients it would have obtained if it had not been in a tie. This is Kendall's rank average method (see remark 2.21). Thus, in our example, we have

$rk(a_{i1}) = k_1$, $rk(a_{i2}) = k_2$, $rk(a_{i3}) = rk(a_{i4}) = (k_3 + k_4)/2$, etc. this definition is a little complicated by the inclusion of ties, but the principle is simple and commonly used.

Example.

We have four alternatives with preferences $a_{i1} \succ a_{i2} \approx a_{i3} \succ a_{i4}$ and the Borda coefficients are $8 > 5 > 3 > 1$; hence $rk(a_{i1}) = 8$, $rk(a_{i2}) = rk(a_{i3}) = (5 + 3)/2 = 4$ and $rk(a_{i4}) = 1$.

Once the function rk has been defined for each criterion j, we write $b(a_i) = \sum (rk_j(a_i))$, each alternative obtaining the sum of the points which it has obtained for each criterion. The aggregate ranking is that defined by the function b, *i.e.* the alternative taken to be the first is the one that has obtained the most points and so on in decreasing order of points. In cases where two alternatives have the same number of points, the alternatives are indifferent for the social order.

The formal expression may be wordy, but the ranking principle of the Borda method is not complicated, and it is widely used, particularly in sports activities; these include the Tour de France and World Motor Racing Championships (with refinements in the latter case). It is also used for some team sports. In the sporting context, each race (in the World Championship) or each stage (in the Tour de France) is a different criterion (complete preorder), and the various rankings in the race or stage are to be aggregated.

Let us now see how the method works with a few examples.

Example 5.1

Consider four candidates A, B, C and D and three criteria C_1, C_2 and C_3 in the following decision matrix:

	C_1	C_2	C_3
A	15	16	03
B	11	13	17
C	08	04	12
D	02	10	09

If we consider the canonical Borda coefficients for four alternatives $4 > 3 > 2 > 1$, we obtain:

	C_1	C_2	C_3	$b(\bullet)$
A	4	4	1	9
B	3	3	4	10
C	2	1	3	6
D	1	2	2	5

The aggregate preorder is therefore: $B \succ A \succ C \succ D$.

Example 5.2 Investment choice (data of section 1.4).

The simplest approach is to take matrix 1.2 and replace each coefficient a_{ij} by the rank obtained by alternative i for criterion j. Once again using the canonical Borda coefficients, $5 > 4 > 3 > 2 > 1$, and using the average ranking method for the ties, we obtain the following coefficients in the Borda matrix.

	pr	gr	ri	ev	pb
P_1	3	1.5	4.5	5	1
P_2	4	1.5	1	2.5	5
P_3	1	3	2.5	1	3.5
P_4	2	4	4.5	4	3.5
P_5	5	5	2.5	2.5	2

Figure 5.1 The Borda matrix

From which we calculate the function b:

	$b(\bullet)$
P_1	15
P_2	14
P_3	11
P_4	18
P_5	17

The aggregate preorder is thus $P_4 \succ P_5 \succ P_1 \succ P_2 \succ P_3$.

Example 5.3 Personnel selection

The reader should perform the same calculation for the personnel selection example (example 3.26). Using the Borda coefficients, $9 > 8 > 7 > 6 > 5 > 4 > 3 > 2 > 1$, the following ranking should be obtained, where the value of $b(\bullet)$ is between brackets.

GEORGIA (33.5) $\succ$ BLANCHE (29.5) $\succ$ CHARLES (27.5) $\succ$ FRANK (25) $\succ$ IRVING (23) $\succ$ EMILY (22.5) $\succ$ HELEN (22.5) $\succ$ DONALD (22) $\succ$ ALBERT (19.5). As we can see, though FRANK is dominated by GEORGIA (section 3.3), he is nevertheless in the top half of the ranking.

Proposition 5.4

The Borda aggregation solves the ranking problem; it generates a complete preorder on the set of alternatives; the Borda aggregation is purely ordinal.

Proof.

Since the Borda ranking is defined on a matrix of ranks, it is clear that it is independent of the utilities, and it is therefore an ordinal method. The function $b(a_i)$ is real and hence induces a total order and is a utility function. Q.E.D.

It might be thought that the Borda ranking has considerable merit, but it does have one major drawback, well-known among sports participants: it allows concert between competitors or teams. We shall now give mathematical backing to this practical observation. For historical reasons we shall call it the 'axiom of independence of irrelevant alternatives', as introduced formally by Arrow (1951); it was, however, already known to Condorcet.

Definition 5.5 The axiom of independence of irrelevant alternatives.

Given a fixed choice set $\mathcal{A}$, n complete preorders (criteria) $\succcurlyeq_j$ and an aggregation procedure giving an aggregate (or social) complete preorder $\succcurlyeq_s$, and a second system of preferences $\succcurlyeq'_j$ giving an aggregate preorder $\succcurlyeq'_s$, the procedure is said to be independent of irrelevant alternatives if:
$\forall\, (a_1, a_2) \in \mathcal{A}^2 \quad \{\forall\, j = 1, 2, ..., n \quad a_1 \succcurlyeq_j a_2 \Longleftrightarrow a_1 \succcurlyeq'_j a_2\}$ implies $\{a_1 \succcurlyeq_s a_2 \Longleftrightarrow a_1 \succcurlyeq'_s a_2\}$.

This axiom is rather esoteric; it means that for any pair of alternatives (a_1, a_2), if a_1 and a_2 are ranked in a certain order for all the criteria (preorders $\succcurlyeq_j$) and if they are ranked in exactly the same order for all the preorders $\succcurlyeq'_j$ then the aggregation procedure must rank the two alternatives in the same order, *i.e.* $a_1 \succcurlyeq_s a_2$ in the first case implies $a_1 \succcurlyeq'_s a_2$ in the second case and vice-versa, which implies $a_1 \succ_s a_2$ if and only if $a_1 \succ'_s a_2$ and $a_1 \approx_s a_2$ if and only if $a_1 \approx'_s a_2$. In other words, the result of the procedure, relative to the two alternatives, only depends on the preorders between the alternatives and not on the alternatives themselves. Note that the exact form that the axiom of independence should take in relation to the independence of irrelevant alternatives is currently a subject of discussion. Several non-equivalent statements of the axiom are possible, and these are set out in Bordes and Tideman (1991).

We can illustrate property 5.5 with an example from the world of sport. Suppose we have five runners A, B, C, D and E competing in three races. A non-sportsman would interpret this as five alternatives which are ranked according to three criteria. The results are:

$$A \succ_1 B \succ_1 C \succ_1 D \succ_1 E$$
$$A \succ_2 B \succ_2 C \succ_2 E \succ_2 D$$
$$C \succ_3 D \succ_3 E \succ_3 A \succ_3 B$$

where $\succ_j$ represents a strict preference. If we give the Borda coefficients $5 > 4 > 3 > 2 > 1$, then Borda ranking gives $A \succ_s C \succ_s B \succ_s D \succ_s E$, since $b(A) = 12$, $b(B) = 9$, $b(C) = 11$, $b(D) = 7$ and $b(E) = 6$.

Suppose now that B is disqualified after a positive dope test, or drops out for team reasons. The new preorders will be:

$$A \succ'_1 C \succ'_1 D \succ'_1 E \succ'_1 B$$
$$A \succ'_2 C \succ'_2 E \succ'_2 D \succ'_2 B$$
$$C \succ'_3 D \succ'_3 E \succ'_3 A \succ'_3 B$$

With the Borda coefficients $5 > 4 > 3 > 2 > 1$, the Borda ranking is now $C \succ'_s A \succ'_s D \succ'_s E$, since $b(A) = 12$, $b(C) = 13$, $b(D) = 9$, $b(E) = 8$. If we consider just the two alternatives A and C, we see that in the first set of three criteria (races) their relative positions are identical to those in the second set, yet on following the Borda procedure, the final result is reversed, with A being strictly preferred first but C being strictly preferred after B dropped out. We have kept to the same notation as in the statement of the axiom of independence of irrelevant alternatives so that the reader can verify for himself that the axiom is not satisfied. The aggregate (or social) ranking between A and C depends not only on the relative order of each criterion, but also on B, *i.e.* on an alternative which, in principle, should not be 'relevant' to the ranking between A and C.

From the point of view of multicriterion decision, this will also mean that introducing or removing alternatives when the Borda procedure is used may change the final ranking. Clearly this is rather embarrassing from an operational point of view.

Note that for m alternatives the canonical Borda coefficients $1 < 2 < 3 < ... < m$ are the most frequently used ones. It is obvious that the Borda ranking will depend on the coefficients chosen. A judicious choice of coefficients can often be used to change the final ranking. For example, in the first set of three races, if we change the coefficients to $10 > 8 > 2 > 1 > 0$, $b(A) = 21$, $b(B) = 16$, $b(C) = 14$, $b(D) = 9$ and $b(E) = 3$, giving the result $B \succ_s C$ in the new ranking compared with $C \succ_s B$ before. In the interests of fair play in sport, it is clearly advisable to fix the coefficients before racing begins, which is actually what is done. The situation is more difficult in multicriterion decision in negotiation problems (see 2.6.4), where these very coefficients could be the subject of discussion among parties who have already anticipated their effect on the final result.

5.2.2 Procedures derived from Borda

5.2.2.1 Borda voting

To introduce the next section, we now show that certain *Borda coefficients* can be given another interpretation. Keeping to our first example 5.1, suppose the criteria C_1, C_2 and C_3 are the preorder of three electors. If an election is fought between A

and B, then A will have two votes (those of the first two electors) and B one vote. If we go on to a vote between A and C, A will have 2 votes and C 1 vote (C_1 and C_2 against C_3 once more). The vote for A versus D gives the same thing. If we add up the votes obtained by A after all these duels we get 6 votes, which is exactly the value of $b(A) - 3$ with the canonical coefficients. This result is also true for B, C and D as the reader will see on enacting all the duels and constructing the square matrix summarizing all the pairwise voting.

	A	B	C	D	$\sum$ *votes for*
A	0	2	2	2	6
B	1	0	3	3	7
C	1	0	0	2	3
D	1	0	1	0	2

Figure 5.2 Pairwise number of votes matrix

Taking the sum of each line, we see that A gets 6 points, B 7 points, C 3 points and D 2 points, which does indeed correspond to the function $b(\bullet) - 3$.

Notation 5.6

Let $V_j(A_i/A_k) = 1$ if $A_i \succ_j A_k$ (victory of A_i over A_k for the criterion, or voting, j); $V_j(A_i/A_k) = 0$ if $A_k \succ_j A_i$ (defeat of A_i) and $V_j(A_i/A_k) = 0.5$ if $A_i \approx_j A_k$. The matrix V defined by $v_{ik} = \sum_j V_j(A_i/A_k)$ if $i \neq k$ and $v_{ii} = 0$ is called the *pairwise number of votes matrix*.

This brings us to a new definition.

Definition 5.7

Borda voting. Given n complete preorders $\succcurlyeq_j$ on m alternatives $A_1, A_2, ..., A_m$, *Borda voting* is the procedure which, for a given alternative A_i, consists of taking the sum of votes (or $\sum_{k \neq i} v_{ik}$) that it obtains in all possible duels of A_i versus A_k. The alternatives are then ranked in order of number of votes.

Proposition 5.8

The number of votes obtained by an alternative in Borda voting is equal to the number of points obtained by the same alternative in the Borda aggregation procedure with the coefficients $m - 1 > m - 2 > ... > 1 > 0$. Consequently, the aggregate preorder produced by the two procedures is identical.

Proof

Mathematically, Borda voting corresponds to $\sum_k v_{ik} = \sum_k \sum_j V_j(A_i/A_k)$. For a given criterion j, suppose that A_i has a ranking r such that there are $m - r - p$ alternatives better than it, p alternatives with the same ranking (ties) and r worse alternatives. With the Kendall ranking method, calculate $rk_j(A_i)$. We have $rk_j(A_i) = [r + (r + 1) + ... + (r + p - 1)]/p = r + (p - 1)/2$. Now calculate $\sum_k V_j(A_i/A_j)$. The alternative A_i obtains one point every time A_i dominates A_k, i.e. r times. For the 0.5 points, there are $p - 1$ scored, since there are p ties including A_i. We therefore obtain $\sum_k V_j(A_i/A_j) = rk_j(A_i) = r + (p - 1)/2$. Since $\sum_k \sum_j V_j(A_i/A_k) = \sum_j \sum_k V_j(A_i/A_k)$, we deduce $\sum_k \sum_j V_j(A_i/A_k) = \sum_j rk_j(A_i)$. Q.E.D.

Remark 5.9

When we used the canonical Borda coefficients $m > m - 1 > ... > 2 > 1$, we let $b(A_i) = \sum_j rk_j(A_i)$ (see 5.2.1). If we use the Borda coefficients $m - 1 > m - 2 > ... > 1 > 0$, we have $\sum_j rk'_j(A_i) = b(A_i) - n$, since we lose one unit for each criterion . As n is constant we deduce that Borda voting also gives the same aggregation as the classical Borda aggregation procedure with the canonical coefficients.

5.2.2.2 Ranking differences

Vansnick (1986), p. 289) has proposed an alternative interpretation of Borda aggregation. We consider the canonical Borda coefficients $m > m - 1 > ... > 2 > 1$, and we consider the function $rk_j(a_i)$ defined in 5.2.1. Ranking differences are obtained by taking, for each pair of alternatives, the ranking difference $rk_j(A_i)$ obtained over the set of criteria by each one. Thus, in our example 5.1, if we perform the calculation for A versus B, we obtain the ranking difference $DR(A/B) = (4 - 3) + (4 - 3) + (1 - 4) = -1$. The completed table of ranking differences is as follows:

	A	B	C	D
A	0	-1	3	4
B		0	4	5
C			0	1
D				0

Figure 5.3 Ranking difference matrix

Since $DR(A_i/A_k) = -DR(A_k/A_i)$, the matrix $dr_{ik} = DR(A_i/A_k)$ is antisymmetric. Note too that $dr_{ik} = \sum_j (rk_j(A_i) - rk_j(A_k))$.

Definition 5.10

In the aggregate preorder according to ranking differences, $A_i \succ A_k \iff dr_{ik} > 0$ and $A_i \approx A_k \iff dr_{ik} = 0$.

Proposition 5.11

The preorders given by ranking differences and the Borda aggregation procedure are identical.

Proof

For the ranking differences, we have $A_i \succ A_k \iff dr_{ik} > 0 \iff \sum_j rk_j(A_i) > \sum_j rk_j(A_k)$, which leads to preference for the Borda method. The same is true for indifference. Q.E.D.

5.2.2.3 'Votes for' less 'votes against'

This method, announced by Luce and Raiffa (1957) is also called as 'The adjusted Borda method' by Black (1958). It consists of considering the sum of the 'votes for' obtained in the duels, less the sum of the 'votes against'. In other words, for a duel A_i/A_k we take the sum of the 'for' criteria less the 'against' criteria (for ties, we do not count anything).

Definition 5.12

'Votes for' less 'votes against' is the name of the method which consists of attributing the real $\sum_{k \neq i} v_{ik} - \sum_{k \neq i} v_{ki}$ to each alternative A_i, i.e. the sum per row less the sum per column of the voting matrix.

With example 5.1 we obtain the matrix:

	A	B	C	D	$\sum$ votes for
A	0	2	2	2	6
B	1	0	3	3	7
C	1	0	0	2	3
D	1	0	1	0	2
$\sum$ votes against	3	2	6	7	

With this matrix, A gets $(6 - 3) = 3$, B gets $(7 - 2) = 5$, C gets $(3 - 6) = -3$ and D gets $(2 - 7) = -5$. We have the classical Borda ranking once more. This property is always true as the following proposition shows.

Proposition 5.13

The 'votes for' less 'votes against' method produces the same aggregation as the Borda aggregation method.

Proof.

Each alternative A_i has assigned to it the real number $\sum_{k \neq i} v_{ik} - \sum_{k \neq i} v_{ki}$. But from 5.6, we have, for $i \neq k$, $v_{ik} + v_{ki} = n$. Therefore $\sum_{k \neq i} v_{ik} - \sum_{k \neq i} v_{ki} = \sum_{k \neq i} (2v_{ik} - n) = 2 \sum_{k \neq i} v_{ik} - (m-1)n$. As $(m-1)n$ is constant, the result of the aggregation will be identical to the Borda vote result (definition 5.7) which produces the same aggregation as the classical Borda aggregation (remark 5.9). Q.E.D.

There exist other, incomparably more complicated methods deriving from the Borda method. These include the work of Cook and Kress (1991). In this method the Borda coefficients attributed to the ranks differ from one criterion to another, and this enables the importance of a criterion to be reflected in a higher coefficient. In the Cook and Kress method, the Borda coefficients are determined, within certain limits, by constraints obtained by solving a linear program. Each alternative is given a score which is the best out of the set of coefficients satisfying these constraints. The alternatives are then ranked according to the scores obtained.

Coming back now to simple methods, all the methods we have just described yield the same aggregate preorder as the Borda method with canonical coefficients. They all obviously suffer from the same drawback: they do not respect the axiom of independence of irrelevant alternatives. On the other hand, they have the advantage of providing a real, total and transitive aggregate preorder which is ordinal and therefore completely independent of the utilities chosen for each criterion. The methods we shall be looking at in the next part of this chapter have the big disadvantage, in certain cases, of yielding transitive social choices.

5.3 The Condorcet method

5.3.1 The Condorcet method

The Marquis Caritat de Condorcet (1743-1794) was a contemporary of Chevalier de Borda. A high thinker, disciple of the physiocrats and self-professed partisan of the Enlightenment and the reforms, he was friend to several encyclopaedists and reformers such as Turgot and Lafayette. At 26 he had already become a member of the Académie des Sciences, to become permanent secretary in 1776. He can be considered as one of the founders, if not *the* founder, of mathematics applied to human science. His book 'Essay on the application of probability analysis to decisions subject to a plurality of voices' is still, two centuries after his death, a work of great interest, and not merely for historical reasons. Then, in the turmoil of the French revolution, as a representative in the Legislative and National Assemblies, he attracted attention through his interest and proposals in the domain of Public Instruction. Classed as a Girondin, too intelligent to be an extremist, he ended up a suspect in the eyes of the radicals, and became a wanted man under the Terror. Arrested, he committed suicide in prison in 1794.

Condorcet was familiar with the work of his fellow-academician Borda and, like him, was concerned with the aggregation of preferences in a voting situation, in his case that of a jury (see 1.2). Since Rousseau, the problem of popular expression within the framework of a social contract had obviously been in the minds of enlightened thinkers.

We can explain the method of Condorcet through example 5.1.

	C_1	C_2	C_3
A	15	16	03
B	11	13	17
C	08	04	12
D	02	10	09

Imagine that Criteria C_1, C_2 and C_3 express the votes among the alternatives. Taking A and B, we see that C_1 and C_2 'vote' for A (through the high marks 15 and 16 compared to 11 and 13), while C_3 votes for B; the result is that A wins by two votes to one, *i.e.* $A \succ_s B$ for the social choice.

What Condorcet proposed doing was to vote over all the possible pairs of candidates. Thus, in the vote A versus B, A gets two votes (criteria 1 and 2) to B's one vote (criterion 2). Similarly, in the vote A versus C, A gets two votes; in B versus C, B wins by three votes to zero. If A and B got the same number of votes, the social relation would be the indifference $A \approx_s B$. The reader will understand why Condorcet's method is sometimes called the *simple majority vote*. We can now introduce the following definition.

Definition 5.14

The *simple majority vote or Condorcet aggregation procedure* is the procedure which, for any pair of alternatives (a_i, a_j) consists of setting $a_i \succ_s a_j$ if and only if the number of criteria for which a_i dominates a_j is strictly greater than the number for which the reverse is true. We deduce that $a_i \approx_s a_j$ when the number of criteria 'for' is equal to the number of criteria 'against'.

Example 5.1 (Sequel)

If we apply the Condorcet procedure to all the pairs in example 5.1, we *obtain* $A \succ_s B$, $A \succ_s C$, $A \succ_s D$, $B \succ_s C$, $B \succ_s D$, $C \succ_s D$. In the social preference relation this can be summarized as:
$$A \succ_s B \succ_s C \succ_s D.$$

In the above form, it is easy to verify that the relation is transitive. In this case, *the Condorcet method leads to a total social preorder. Candidate A is called 'the Condorcet winner'.*

We recall that for this example, Borda Aggregation gave $B \succ_s A \succ_s C \succ_s D$: the two methods do not yield the same ranking.

Remark 5.15

If we use the matrix $V = (v_{ik})$ giving the number of votes (figure 5.2), we can say that the Borda vote is based on the sum $\sum_{k \neq i} v_{ik}$, whereas Condorcet's method is based on the difference $d_{ik} = v_{ik} - v_{ki}$, since:

$$d_{ik} > 0 \Longleftrightarrow A_i \succ_s A_i \text{ and } d_{ik} = 0 \Longleftrightarrow A_i \approx_s A_k.$$

Definition 5.16

A *Condorcet winner* is an alternative which, in the social relation resulting from the Condorcet method, dominates all the others.

Is the Condorcet method the ideal one? Unfortunately not; an unpleasant surprise is in store, one of which Condorcet was quite aware since the following example is taken from his book. Consider once more three candidates, but this time with sixty voters whose preferences are as follows:

there are	10 voters for	$C \succ_1 A \succ_1 B$
there are	8 voters for	$C \succ_2 B \succ_2 A$
there are	23 voters for	$A \succ_3 B \succ_3 C$
there are	17 voters for	$B \succ_4 C \succ_4 A$
there are	2 voters for	$B \succ_5 A \succ_5 C$

After the voting we see that:

$A \succ_s B$ (33 votes out of 60 for A)
$B \succ_s C$ (42 votes out of 60 for B)
$C \succ_s A$ (35 votes out of 60 for C).

So what is the problem raised by this example from Condorcet? Just this: the social relation $\succ_s$ is not transitive. Here $A \succ_s B$ and $B \succ_s C$ which should mean $A \succ_s C$ – but we have the opposite!

Here is a statement of the property which is often called *Condorcet's paradox:* *Even if the voters' preferences (criteria) are complete preorders, for certain configurations of these preorders, Condorcet's method yields a non-transitive social relation.*

Example 5.17

The Condorcet triplet. It is easy to show Condorcet's paradox by a simple example called *the Condorcet triplet.* We just have to consider three voters with the following preferences for three candidates:

$$A \succ_1 B \succ_1 C$$
$$B \succ_2 C \succ_2 A$$
$$C \succ_3 A \succ_3 B$$

The social relation yielded by simple majority voting is:

$A \succ_s B$, $B \succ_s C$ and $C \succ_s A$, which is the above paradox. A, B and C are said to form a circuit, which can be represented as follows, using the conventions of figure 2.6:

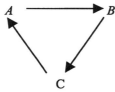

We can now summarize the properties of Condorcet's method.

Proposition 5.18

Condorcet's method is an ordinal method which yields a total social relation satisfying the axiom of independence of irrelevant alternatives, but which is not necessarily transitive.

Proof

Condorcet's method is ordinal as it only involves the preorders associated with the criteria, and not the utilities. The social relation is total since any two alternatives are obviously ranked. It satisfies the axiom of independence of irrelevant alternatives since the ranking of two alternatives only depends on the relative ranking of the two alternatives in question for each of the criteria j, and nothing else. Condorcet's method *does not necessarily yield a transitive social relation*, as is shown by the Condorcet triplet in example 5.17. Q.E.D.

Are we to conclude that Borda's method is better than Condorcet's, at least when we wish to obtain a rational social relation? Condorcet himself did not think so; though he did not explicitly reveal the axiom of independence of irrelevant alternatives, he did point out another disadvantage: a Condorcet winner is not a

Borda winner which, as far as Condorcet was concerned, was sufficient reason to disqualify Borda voting. To illustrate this, take example 5.1. We have $A \succ_s B \succ_s C \succ_s D$ for Condorcet and $B \succ_s A \succ_s C \succ_s D$ for Borda. This effect stems from B being more 'average' than A.

Remark 5.19

To illustrate the above property, Condorcet, in his book, gave the following example. Suppose we have 81 voters with the following preferences:

there are	30 voters for	$A \succ_1 B \succ_1 C$
there is	1 voter for	$A \succ_2 C \succ_2 B$
there are	10 voters for	$C \succ_3 A \succ_3 B$
there are	29 voters for	$B \succ_4 A \succ_4 C$
there are	10 voters for	$C \succ_5 B \succ_5 A$
there is	1 voter for	$C \succ_6 B \succ_6 A$

By simple majority voting we obtain:

$A \succ_s B$ (by 41 votes against 40)
$B \succ_s C$ (by 69 votes against 12)
$A \succ_s C$ (by 60 votes against 21).

This leads to the complete preorder have $A \succ_s B \succ_s C$. Strict democrat that he was, Condorcet had no doubt that A was the best choice. And yet ... if we now apply Borda voting, we get the complete preorder have $B \succ_s A \succ_s C$, B obtaining 109 votes, A 101 and C 33.

We must leave the two great thinkers arguing their cases eyebrow to eyebrow and conclude that no method is perfect!

Before going on, let us see what Condorcet's method does with our examples of investment (section 1.4) and personnel selection (3.26). In these example, a disadvantage of Condorcet's method becomes apparent, and this is the time it takes to work out the binary relation permutations. The reader with enough courage can check he gets the same results as us. We have summarized them according to the chains for the transitive parts.

Investment example.

We obtain: $P_4 \succ_s P_5 \succ_s P_2 \succ_s P_3$ and $P_5 \succ_s P_1 \approx_s P_2 \succ_s P_3$ and $P_4 \approx_s P_1$. We do not have transitivity since, in our example, we have $P_4 \succ_s P_5$, $P_5 \succ_s P_1$ and $P_4 \approx_s P_1$. P_4 is both the Condorcet winner and the Borda winner.

Personnel selection example.

Using candidate's initials, we obtain:
$G \approx_s B \succ_s F \succ_s E \succ_s D \succ_s A$ and $G \succ_s I \succ_s C \approx_s D \succ_s A$ and $G \approx_s B \succ_s F \succ_s H \succ_s I \succ_s A$ and $C \succ_s B \succ_s H \succ_s E$ and $C \approx_s F$ and $E \approx_s I$ and $H \approx_s D$. The combinatorics, as we can see, rapidly get heavy. We do not have transitivity here; for example, $I \succ_s C$, $C \succ_s B$ and $B \succ_s I$. On the other hand, there is a Condorcet winner, namely Georgia.

5.3.2 Compensation between criteria.

In addition to being ordinal, the Condorcet method has the useful property that the criteria (or voters) are independent. Let us try to express this idea formally. For the sake of simplicity, we shall in this section only consider complete preorders $\succcurlyeq_j$ for the criteria.

Definition and notation 5.20

We shall say that a pair of alternatives (a_i, a_k) is equivalent to (a_r, a_s), denoted '(a_i, a_k) has the same criteria as (a_r, a_s)', abbreviated to (a_i, a_k) SC (a_r, a_s), if and only if they have the same criteria 'for' and the same criteria 'against', formally expressed $\{j \mid a_i \succ_j a_k\} = \{j \mid a_r \succ_j a_s\}$ and $\{j \mid a_k \succ_j a_j\} = \{j \mid a_s \succ_j a_r\}$. It can easily be verified that SC is a relation of equivalence and that (a_i, a_k) SC (a_r, a_s) <===> (a_k, a_i) SC (a_s, a_r).

Definition 5.21

An aggregation procedure is said to be *non-compensatory* if and only if it satisfies:
$\forall$ the alternatives a_i, a_k, a_r, a_s (a_i, a_k) SC (a_r, a_s) ==> $[a_i \succ_s a_k$ <===> $a_r \succ_s a_s]$. If this is not the case, it is said to be *compensatory*.

The definition indicates that there can be no compensation for what would be lost on one criterion, by what is won on another. Adapted from Bouyssou (1986), it is one of the simplest possible definitions of compensation. Numerous variations on the notion of compensation are to be found in Fishburn (1976) and Bouyssou (1986). We shall see that the weighted sum, for comparable weights (see proposition 5.52), and the Borda procedure are obviously compensatory.

Example 5.22

Consider the following decision matrix:

	C_1	C_2	C_3	C_4
A	15	15	0	2
B	3	3	15	5
C	12	12	9	15
Weights	1	1	1	1

The weighted sums are: 32 points for A, 26 for B and 48 for C, giving the ranking $C \succ_s A \succ_s B$. However, (C,A) SC (B,A), proving that definition 5.21 is not satisfied; if it were, B would dominate A. The situation is the same for the Borda procedure with the canonical weights $3 > 2 > 1$: A gets 8 points, B gets 7 points and C 9 points, i.e. the same social preorder as with the weighted sum.

We have just seen that Borda procedure is compensatory; the Condorcet procedure, on the other hand, is not whereas for the weighted sum it depends on the weights (see 5.52).

Proposition 5.23

The Condorcet procedure is non-compensatory.

Proof.

Suppose that for any four alternatives we have (a_i, a_k) SC (a_r, a_s); this means that $\{j \mid a_i \succ_j a_k\} = \{j \mid a_r \succ_j a_s\}$ and $\{j \mid a_k \succ_j a_i\} = \{j \mid a_s \succ_j a_r\}$. If $a_i \succ_s a_k$ for Condorcet, this means that the number of criteria such that $\{j \mid a_i \succ_j a_k\}$ is strictly greater than those such that $\{j \mid a_k \succ_j a_i\}$. Through the SC relation this applies also to the pair (a_r, a_s) and therefore $a_r \succ_s a_s$. As the reasoning can also be applied in the opposite sense we have equivalence. Q.E.D.

To summarize, whenever we have *summation over criteria with 'comparable' weights, the aggregation method is compensatory. Furthermore, when the sum is over the rankings, the procedure does not respect the axiom of independence of irrelevant alternatives.*

5.3.3 Matrices of binary relations

Definition and notation 5.24

Consider any *complete* binary relation, $\succeq$, of asymmetric ($\succ$) and symmetric ($\approx$) parts, defined on the finite choice set $\mathcal{A} = \{a_1, a_2, ..., a_m\}$. This relation can be represented by a square matrix $C = (c_{ik})$ where:

$c_{ik} = 1$ if $a_i \succ a_k$, $c_{ik} = 0.5$ if $a_i \approx a_k$ and $c_{ik} = 0$ otherwise.
The matrix C is called the *binary relation matrix*.

The properties of the matrix C of a binary relation $\succeq$ obviously depend on the properties of the relation $\succeq$. For example, $\succeq$ is reflexive if and only if the matrix diagonal is formed of values of 0.5.
For reasons that will become clear, we actually prefer another matrix which is easier to handle.

Definition and notation 5.25

Consider the matrix $C' = (c'_{ik})$ defined by:
$c'_{ik} = 1$ if $a_i \succ a_k$, $c'_{ik} = 0$ if $a_i \approx a_k$ and $c'_{ik} = -1$ otherwise.
The matrix C' is called the *antisymmetric matrix of the binary relation* (see proposition 5.26).

Proposition 5.26

i) The two matrices C and C' are related as follows:
$C' = 2C - U1$ where $U1$ is the matrix formed of ones.
ii) The matrix C' is antisymmetric.
iii) The sum of the coefficients of a row of the antisymmetric matrix C' of the Condorcet relation is equal to the number of victories less the number of defeats of the alternative in the row.

Proof

i) is immediately verified. For ii), if $a_i \succ a_k$ then $c'_{ik} = 1$ and we have neither $a_k \succ a_i$ nor $a_k \approx a_i$, hence $c'_{ki} = -1$. If $a_i \approx a_k$ we have $a_k \approx a_i$ and hence $c'_{ik} \approx c'_{ki} = 0$. To verify iii), consider the sum $\sum_k c'_{ik}$; each victory of a_i against a_k scores a point while each defeat counts -1, and in the case of ties there is no score. Q.E.D.

In example 5.1, the antisymmetric matrix of the Condorcet relation is as follows:

	A	B	C	D	$\sum$ (victories – defeats)
A	0	1	1	1	3
B	−1	0	1	1	1
C	−1	−1	0	1	−1
D	−1	−1	−1	0	−3

5.3.4 Copeland's method

We finish this section with a method derived from Condorcet. According to Luce and Raiffa (1957) it was proposed by the American researcher Copeland (1951).

Definition 5.27

Copeland's preference aggregation method consists of taking the sum of the victories less the defeats in a simple majority vote and ranking the alternatives in terms of this result. In other words, Copeland's method ranks the alternatives according to their score in the sum of rows in the antisymmetric matrix of the Condorcet relation.

In example 5.1, A thus gets 3 points, B 1 point, $C-1$ points and D -3 points. The 'Copeland winner' is therefore the same as the Condorcet winner. The Copeland social preorder is $A \succ_s B \succ_s C \succ_s D$, identical to that of Condorcet; this is not by chance, as we shall now see.

Proposition 5.28

The Copeland method produces a complete social preorder. Suppose furthermore that the Condorcet method generates a complete preorder; Copeland's method will generate the same preorder and this preorder is similar to that obtained by summing the victories of each alternative.

Proof

The relation is complete and transitive since the 'sum of the points', a utility function of the social preorder, is a real function.

If the Condorcet method generates a complete preorder, the alternatives can, following this preorder, be ranked along a path (corollary 1 of theorem 2.19). Let A_i be a given alternative and r_i its rank counting from the end, i.e. we have $q_i = (m - 1) - r_i - p_i$ better alternatives, p_i alternatives of the same rank (ties, not including A_i) and r_i worse alternatives. We note that with our notation, the number of victories of any alternative A_i is equal to r_i, the number of defeats is equal to q_i, while the Copeland number, Cop_i, satisfies $Cop_i = r_i - q_i = 2r_i + p_i - m + 1$.

Consider two alternatives such that $A_i \succ_s A_k$ for Condorcet; then the number of victories satisfies $r_i > r_k$. The number of defeats of A_k, q_k, satisfies $q_k > q_i + p_i$. From these two inequalities we deduce $r_i - q_i - p_i > r_k - q_k$; since $p_i \geq 0$, this implies $Cop_i > Cop_k$ and hence $A_i \succ_s A_k$ for Copeland.

Inversely, consider two alternatives $A_i \succ_s A_k$ for Copeland; then we have $Cop_i > Cop_k$. Suppose that $r_i = r_k$, then A_i and A_k are tied and $q_i = q_k$, which is absurd. Suppose that $r_i < r_k$; the winners of A_i include those of A_k and ties, i.e. $q_i > p_k + q_k$. Combining the two inequalities, we have $r_i > r_k + p_k$, which is absurd since $p_k \geq 0$. There remains the possibility $r_i > r_k$, leading to $A_i \succ A_k$ for Condorcet.

Thus, if the Condorcet method leads to a preorder, the methods of Condorcet and Copeland will both give the same result, and it is sufficient to count the victories r_k. Q.E.D.

Corollary to proposition 5.28

The methods of Copeland and Condorcet can only give different results when Condorcet's method leads to intransitivity.

Remark 5.29

The reader should take care not to confuse the method 'votes for' less 'votes against', which, as its name indicates, counts the votes, and Copeland's method, which counts the victories less the defeats. The former is close to the Borda method and produces the same aggregation, whereas the latter is, as we have just seen, close to the Condorcet method.

Proposition 5.28 suggests that, if Condorcet had thought of Copeland's method, he would have found it superior to that of Borda. However, it suffers from the same drawback as the latter: it is not independent of irrelevant alternatives. We shall show this with an example. Suppose we have the following preferences of four voters for three candidates:

$$A \succ_1 B \succ_1 C$$

$$B \succ_2 C \succ_2 A$$

$$C \succ_3 A \succ_3 B$$

$$C \succ_4 B \succ_4 A$$

Simple majority voting will give $A \approx_s B$, $B \approx_s C$ and $C \succ_s A$; note the non-transitivity of indifference in this case. Borda voting will give $C \succ_s B \succ_s A$; Copeland voting will give the same result as Borda. Suppose now that C is eliminated, the rest remaining unchanged. We obtain:

$$A \succ'_1 B \succ'_1 C$$

$$B \succ'_2 A \succ'_2 C$$

$$A \succ'_3 B \succ'_3 C$$

$$B \succ'_4 A \succ'_4 C$$

Condorcet aggregation still gives $A \approx'_s B \succ'_s C$ which has obviously become a complete preorder, and Copeland gives the same result as Condorcet (proposition 5.28), i.e. $A \approx'_s B \succ'_s C$. Before, we had $B \succ_s A$, now we have $A \approx'_s B$ and the axiom of independence of irrelevant alternatives is therefore violated. Incidentally, Borda also gives $A \approx'_s B \succ'_s C$; for once all three agree!

Finally, it should be said that there are no miracles, and that when the Condorcet paradox appears, the other two methods will either not give the same results, as in example 5.1, or they will avoid the problem by introducing indifference. This is seen clearly in the Condorcet triplet (example 5.17), for there, as the reader can confirm, Borda and Copeland find $A \approx_s B \approx_s C$ while Condorcet gives $A \succ_s B$, $B \succ_s C$ and $C \succ_s A$. Indifference is perhaps an appropriate response here, as we shall see (section 7.3). Note that the introduction of indifference appears more likely than a divergence between Borda and Copeland, Fishburn (1973) having shown that the latter two methods are in agreement in over 80% of randomly generated cases.

Unfortunately, when the Condorcet method is used the probability of ending up with intransitivity of the Condorcet triplet type is by no means negligible. Assuming that each preorder has the same probability $1/m!$ of being chosen by the voters, it can be shown that the probability of ending up with a preference circuit rises rapidly, both with a rising number of alternatives and with a rising number of voters. Only for a small number of alternatives (3 to 5) does this probability remain small: less than 20% (Gehrlein, 1983).

Before going further into the subject, it is important to understand that no ordinal method possess all the qualities, and to show this, we go a little further into social choice in the next section. The reader who is in a hurry can skip the section.

5.4 Social choice and Arrow's theorem

As we said above, this section is not essential to multicriterion decision, though it will enable certain points to be clarified; the results, however, are so fundamental from the point of view of economics, philosophy and politics that we considered we should at least include a brief presentation.

First we have to formalize a few concepts to make the rest of the section simpler to follow. Suppose there are n voters (or n criteria) possessing n complete preorders $\succcurlyeq_j$ on a choice set $\mathcal{A}$. If we call $\mathcal{PR}$ the set of all the possible complete preorders on $\mathcal{A}$, there will be m! of these if $\mathcal{A}$ has m elements and if we only consider strict preorders (*i.e.* no ties). The problem of social choice, or the problem of aggregation of criteria, is mathematically the search for a mapping of $\mathcal{PR}^n = \mathcal{PR} \times \mathcal{PR} \times ... \times \mathcal{PR}$ in $\mathcal{PR}$; the image of the function in $\mathcal{PR}$ defines the aggregate preorder or social preorder $\succcurlyeq_s$. We can ask whether this function exists and what properties it should have. It was Kenneth Arrow who had the idea of posing the problem in these terms in the fifties; in 1972 he was awarded the Nobel prize for economics for the sum of his contributions to economic analysis and in particular to social choice.

The axiom of universality

The domain of definition of the social choice function is the whole of $\mathcal{PR}^n$.

This axiom states that whatever the preorders of the voters, or whatever the rankings adopted by the criteria, the social choice function is able to function; voters' choices are not subject to any restrictions.

The axiom of unanimity or the Pareto axiom.

If for a pair of alternatives (a,b) everybody prefers a to b (*i.e.* whatever $i = 1, 2, ..., n$, $a \succcurlyeq_j b$), then the society prefers a to b ($a \succ_s b$).

This axiom is so natural that it hardly merits any commentary. When a choice is unanimous, *i.e.* when a dominates *b* for the product preorder, then society must ratify the choice in the interests of rational behavior.

The axiom of independence of irrelevant alternatives

We have already looked at this axiom in section 5.2. Remember that it means that the social choice between two alternatives *a* and *b* depends only on what the voters think of *a* and *b*, and not at all on what they think of the other alternatives.

In mentioning above the mathematical search for a social choice function whose image belongs to $\mathcal{PR}$ we were anticipating on the next two axioms.

The axiom of transitivity.

The social relation $\succcurlyeq_s$ of preference is transitive (we also assume, without noting this in any axiom, that it is reflexive).

The axiom of transitivity of social preference requires the same rationality from society as from each member of the society – an assumption which is of course open to discussion (see section 2.2.3 *et seq.*) The axiom allows us to say that the social relation is a preorder; we require this preorder to be complete, which is the fifth and last axiom.

The axiom of completeness.

The social relation is complete, *i.e.* any pair of alternatives (a, b) is either in the relation $a \succcurlyeq_s b$ or $b \succcurlyeq_s a$, or both $(a \approx_s b)$.

We have seen that Borda and Copeland ranking satisfy four of the above axioms but violate the axiom of independence of irrelevant alternatives. Condorcet voting also respects four axioms, but does not always satisfy transitivity. The obvious question, then, is: 'Is there a single procedure which satisfies all five axioms?' The answer is 'Yes'. We choose any voter or criterion j_0 which we shall call the dictator for reasons that will become immediately apparent. For all pairs of alternatives (a, b) let $a \succ_{j_0} b \Longrightarrow a \succ_s b$. In other words, if the agent j_0 strictly prefers *a* to *b*, then society must do the same, hence the name *dictator*. It is easy to complete this procedure of social choice so that it satisfies the five axioms above; it is sufficient, for example, to add $a \approx_{j_0} b \Longrightarrow a \succcurlyeq_s$ as in the lexicographic procedures that we shall be looking at in section 5.6. Finally, as the reader will probably have noted, since each agent has the power of a dictator, there are *n* functions of dictatorial social choice.

The question of existence is solved, but Arrow's theory shows that there is unfortunately not a plethora of solutions.

Theorem 5.30 (Arrow, 1951)

There is no other social choice function other than dictatorial procedures which satisfies the five axioms above.

The proof is both instructive and mathematically original, but space prevents us from including it here. The interested reader will find it in the books by Black (1958), Kelly (1978) and Ekeland (1979).

The result of Arrow's theorem, which was unexpected among partisans of democracy, gave rise to much discussion and work, some of which is still continuing today. We shall begin by taking a look at these discussions and we shall then examine the consequences from the standpoint of multicriterion decision.

People's first shock reaction to Arrow's theorem was to go over the five axioms with a view to weakening them. However, it quickly became apparent that only two could really be questioned, the others being too basically naturally to be attacked. Those two axioms are universality and transitivity. The axiom of independence of irrelevant alternatives cannot seriously be questioned if we wish to end up with a procedure that is safe from manipulation, which does not mean that the methods of Borda and Copeland are without interest, especially if the alternatives are fixed and intangible.

To begin with the axiom of universality: in the example of the Condorcet triplet (5.17), the difficulty stems from the fact that each of the three voters has a preorder which is in strong contradiction with that of the other two, hence the idea that the agents should not be allowed a completely free choice of preorder. If the agents are restricted to certain preorders, then there is a chance that the Condorcet method will be transitive. Some useful results have been obtained in this way, results which prove that with certain restrictions in the domain of the social choice function, simple majority voting is transitive and that the Condorcet procedure therefore satisfies the four axioms other than universality. Among the most interesting results is the notion of unimodal preference introduced by Black (1948a, b and c) and developed by Sen (1970). Basically, preferences are unimodal if there exists a ranking of alternatives along an axis, for example a_1, a_2, a_3, ..., a_m, and if the voter chooses a preferred alternative lying on this axis, for example a_p (the mode), subsequently ranking his preferences to be compatible with the two paths:

$$a_p \succ a_{p-1} \succ a_{p-2} \succ ... a_1 \text{ and } a_p \succ a_{p+1} \succ a_{p+2} \succ ... \succ a_m.$$

This type of preference reflects traditional political preferences quite well: the elector chooses a party on an axis extending from the extreme left to the extreme right, then positions his choice to the right or left according to distance from his preferred party (see figure 5.4).

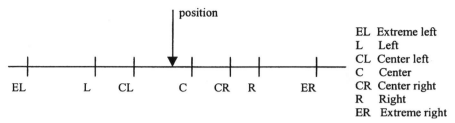

Preferences: $C \succ CL \succ CR \succ L \succ R \succ EL \succ ER$

Figure 5.4 Unimodal preferences

Remember that, if there exists a ranking of alternatives such that the preferences of all the voters are, relative to this ranking, unimodal, then simple majority voting is transitive. Consider three alternatives a, b and c; there exist six possible preorders, omitting indifference for the sake of simplicity:

$a \succ b \succ c$, $a \succ c \succ b$, $b \succ a \succ c$, $b \succ c \succ a$, $c \succ a \succ b$ and $c \succ b \succ a$.

Suppose the alternatives are ranked in the order a, b and c. The unimodal preferences will be: $a \succ b \succ c$, $b \succ a \succ c$, $b \succ c \succ a$ and $c \succ b \succ a$.

We can thus show that in a society, if the voters only have a choice between the four preorders above (instead of six), then the Condorcet procedure is transitive and satisfies the last four axioms. As far as the criteria are concerned, we are saying that they should not be too contradictory. In the example above, $a \succ b \succ c$ and $b \succ c \succ a$ are possible but it is particularly vital to eliminate $c \succ a \succ b$ which would complete the Condorcet triplet and lead to intransitivity. The proofs are lengthy, but the principle is simple. To sum up, *if the criteria are not obviously contradictory, the Condorcet procedure leads to a social preorder and becomes a criterion aggregation procedure satisfying the last four axioms*. In fact, to obtain transitivity, it is sufficient to impose conditions such as forbidding the Condorcet triplet (see Arrow and Raynaud, 1986).

We now have an idea of the main results from the axiom of universality; a more formalized account of the results is given in Villar (1988), and research is still continuing on the subject. Now we examine the axiom of transitivity. Section 2.2.3 is relevant here. Transitivity of preference is probably too strong a requirement, and transitivity of strict preference should suffice. Sen (1970), who in 1998, got as Arrow, the Nobel prize of economics, showed that if the axiom of transitivity is replaced by the axiom of acyclicity of strict preference, there exists a procedure other than dictatorship and which we shall call Sen's rule, which respects all the axioms. First we define acyclicity, then we shall state this rule.

Definition 5.31

The preorder $\succeq$ is *acyclic* if and only if there exists no circuit of any length for strict preference, of the type:
$a_1 \succ a_2 \succ a_3 \succ ... \succ a_q \succ a_1$.

Thus, circuits of the Condorcet triplet type are forbidden.

Definition 5.32

In the Sen procedure, $a \succ_s b$ if and only if a strictly dominates b for the product preorder; in the other cases $a \approx_s b$. In other words, if everybody prefers a to b then $a \succ_s b$, whereas if at least one of the voters strictly prefers b to a, then $a \approx_s b$.

The reader will almost certainly have realized that this procedure is hardly more attractive than dictatorship, since as soon as there is any disagreement between the voters, the social choice ends in indifference! Sen's procedure is therefore important. From the point of view of multicriterion theory, this means that all the Pareto-optimum alternatives are indifferent, which does not advance matters a lot!

Gibbard (1974) went further, showing that any social choice function satisfying the first four axioms and the transitivity of strict preference is necessarily an oligarchic rule, *i.e.* there exists a sub-group of voters, called the oligarchy, whose member agree between themselves on the choice between two alternatives; this choice becomes the social choice, and if the oligarchs do not agree between themselves the social choice becomes that of indifference. Note that a dictatorship is an oligarchy of one person whereas Sen's rule is an oligarchy of the whole society. As Blair and Pollack (1983) remarked, we have the choice between impotence if there are many members of the oligarchy and inequality if there are few members.

Acyclicity is a weaker property than transitivity of strict preference, which is in turn weaker than the semiorder (*cf.* section 2.5). It is therefore not surprising that there is still only dictatorship which satisfies the first four axioms plus semiorder. For an excellent review of all the properties, accompanied by a good bibliography, see Villar (1988).

What are the consequences for multicriterion analysis of the results above? The first thing to note is that there exists no purely ordinal method of aggregating criteria which is perfect. According to Arrow's theorem we have to choose the properties we want to favor from among the alternatives. When the criteria are not too contradictory, Condorcet's simple majority voting is unlikely to have too much untransitivity and hence is likely to be a good solution, but it has the drawback of being tedious to use. We shall see in the next section how to go from near-transitivity to complete transitivity. When there is only a small risk of manipulation related to non-independence of irrelevant alternatives, the Borda method has the immense advantage of being simple to use and giving acceptable results.

In multicriterion decision, we often have to resort to the logic of the least bad. Given that no method is likely to be perfect with regard to Arrow's theorems, we might ask if 'on average', some methods of aggregation are better than others. Here we can cite the work of Pérez and one of the authors of this book (Pérez, 1991 and 1994, Pérez and Barba-Romero, 1995) who defined the formal properties of consistency that can reasonably be expected to be fulfilled, always or in most cases, by aggregation methods. These authors define *'degree of inconsistency'*, for example, as being the probability that an irrelevant alternative will affect the ranking of another pair, and, by using either formal tools or statistical simulations, they manage to make significant comparisons between various aggregation procedures. Thus they have shown that Blin's method (see 8.3.3) is less good than that of Borda, and is more complicated to boot (see also chapter 12).

This section emphasizes once again that the choice between efficient alternatives is a matter of taste, each one defensible according to the importance attached to one or other of the criteria rather than being anything to do with the rationality of the model. As far as multicriterion analysis is concerned, this means that the discipline is basically there to aid decisions, not to make them since behind every multicriterion decision and beyond the logic of the model there is a decider or a group of deciders who will either decide on a method for aggregating the criteria or who will make the decision themselves from among a small number of admissible alternatives between which the method would have been incapable of discriminating. To end this section on a more positive note, multicriterion models are generally richer than voting models, if only through the introduction of the notion of the importance of the criteria (the weights), which is obviously impossible in the voting method since each voter has, in principle, the same importance.

5.5 The method of Bowman and Colantoni

We include this method here for two reasons: it is an ordinal method and, more importantly, it throws a new light on the binary relations and the methods of the Condorcet school. The method was first published by Bowman and Colantoni (1973) and widely popularized in France towards the end of the seventies by the work of two IBM engineers, F. Marcotorchino and P. Michaud (1979). In fact, the idea of transforming the matrix of any binary relation into a 'near' matrix, in a sense to be defined, representing a transitive relation goes back to Slater (1961) and Kemeny and Snell (1962).

In section 5.3 we introduced the notion of a matrix (non-antisymmetric) associated with a binary relation (definition 5.24). We keep the same notation here and call the associated antisymmetric matrix CD: $CD = 2C - U1$ (proposition 5.26).

Consider now n criteria, hence n complete preorders, $\succcurlyeq_j$ defined on $\mathcal{A}$. For each of the criteria j we consider the associated antisymmetric matrix CD_j. Let $CD = \sum_j CD_j$. The coefficient cd_{ik} of this matrix sums the number of criteria j such that $a_i \succ_j a_k$ (a_i is preferred to a_k) less the number of criteria such that not($a_i \succcurlyeq_j a_k$) or $a_k \succ_j a_i$ (proposition 2.15). In other words (remark 5.15), we have: $cd_{ik} > 0$ if and only if in Condorcet's method $a_i \succ_s a_k$ and $cd_{ik} = 0$ if and only if in Condorcet's method $a_i \approx_s a_k$ (indifference for the social relation).

The search for an aggregate relation is interpreted as the search for a relation whose matrix $m \times m$ is unknown. Call $X = (x_{ik})$ the ordinary matrix (non-antisymmetric, definition 5.24) of this unknown binary relation $\succcurlyeq$ with $x_{ik} = 1$ if $a_i \succ a_k$ ($a_i \succcurlyeq a_k$ and not($a_k \succcurlyeq a_i$)), $x_{ik} = 0.5$ if $a_i \approx a_k$ ($a_i \succcurlyeq a_k$ and $a_k \succcurlyeq a_i$) and $x_{ik} = 0$ otherwise.

Proposition 5.33

The binary relation $\succcurlyeq$, whose matrix is X, is complete if and only if:
$$\forall i \text{ and } k \qquad x_{ik} + x_{ki} \geq 0.5.$$

Proof

Our relation is equivalent to $x_{ik} \neq 0$ or $x_{ki} \neq 0$. Q.E.D.

Proposition 5.34

The complete binary relation $\succcurlyeq$ whose matrix is X satisfies:
i) $\approx$ is reflexive if and only if $\forall k \quad x_{kk} = 0.5$;
ii) $\approx$ is symmetric if and only if $\forall i \text{ and } k \quad x_{ik} + x_{ki} \geq 1$;
iii) $\succ$ is asymmetric if and only if $\forall i \text{ and } k \quad x_{ik} + x_{ki} \leq 1$.

Proof

The first point is obvious. For point ii) we have $x_{ik} = 1$, 0 or 0.5. Since $\succcurlyeq$ is complete, we have $x_{ik} + x_{ki} \geq 0.5$ (proposition 5.33). If $x_{ik} = 1$ or $x_{ki} = 1$, relation ii) tells us nothing and it is therefore equivalent to $x_{ik} + x_{ki} \geq 1$ for $x_{ik} = 0.5$, which is indeed equivalent to the symmetry of $\approx$.

Given the definitions, relation iii) is equivalent to $x_{ik} = 1 \Longrightarrow x_{ki} = 0$ and $x_{ki} = 1 \Longrightarrow x_{ik} = 0$, which is indeed the definition of the asymmetry of $\succ$. Q.E.D.

Corollary

The binary relation $\succcurlyeq$ satisfies the following conditions: i) $\succcurlyeq$ is complete and ii) $\approx$ is reflexive and symmetric and iii) $\succ$ is asymmetric if and only if its matrix X is such that:
$$\forall i \text{ and } k \qquad x_{ik} + x_{ki} = 1. \tag{1}$$

Proof

This results from the union of the conditions of proposition 5.34. Q.E.D.

In what follows we shall assume that we are only interested in complete relations $\succcurlyeq$ such that $\succ$ is antisymmetric and $\approx$ symmetric and reflexive (weak assumptions on rationality, see section 2.2), in other words the relations $\succcurlyeq$ for which the associated matrix X satisfies (1).

Let $F(X) = \sum_{ik} cd_{ik} x_{ik}$ where cd_{ik} is the previously defined coefficient of the antisymmetric Condorcet matrix.

Proposition 5.35

Among the solutions of the mathematical program:
Maximize $F(X)$ under the constraints $\qquad\qquad$ (2)
$\forall\ i$ and k $\quad x_{ik} + x_{ki} = 1$ and $\quad x_{ik} = 0$ or 0.5 or 1
we have the Condorcet relation.

Proof

We have $cd_{kk} = 0$ and $cd_{ik} = -cd_{ki}$ since CD is antisymmetric (proposition 5.26). Therefore, we obtain $\sum_{ik} cd_{ik} x_{ik} = \sum_{i<k} cd_{ik}(x_{ik} - x_{ki})$. Since $x_{ik} + x_{ki} = 1$, we have $x_{ik} - x_{ki} = 2x_{ik} - 1$. Let $y_{ik} = 2x_{ik} - 1$; y_{ik} only takes the values 1, 0 and -1. The matrix $Y = (y_{ik})$ is the antisymmetric matrix associated with X (proposition 5.26 i). After change of variable, the program in x is therefore equivalent to:

Maximize $\sum_{i<k} cd_{ik} y_{ik}$ under the constraints
$y_{ik} + y_{ki} = 0$ and $y_{ik} = 1$ or 0 or -1.

The maximum of the latter program is obviously reached by those solutions which satisfy $y_{ik} = 1$ when $cd_{ik} > 0$, $y_{ik} = -1$ when $cd_{ik} < 0$ and y_{ik} indeterminate when $cd_{ik} = 0$. Going back to x, we see that this implies $a_i \succ a_k$ when $cd_{ik} > 0$, $a_k \succ a_i$ when $cd_{ik} < 0$ and anything at all if $cd_{ik} = 0$. When $cd_{ik} = 0$ we can take $y_{ik} = 0$, in other words $a_i \approx a_k$ and we obtain the Condorcet relation. Q.E.D.

Remark 5.36

From the above proof, we see that the non-uniqueness of the solution to program (2) stems from the coefficients $cd_{ik} = 0$. In other words, if there is no indifference in the Condorcet relation, then it is the only solution to program (2).

The idea of Bowman and Colantoni is to search for a social relation 'close' to that of Condorcet but which is constrained to be transitive.

Proposition 5.37

Consider the complete binary relation $\succcurlyeq$ whose matrix is X. The relation $\succ$ will be transitive if and only if:
$\forall\ i$, k and h, different among themselves $\quad x_{ik} + x_{kh} - x_{ih} \le 1$. $\qquad\qquad$ (3)

Proof.

The condition is necessary, since if $a_i \succ a_k$ $(x_{ik} = 1)$ and $a_k \succ a_h$ $(x_{kh} = 1)$, transitivity implies $a_i \succ a_h$, i.e. $x_{ih} = 1$, which satisfies (3). The condition is sufficient since if we have $x_{ik} = 1$ $(a_i \succ a_k)$ and $x_{kh} = 1$ $(a_k \succ a_h)$, relation (3) does lead to $x_{ih} \ge 1$ i.e. $x_{ih} = 1$ and consequently $a_i \succ a_h$. Q.E.D.

Consider the mathematical program :

> Maximize F(X) under the constraints   (4)
> $x_{ik} + x_{kh} - x_{ih} \leq 1$ whatever i, k and h, different among themselves
> $x_{ik} + x_{ki} = 1$ whatever i and k
> $x_{ik} = 0$ or 0.5 or 1.

Proposition 5.38

Solving program (4) produces one or several aggregate binary relations which satisfy the weak assumptions on the rationality of the decider.

Proof

This results from the corollary to proposition 5.34 and from proposition 5.37. Q.E.D.

Definition 5.39

The method which consists of adopting a binary relation which is a solution of (4) as aggregate relation will be called *Bowman-Colantoni aggregation* or B-C for short. All B-C relations satisfy the weak assumptions on rationality.

Proposition 5.40

Program (4) is equivalent to the following program:

> Max $\sum_{i<k} cd_{ik} y_{ik}$ under the constraints   (5)
> $y_{ik} + y_{kh} - y_{ih} \leq 1$ whatever i, k and h, different among themselves
> $y_{ik} + y_{ki} = 0$ whatever i and k
> $y_{ik} = 1$ or 0 or -1.

Proof

With the same change of variable as in 5.35 let $y_{ik} = 2x_{ik} - 1$. We have seen (proposition 5.35) that F(x) = $\sum_{i<k} cd_{ik}(x_{ik} - x_{ki})$ which is therefore equal to $\sum_{i<k} cd_{ik} y_{ik}$. Replacing x_{ik} by its value in terms of y_{ik}, the constraints are shown to be those given. Q.E.D.

The reader will easily appreciate that the above program is not easy to solve, either by hand or machine, because firstly it is a program in integers and secondly it possesses a large number of constraints. The number of inequalities associated with the constraint of transitivity is equal to the number of arrangements possible according to $A^3_m = m(m - 1)(m - 2)$; if $m = 10$ there will be 720 transitivity constraints. Considerable computing power is necessary to solve this program, and it may perhaps be understood why the method has been mainly promoted by IBM!
The reader will by now be familiar enough with ordinal multicriterion methods to suspect that the B-C method will not perform miracles.

We now apply the B-C method to the Condorcet triplet (5.17). The anti-symmetric matrices corresponding to the three preorders P_1, P_2 and P_3 are, in that order:

	A	B	C
A	0	1	1
B	-1	0	1
C	-1	-1	0

	A	B	C
A	0	-1	-1
B	1	0	0
C	1	-1	1

	A	B	C
A	0	1	-1
B	-1	0	-1
C	1	1	0

The *CD* matrix is equal to the sum of all three:

	A	B	C
A	0	1	-1
B	-1	0	1
C	1	-1	0

Program (5) is written:

Max $(y_{12} + y_{23} - y_{13})$ under the constraints
$y_{12} + y_{23} - y_{13} \leq 1, y_{23} + y_{31} - y_{21} \leq 1, y_{31} + y_{12} - y_{32} \leq 1,$
$y_{13} + y_{32} - y_{12} \leq 1, y_{21} + y_{13} - y_{23} \leq 1, y_{32} + y_{21} - y_{31} \leq 1$ and
$y_{ik} + y_{ki} = 0$ for all i and k and $y_{ik} = 1$ or 0 or -1.

Note first of all that if we exclude the transitivity constraints, this program has the unique solution $y_{12} = 1$, $y_{23} = 1$, $y_{13} = -1$ which does indeed correspond to the Condorcet relation $A \succ B$, $B \succ C$ and $C \succ A$ (remark 5.36) and to an optimum of 3 for the 'economic function'.

If we consider the transitivity constraints we can, through the relation $y_{ik} + y_{ki} = 0$, limit ourselves to y_{ik} such that $i < k$. Under these conditions the program reduces to:

Max $(y_{12} + y_{23} - y_{13})$ under the constraints
$-1 \leq y_{12} + y_{23} - y_{13} \leq 1$ and
$y_{ik} + y_{ki} = 0$ for all i and k and $y_{ik} = 1$ or 0 or -1.

The optimum for the 'economic function' is now just 1 and we obtain three possible optimum solutions without indifference, *i.e.* $y_{12} = 1$ $y_{23} = 1$ $y_{13} = 1$ which is, in terms of the B-C social relation, $A \succ B \succ C$; or $y_{12} = 1$ $y_{23} = -1$ $y_{13} = -1$ which is, in terms of the B-C social relation, $C \succ A \succ B$; or $y_{12} = -1$ $y_{23} = 1$ $y_{13} = -1$ which is, in terms of the B-C social relation, $B \succ C \succ A$. The *decision maker* is none the wiser! *The trouble with the method is the non-uniqueness of the solution.* And in our simple example we have not considered solutions of maximization with indifference! For $y_{12} = 1$ $y_{23} = 0$ $y_{13} = 0$, which is, in terms of the B-C social relation, $A \succ B$ and $B \approx C$ and $A \approx C$ is also a solution – and there are two more of this type!

We can state this state of affairs as follows:

Remark 5.41

The B-C aggregation procedure can very easily lead to several non-equivalent social relations all of which are solutions to the maximization program (4).

It is however of interest to note that with the B-C method we do in a way obtain the 'closest' transitive social relations to those of Condorcet.

To explain this, we have to introduce the notion of distance and to simplify the notion of matrix of a binary relation.

Definition 5.42

The distance between any two matrices X and C is defined by:
$$d(X,C) = \sum_{ik} |x_{ik} - c_{ik}|.$$

The distance is a function such that for all matrices X and C, $d(X,C) \geq 0$, $d(X,C) = d(C,X)$ and whatever the matrix Y, $d(X,C) \leq d(X,Y) + d(Y,C)$. When this distance is applied to the antisymmetric matrices of binary relations, Kemeny and Snell (1962, p.18) showed that this distance is the only one which is consistent with a reasonable notion of distance between preorders (a distance depending only on the preorders, and not on the 'names' of the alternatives and on independence with regard to the parts of the preorders which are identical).

Definition 5.43

The *simplified matrix of a binary relation* $\succcurlyeq$ is the matrix $D = (d_{ik})$ such that $d_{ik} = 1$ if $a_i \succcurlyeq a_k$ and $d_{ik} = 0$ otherwise. Compared to the binary relation matrix defined in 5.24, we have $d_{ik} = 1$ if and only if $c_{ik} > 0$. The *simplified antisymmetric matrix* $2D - U1$ only contains ones and minus ones; the zeros of the normal antisymmetric matrix have been transformed into ones.

Let C be the matrix of the Condorcet relation. Consider an unknown relation $\succcurlyeq$ whose simplified matrix is equal to X. We are concerned with the distance $d(X,C) = \sum_{ik} |x_{ik} - c_{ik}|$. Since x_{ik} is equal to zero or one and $c_{ik} = 0$ or 0.5 or 1, we can verify that $|x_{ik} - c_{ik}| = x_{ik} + c_{ik} - 2x_{ik}c_{ik}$. Searching for a matrix X as close as possible to the matrix C, we seek to minimize $G(X) = \sum_{ik} [c_{ik} - x_{ik}(2c_{ik} - 1)]$. But we know that $2c_{ik} - 1 = cd_{ik}$ (the antisymmetric matrix of the Condorcet relation), and since c_{ik} is fixed, we deduce that minimizing G is equivalent to maximizing $\sum_{ik} cd_{ik} x_{ik}$ (note the change of sign).

Proposition 5.44

The relations $\succcurlyeq$ from program:

$$\text{Maximize } \sum_{ik} cd_{ik} x_{ik} \text{ under the constraints} \tag{6}$$

$x_{ik} + x_{kh} - x_{ih} \leq 1$ whatever i, k and h, different among themselves
$x_{ik} + x_{ki} \geq 1$ whatever i and k different among themselves
$x_{ik} = 0$ or 1

are complete preorders which lie at a minimum distance (in the above sense) from the Condorcet relation.

Proof

We have just seen that maximizing $\sum_{ik} cd_{ik} x_{ik}$ minimizes the distance. The relation $x_{ik} + x_{ki} \geq 1$ is equivalent to $x_{ik} + x_{ki} \neq 0$, proving that $\succcurlyeq$ is complete and reflexive (for $i = k$). Transitivity is shown as in proposition 5.37. Q.E.D.

Remark 5.45

Since $cd_{ik} = -cd_{ik}$, the function to be maximized in (6) is written:

$$\sum_{i<k} cd_{ik}(x_{ik} - x_{ki}) .$$

Given remark 5.45, if we solve program (6) for the example of the Condorcet triplet (see above 5.17), we find:

$$\text{Maximize } (x_{12} - x_{21}) + (x_{23} - x_{32}) - (x_{13} - x_{31}) \quad \text{under the constraints}$$
$$x_{12} + x_{23} - x_{13} \leq 1, \quad x_{23} + x_{31} - x_{21} \leq 1, \quad x_{31} + x_{12} - x_{32} \leq 1,$$
$$x_{13} + x_{32} - x_{12} \leq 1, \quad x_{21} + x_{13} - x_{23} \leq 1, \quad x_{32} + x_{21} - x_{31} \leq 1 \text{ and}$$
$$x_{ik} + x_{ki} \geq 1 \text{ for all } i \text{ and } k, x_{ik} = 0 \text{ or } 1.$$

Solving this program by hand is tedious: it has four solutions $x_{kk} = 1$ and $x_{12} = x_{21} = x_{23} = x_{32} = x_{13} = x_{31} = 1$, *i.e.* in terms of the relation, $A \approx B \approx C$, the other solutions being $x_{12} = 1$, $x_{21} = 0$, $x_{23} = 0$, $x_{32} = 1$, $x_{13} = x_{31} = 1$, *i.e.* in terms of the relation, $A \approx C \succ B$ and by circular permutation $A \approx B \succ C$ and finally $B \approx C \succ A$.

The result of proposition 5.44 is very interesting theoretically because it gives a practical way of finding a transitive solution close to Condorcet. However, it is to be feared that when there are many circuits of the Condorcet triplet type ($A \succ B$ and $B \succ C$ and $C \succ A$), there may also be many possible solutions. When there are few circuits, the method will almost certainly produce valid results. Note too that for matrices of more than about ten alternatives, the calculations rapidly become prohibitively long, and program (6) in integers is then solved in continuum by traditional powerful methods of the simplex type. According to Marcotorchino and Michaud (1979), solving the continuous program does give a solution in 0 or 1. This experimental result remains without theoretical foundation, however.

Remarks 5.46

a) Program (6) is not unique; several variants are possible. Thus, to aggregate the preorders $\succ_j$, $j = 1, ..., n$, Kemeny and Snell (1962) proposed searching for the preorder $\succ_s$ which achieves either the minimum in $\sum_j d(AP_j, AP_s)$ (the median preorder), or the minimum in $\sum_j [d(AP_j, AP_s)]^2$ (the average preorder), where AP_j denotes the antisymmetric matrix of the binary relation $\succ_j$. A complete review of the properties of median preorders is to be found in Barthélemy and Monjardet (1981).

b) The distance that we have just been considering, though it is the most common as far as matrices are concerned, is not the only one possible. Slater's (1961) original idea was to consider the simplified matrix of the Condorcet relation. and to search for the minimum number of arcs to invert in the corresponding graph. This is equivalent to minimizing the number of changes from 1 to 0 and from 0 to 1, to transform the simplified Condorcet matrix into the matrix of a transitive relation, *i.e.* one satisfying the constraints that we have seen. This method has the major disadvantage of not considering the arcs that are inversed; thus with arcs in high majority, say A for example, which win against B for nearly all criteria, we may very well end up with $B \succ A$ in the final relation. To overcome this difficulty, Vidal and Yehia Alcoutlabi (1990) propose weighting the arcs by the difference in rankings between the two alternatives defining an arc. The difference in rankings is, for example, measured on the matrix of ranking differences (figure 5.3) or by the difference between 'votes for' and 'votes against' (definition 5.12). Vidal and Yehia Alcoutlabi show that the program which achieves this type of minimization is equivalent to a quadratic assignment program for whose solution algorithms exist. It is also possible to search for the nearest semiorder to a given binary relation; see Lavialle and Vidal (1992). The disadvantage as usual is that there are generally many possible solutions to these programs which are not equivalent from the standpoint of preferences, though some regular features may appear in the preorders found.

To remain close to Condorcet, a simpler approach is to take the Condorcet social relation and eliminate from it the circuits caused by intransitivity by stipulating that the alternatives forming circuits are equivalent. This is the recommended method for outranking methods (see section 7.3). The B-C method has nevertheless demonstrated the links between multicriterion analysis and optimization through the very useful matrix representation of binary relations.

5.6 Lexicographic methods

5.6.1. The basic lexicographic method

The lexicographic methods are among those ordinal methods having specific properties. They satisfy the five axioms in Arrow's theorem, so they are dictatorial:

a criterion acts as a dictator. Later we shall understand why this is a correct interpretation of these methods, but first, a definition is needed.

Definition 5.47

Given a finite set of vectors $|R^n$ and a preorder $\succcurlyeq_j$ corresponding to each of the coordinates $j = 1, ..., n$, we shall say that the vector $x = (x_1, x_2, ..., x_n)$ strictly dominates $y = (y_1, y_2, ..., y_n)$ for the lexicographic preorder $\succcurlyeq_{lex}$ ($c \succ_{lex} y$) if and only if there exists j such that $x_j \succ_j y_j$ and there does not exist $h < j$ such that $y_h \succ_h x_h$. In other words, for the first coordinate j such that x_j is not equal to y_j, x_j dominates y_j.

The above definition is an adaptation of a more general one given by Fishburn (1974) in a publication on the systematic analysis of lexicographic order relations and associated utility functions. These concepts first appeared at the beginning of the century in the mathematical literature stemming from the ideas of Hausdorff, and were subsequently taken up in economics and decision theory (Debreu, 1954, Georgescu-Roegen, 1954), though still earlier and illustrious predecessors do exist, such as Carl Menger in the second half of the nineteenth century, who was already using the concept of lexicographic ranking. Fishburn (1974) and Chipman (1971) provide further interesting historical details.

The simplest example of the use of a lexicographic preorder, and the one which gives it its name, is the method of entering words in a dictionary. Each word can be thought of as a vector formed from letters in a n-dimensional space (n being the length of the longest word). Words with fewer letters are completed with the necessary numbers of 'zeros', and for each coordinate the preorder of the letters is alphabetical order, the character 'zero' preceding all other letters. Lexicographic order is also used in some voting procedures for deciding between ties by introducing a second criterion such as age or a casting vote.

Within the multicriterion context it is a simple matter to define a method of multicriterion choice based on lexicographic preorder.

Definition 5.48 *The lexicographic multicriterion method.*

Consider the decision matrix (a_{ij}) $i = 1, 2, ..., n$ and the weights $w = (w_1, w_2, ..., w_n)$; the process is as follows:

1) Rank the criteria in order of weights w_j from the highest to the lowest. Re-number the criteria so that the highest is #1 and so on down to the lowest, #n.

2) For the aggregate preorder $\succ_{lex}$, let $a_i \succ_{lex} a_k$ if and only if there exists j such that $a_{ij} \succ_j a_{kj}$ and there does not exist $h < j$ such that $a_{kh} \succ_h a_{ih}$. (This is definition 5.47 adapted for vectors $a_i = (a_{i1}, a_{i2}, ..., a_{in})$).

Thus we take the highest criterion and we rank the alternatives according to this criterion. When there are ties, we resolve them using criterion #2; if there are still ties, we use criterion #3 and so on. In the sense of Arrow's theorem criterion #1 is a dictator since $a_i \succ_1 a_k \Longrightarrow a_i \succ_{lex} a_k$.

Proposition 5.49

The lexicographic method leads to a complete preorder, it is ordinal and it is non-compensatory.

Proof

These properties are the direct result of the definitions. Q.E.D.

Example 5.50

We shall apply the lexicographic method to the example of personnel selection; the data are the same as in figure 4.1 in the previous chapter, reproduced below. The criteria have already been ranked according to their weights, and tied criteria have been left in their initial order.

Alternatives	Criteria				
	C_1 MAX	C_2 MAX	C_4 MAX	C_5 MAX	C_3 MIN
Albert	6	5	5	5	28
Blanche	4	2	10	9	25
Donald	5	7	9	6	35
Emily	6	1	6	7	27
Georgia	6	8	7	9	30
Helen	5	6	4	8	26
Weights:	5	5	4	4	2

Figure 5.5 Data for the candidate selection example with weights

According to method 5.48, we begin with criterion #1. The result is then:

Albert	6
Emily	6
Georgia	6
Donald	5
Helen	5
Blanche	4

There are ties which we now resolve with the second criterion, giving the final ranking:

	C_1	C_2
#1 Georgia	6	8
#2 Albert	6	5
#3 Emily	6	1
#4 Donald	5	7
#5 Helen	5	6
#6 Blanche	4	

The advantages and disadvantages of the lexicographic method are easy to see. The first feature of the method is that it does not require much information. In our example, the data on criteria #3 to #5 are not used. Also, the method is ultra-simple and requires no calculations.

The second feature is that the weights, like the evaluations of alternatives, only need to be ordinal; at no time does the method use any information apart from order. This recalls the first feature: a low information requirement. The third feature is that the weights must have an order rather than a simple preorder; otherwise we have to distribute the equal weights in a more or less arbitrary way, just as we did in our example. In example 5.50 we could just as well have considered criterion C_2 as the most important criterion instead of C_1 which has the same weight of 5. In that case the ranking would have been rather different:

	C_2
#1 Georgia	8
#2 Donald	7
#3 Helen	6
#4 Albert	5
#5 Blanche	2
#6 Emily	1

There would have been no need to consider a second criterion because there would have been no ties.

The fourth feature of this method is obviously a big disadvantage: it has a very high dependence on weights (or the order of the criteria). Any change, however small, in the weights which induces a change in the order will lead to large modifications to the final order of the alternatives.

The fifth characteristic concerns the dictatorial nature of the method, the 'dictator' being the criterion with the highest weight; the other criteria cannot alter the strict preference of the dictator – they can only have any effect if he is indifferent. He chooses first and the others cannot question his judgement.

The last point to note is the non-compensatory character of the method. This can be considered as a weakness because it does not make the formalized search for a compromise any easier, which some parties may find. But it can also be an advantage because in this way we do not have to provide cardinal weights with a background of substitution rates (definition 4.2) between criteria which may have little connection with each other. The lexicographic method is really the prototype of the non-compensatory methods, and we can state this in the form of a remark.

Remark 5.51

Fishburn (1974, theorem 1) showed that among all non-compensatory aggregation methods leading to a non-trivial aggregate preorder (*i.e.* such that there exist alternatives a, b and c such that $a \succ b$ and $b \succ c$), only lexicographic methods are to be found.

Given the well-affirmed non-compensatory character of the lexicographic method (*cf.* the above remark), it would seem to be diametrically opposed to the weighted sum method. The reader will therefore be surprised to learn the following counter-intuitive result, which will be useful in the next chapter.

Proposition 5.52

Utilities and weights can be chosen in such a way that the lexicographic preorder can be calculated through a weighted sum.

Proof

We renumber the criteria and assume the weights are $w_1 > w_2 > ... > w_n$. Using a normalization of the type found in procedure 2 (figure 2.8), we assume that the values of the criteria are all within the interval [1,2]; it is sufficient to take procedure 2 and to add 1 to all the values, this being a strictly increasing transformation. Now we write $U(a_i) = \sum_j 10^{n-j} a_{ij}$; the function U represents the lexicographic preorder, as we can see if we consider $\sum_{k>j} 10^{n-k} a_{ik}$: for every j < n we have $(2*10^{n-j})/9 > 2[(10^{n-j} - 1)/9] = 2\sum_{k>j} 10^{n-k} \geq \sum_{k>j} 10^{n-k} a_{ik}$. Since $10^{n-j} a_{ij} \geq 10^{n-j} > (2*10^{n-j})/9$, the contribution of the criteria $k > j$ to the sum cannot change the order induced by the criteria $i \leq j$. Q.E.D.

The above proposition serves as yet another warning to consider weights cautiously, since they can even be made to produce lexicographic aggregation!

To end this section we return to the lexicographic method itself. Probably because of the drawbacks we have seen, it does not seem to be very widely used in multicriterion decision aids, yet it is used daily in everyday life. For example Chankong and Haimes (1983) quote the purchase of household electrical goods by individuals, which seems to be based on lexicographic choice; is it not price that dictates in the purchase of a great many consumer durables? We cannot therefore afford to ignore the consequences of this method of making decisions.

These practical scenarios illustrate the low level of information required, a feature which has encouraged the search for variants which do not share the disadvantages of the basic lexicographic method.

5.6.2. The lexicographic 'semiorder'

We shall see below the reason for the inverted commas. The notion of semiorder was introduced in section 2.5.

Definition 5.53 *Lexicographic 'semiorder'.*

1) An indifference threshold s_j is defined for each criterion such that two alternatives a_i and a_k are considered as indifferent or tied relative to j if and only if $|a_{ij} - a_{kj}| \le s_j$.
2) We then apply the lexicographic method with these ties in the wide sense defined in 1).

Two points must be made: the evaluation of the alternatives a_{ij} must be cardinal, and there will clearly be more ties than in the basic lexicographic method. It will probably be necessary to use more criteria.

It has been shown (Tversky, 1969) that the lexicographic 'semiorder' is liable to lead to intransitivity in the final relation $\succeq$ and even in $\succ$; we shall see an example, and it is in this sense that the lexicographic 'semiorder' is not actually a semiorder (*cf.* proposition 2.32, hence the inverted commas). Nevertheless, Pirlot and Vincke (1992) give conditions for the lexicographic 'semiorder' to become a real semiorder. But in spite of the lack of transitivity in the general case, it should not be concluded that this method cannot be adapted to real situations; quite the reverse, as for above mentioned problem of the purchase of consumer durables. Let us see what the method gives in our personnel selection example.

Example 5.54

Application of the lexicographic 'semiorder' method to personnel selection.
We use the data of figure 5.5 and assume in addition that we have defined the following indifference thresholds:

Criteria	Indifference threshold
C_1: Higher studies	1 year
C_2: Experience	2 years
C_3: Age	3 years
C_4: Interview	3 points
C_5: Tests	3 points

The lexicographic 'semiorder' method is applied to the alternatives pair by pair, since it is too difficult to get the overall ranking in one operation. The result is given as an antisymmetric matrix for the final aggregate relation (definition 5.25). We obtain:

	A	B	D	E	G	H
Albert	0	1	−1	1	−1	0
Blanche		0	−1	−1	−1	−1
Donald			0	1	−1	1
Emily				0	−1	−1
Georgia					0	−1
Helen						0

For example, to compare Albert with Donald we examine the criteria in order of weights:

Criterion C_1: $|6 - 5| = 1 \leq s_1$

Criterion C_2: $|5 - 7| = 2 \leq s_2$

Criterion C_4: $|5 - 9| = 4 > s_4$, hence Donald dominates Albert.

The final relation may not be transitive. Thus, from the above matrix we have Georgia $\succ$ Donald $\succ$ Helen $\succ$ Georgia.

In section 3.4 we referred to the *elimination by aspects* (EBA) *method* which combines pre-analysis of satisfaction with lexicographic elimination. This method can be adapted to search for an aggregate preorder by comparing alternatives as in the lexicographic 'semiorder' but combining this with generalized elimination by aspect. Let $[L, M, N]$ be a partition of the set of criteria (to be maximized); according to Barthélemy and Mullet (1986, 1989 and 1992) we shall, for example, say that alternative a is preferred to alternative b if and only if:

$\forall j \in L \quad (a_i - b_j) \geq 2* th$ (*th* stands for threshold)

or else if $\forall j \in L \quad (a_i - b_j) \geq 1* th$ and $\forall j \in M \quad (a_i - b_j) \geq 1* th$

or else if $\forall j \in M \quad (a_i - b_j) \geq 3* th$ and $\forall j \in N \quad (a_i - b_j) \geq 2* th$.

The thresholds and parameters are defined by the decision maker. This method, called *the mobile base method*, can be schematized, in our example, by the formal expression $L^2 + LM + M^3 N^2$ where a product indicates conjunction and a sum disjunction, while the indices correspond to thresholds. Barthélemy and Mullet give both practical and theoretical accounts of this mobile base method, the theory being based on the above 'polynomial' representation.

5.6.3. Other lexicographic methods

The second variant, which may be called the *lexicographic permutation method* (not to be confused with permutation methods of section 8.4), was introduced by Massam and Askew (1982) to solve a problem of choice of urban amenities. In order to lessen the very great sensitivity of the lexicographic method to weights, they proposed forgetting entirely about weights and applying the lexicographic method for each of the $n!$ possible rankings of the criteria. For each of these we note the best alternative and then calculate the number of times each alternative has been the best.

This method cannot be used without a computer because the lexicographic method must be applied $n!$ times. In our example there will be $5! = 120$ possible rankings of criteria. Doing the calculations with the help of the program DMDI (see chapter 10), we obtain the following totals:

Alternative	Total number of first rankings
#1 Blanche	60
#2 Georgia	36
#3 Emily	18
#4 Albert	6
#5 Donald	0
#6 Helen	0

The results are rather different from those we have obtained up to now in the same example. The first rank of Blanche is surprising because with the basic lexicographic methods she always trailed in the bottom ranks. It would seem hard to justify this method except when we can say that the criteria have equal importance and the decision maker is incapable of ranking them.

The last variant, which we shall only describe briefly, is *lexicographic ordering with aspiration levels* mentioned by Keeney and Raiffa in their book (1976, page 78). For each criterion C_j (supposing that we wish to maximize), we set an aspiration level asp_j. To be able to say that an alternative a_i is lexicographically better than another alternative a_k for criterion j, we require that $a_{kj} < asp_j$ and that a_i dominates a_k $(a_{ij} > a_{kj})$. If this is not the case the alternatives a_i and a_k are considered as tied and we proceed to the next criterion following the order of importance. Here, many indifferences are likely to appear between the alternatives but the intransitivity that appears in lexicographic 'semiorder' will be absent. Despite this advantage, Keeney and Raiffa themselves state that the rationality of the method remains hard to prove.

6 ADDITIVE UTILITY FUNCTIONS AND ASSOCIATED METHODS

6.1 Introduction

In chapter 4 we saw that the weighted sum is a simple method of aggregating preferences. However, this type of aggregation does pre-suppose a certain number of assumptions on the nature of the decision maker's preferences; these assumptions are quite strong and by no means simple to explain, as we shall see in this chapter. The key question is: what properties must the decision maker's preferences satisfy if they are to be expressed in the form of a sum of utility functions?

In the first part of this chapter we shall give the main results concerning this question and within the scope of this book, and we shall try to do this in a way which is both rigorous and accessible. Sections 6.2, 6.3 and 6.4 are straightforward, and the reader who is not mathematically inclined can skip section 6.5. For the practical aspects, the reader can go directly to sections 6.6 and 6.7; each of these two sections deals with an aggregation method stemming directly from research on the additivity of preferences. The first Multi-Attribute Utility Theory (MAUT) is described in detail in Keeney and Raiffa (1976) and in Keeney (1992). The method, which has among other advantages the ability to be adapted to choice in the face of uncertainty (an aspect we shall not be dealing with in this book) is very widely used, especially in the United States. The other method, UTA, enables a partial choice in a subset of $\mathcal{A}$ to be extended to the whole choice set.

6.2 The problem of comparing utilities

Suppose we have two criteria in a choice set $\mathcal{A}$. Each of the two associated preorders will be represented by a utility function, U_1 and U_2 respectively. Now let a_1 and a_2 be two alternatives to be compared. We define $U \equiv (U_1, U_2)$, U being a mapping of $\mathcal{A}$ in $|R^2$, and we apply our reasoning in the set $U(\mathcal{A})$. Let $U(a_1) = (x_1, y_1)$ and $U(a_2) = (x_2, y_2)$. Suppose we have neither $(x_1, y_1) \succcurlyeq (x_2, y_2)$ nor $(x_2, y_2) \succcurlyeq (x_1, y_1)$ where $\succcurlyeq$ is the partial canonical preorder of $|R^2$. The decision maker may however think that (x_1,y_1) $\succ_d (x_2,y_2)$, where $\succ_d$ denotes the decision maker's preorder. The decision maker's stance can very well be expressed by:

$$(x_1,y_1) \succ_d (x_2,y_2) <==> [x_1 \geq x_2 \text{ and } x_1 - x_2 \geq y_2 - y_1]$$
$$\text{or } [x_1 \leq x_2 \text{ and } x_2 - x_1 \leq y_1 - y_2] \tag{1}$$

which is further equivalent to:

$$(x_1,y_1) \succ_d (x_2,y_2) <==> x_1 + y_1 \geq x_2 + y_2 \tag{2}$$

The above expression is of course only an example to show that, for the decision maker, what is lost on one criterion is compensated by what is gained on the other. Here, one unit lost on one criterion is exactly compensated by one unit gained on the other.

Suppose that the decision maker has a complete preorder in $\mathcal{A}$ which can be represented by a utility function V_d. From the definition of utility we have:

$$a_1 \succ_d a_2 <==> V_d(a_1) \geq V_d(a_2)$$

Taking the example above, the decision maker had:

$$a_1 \succ_d a_2 <==> U_1(a_1) + U_2(a_1) \geq U_1(a_2) + U_2(a_2) \tag{2'}$$

If we compare the above two expressions we see that the wishes of a decision maker skilled in the art of compensation between criteria are fulfilled if $V_d(a) = U_1(a) + U_2(a)$.

What we have just seen for two criteria can be generalized to several, and we therefore have to ask about the existence of utility functions capable of representing the decision maker's preorder, and of the form $U_1 + U_2$.

For the sake of simplicity, and also to work within the context usually found in the literature, we shall modify our model slightly. In all that follows in this section we shall assume that the alternatives in $\mathcal{A}$ are described by their values relative to the n attributes. We therefore assume the existence of criteria that can be represented by utility functions U_j with the numerical values of the attributes belonging to the sets $X_j = U_j(\mathcal{A})$, $j = 1, 2, ..., n$. Instead of considering the alternatives a_i in $\mathcal{A}$, the decision maker will choose directly from the vectors $U(a_i) = (x_{i1}, x_{i2}, ..., x_{in}) \in X_1 \times X_2 \times ... \times X_n = \pi_j X_j$ where $x_{ij} = U_j(a_i)$. We shall therefore assume the existence of a decision maker's preorder, $\succ_d$, in the set $\pi_j X_j$ or else in a subset $X \subset \pi_j X_j$.

6.3 Definition and cardinality of additive utility functions

We next consider a Cartesian product $\pi_j X_j$ where each Xj is finite or infinite and contained in $|R$. The choice set $X \subset \pi_j X_j$ is endowed with a complete preorder $\succ_d$. In this book we restrict ourselves to finite sets of alternatives; however, we shall be unable to avoid considering the case where X is infinite in the following sections. This is necessary in order to reveal the limiting and restrictive nature of the existence assumptions on additive utility functions. In section 6.6 we shall show how to use the results involving infinite X in the discrete situation (example 6.44).

Definition 6.1

We shall say that a *complete preorder* $\succcurlyeq_d$ in $X \subset \pi_j X_j$ is *additively separable* if there exists a utility function V_d representing the preorder $\succcurlyeq_d$ in X and functions U_j defined on X_j such that for all $x = (x_1, x_2, ..., x_n)$ we have:
$V_d(x) = U_1(x_1) + U_2(x_2) + ... + U_n(x_n)$
We then say that the *utility function* V_d *representing* $\succcurlyeq_d$ *is additively separable or additive.*

Remark 6.2

In the above definition nothing is said about the functions U_j; they do not necessarily have any relation to the utility functions which may have been used to construct X_j. However, since $\succcurlyeq_d$ is complete if V_d and U_j together with W_d and T_j fulfil definition 6.1, we can consider two elements of X, $x = (x_1, x_2, ., x_j, .., x_n)$ and x' $= (x_1, x_2, ., x'_j, .., x_n)$ which only differ in the jth. co-ordinate, and say for example that $x \succcurlyeq_d x'$. From the additive form we deduce, after simplification, $U_j(x_j) \geq U_j(x'_j)$. Conversely, $U_j(x_j) \geq U_j(x'_j)$ implies $V_d(x) \geq V_d(x')$ and $W_d(x) \geq W_d(x')$ hence $T_j(x_j) \geq T_j(x'_j)$. As the reverse is also true we conclude that:

$$U_j(x_j) \geq U_j(x'_j) \Longleftrightarrow T_j(x_j) \geq T_j(x'_j).$$

Thus the functions U_j and T_j as utility functions are necessarily strictly increasing transforms of each other (proposition 2.24). We shall see below that this mapping is necessarily affine.

Remark 6.3

Multiplicative form. The function $\exp(V_d)$ is a utility function representing the same preorder as V_d; it is clearly multiplicative since $\exp(V_d) = \exp(U_1)*\exp(U_2)* ...$ $*\exp(U_n)$. Conversely if $V_d = W_1*W_2 ... *W_n$ with $W_j \geq 0$, it is sufficient to take logs to show that V_d is additive. The two expressions 'additive' and 'multiplicative' are almost equivalent (and are exactly so for positive values). Either form can be used depending on practical considerations (see section 4.1.4).

As we have seen above, when there is an additive utility function, it is as though the decision maker were capable of comparing utility ranges; in (1), for example, he estimates that $U_1(x_1) - U_1(x_2) \geq U_2(y_2) - U_2(y_1)$. The reader will remember that this expression has a whiff of cardinality about it (section 2.4.2), and we shall show that additivity is indeed a cardinal property. The following proposition is inspired from Jaffray (1982).

Proposition 6.4 (Jaffray, 1982)

If $V_d(X)$ contains a neighborhood of 0 in $|R$, which requires X to be infinite, and if the preorder $\succeq_d$ is additively separable and is represented by the utility V_d, then a necessary and sufficient condition for another utility function W_d, additively separable, to represent the same preorder is that $W_d = k_1 V_d + k_2$ (where $k_1 > 0$).

Proof

For the sake of simplicity we shall set out the proof for $n = 2$; the result is easily generalized.

i) The condition is sufficient. Let $V_d = U_1 + U_2$ and suppose $W_d = k_1 V_d + k_2$ (where $k_1 > 0$); V_d and W_d obviously represent the same preorder. Furthermore, W_d is additive because the functions $k_1 U_1 + k'_2$ and $k_1 U_2 + k''_2$ satisfy the additivity condition for all k'_2 and k''_2 such that $k'_2 + k''_2 = k_2$.

ii) The condition is necessary. Let V_d and W_d be two functions satisfying the above definition. They are additively separable and represent the same preorder; we therefore have:

$$V_d = U_1 + U_2 \tag{3}$$
$$\text{and } W_d = T_1 + T_2 \tag{4}$$

and moreover there exists S strictly increasing (proposition 2.24) such that:

$$V_d = S(W_d) \tag{5}$$

Considering once more remark 6.2 above, there exist S_1 and S_2 strictly increasing such that $U_1 = S_1(T_1)$ and $U_2 = S_2(T_2)$. Let $A_1 = T_1(a)$ and $A_2 = T_2(a)$; for all $a \in \mathcal{A}$, equation (5) becomes:

$$S_1(A_1) + S_2(A_2) = S(A_1 + A_2) \tag{6}$$

If we set two particular values A_{01} and A_{02}, we see that $S_1(A_1) = A(A_1 + A_{02}) - S_2(A_{02})$ and $S_2(A_2) = S(A_{01} + A_2) - S_1(A_{01})$, which on rearranging gives:

$$S(A_1 + A_2) = S(A_1 + A_{02}) + S(A_{01} + A_2) - S(A_{01} + A_{02}) \tag{7}$$

On introducing the change of variable $A'_1 = A_1 - A_{01}$ and $A'_2 = A_2 - A_{02}$, and introducing the function $f(t) = S(t + A_{01} + A_{02}) - S(A_{01} + A_{02})$, equation (7) becomes the Cauchy functional equation:

$$f(A'_1 + A'_2) = f(A'_1) + f(A'_2)$$

It is known that this equation in $|R$ or in a non-void interval of $|R$ whose interior is non empty, only has the linear solution $f(A') = k_1 A'$ (Hölder's theorem, see Fishburn, 1970, p.55). We deduce that S is affine and that k_1 is strictly positive since S is strictly increasing. Q.E.D.

Remark 6.5

When X is finite, Cauchy's equation can have solutions other than the linear function, and hence it cannot be rigorously proved, in this case, that all additive functions are necessarily transforms of each other through an affine transformation.

We shall nevertheless assume that, even when X is finite, and unless exotic solutions are sought, *the utility functions V_d are unique up to a strictly positive affine transformation,* and are therefore cardinal utilities (proposition 2.28).

Assuming then that V_d is unique up to a strictly positive affine transformation, we now specify the links between V_d and the sum $U_1 + U_2$. We begin by noting that V_d, U_1 and U_2 cannot be considered as independent utilities.

Remark 6.6

If V_d is an additive utility function such that $V_d = U_1 + U_2$, then the functions V_d, U_1 and U_2 are not mutually independent.

For example, suppose that we have $V(x) = U_1(x_1) + U_2(x_2)$. Consider $V(x_1) = U_1(x_{11}) + U_2(x_{12}) = 2 + 0$, $V(x_2) = U_1(x_{21}) + U_2(x_{22}) = 0 + 2$ and $V(x_3) = U_1(x_{31}) + U_2(x_{32}) = 0 + 3$; for the decision maker, we have $x_3 \succ_d x_1 \approx_d x_2$. Replace U_1 by U_1^2, which is an allowed strictly increasing transformation, at least if we assume, as in the present case, that $U_1 > 0$. If we now consider $W = U_1^2 + U_2$, we will have $W(x_1) = 4$, $W(x_2) = 2$ and $W(x_3) = 3$, and hence $x_1 \succ_d x_3 \succ_d x_2$. Thus, this does not represent the same preorder. The reader will readily see that the situation is no better if we transform U_1 into $k_1 U_1 + k_2$ and U_2 into $k'_1 U_2 + k'_2$ where $k_1 \neq k'_1$.

The following proposition summarizes the links between the additive utility functions V_d, U_1 and U_2 such that $V_d = U_1 + U_2$.

Proposition 6.7

Under the assumptions of proposition 6.4, when a utility function V_d represents an additively separable preorder with $V_d = U_1 + U_2$, the functions V_d, U_1 and U_2 are defined jointly. A necessary and sufficient condition for an alternative additive function $W_d = T_1 + T_2$ to represent the same preorder is that W_d, T_1 and T_2 differ from V_d, U_1 and U_2 up to a *unique* strictly positive affine transformation (*i.e.* with the same slope $k > 0$).

Proof

i) The condition is sufficient. Clearly, if we write $U'_1 = kU_1 + k_2$ and $U'_2 = kU_2 + k'_2$ ($k > 0$), we obtain $V'_d = k(U_1 + U_2) + k_2 + k'_2$, which does represent the same preorder as V_d.

ii) The condition is necessary. Let $V_d = U_1 + U_2$ and $W_d = T_1 + T_2$ be two additive utilities representing the same preorder. Let $x = (x_1, x_2)$, and for all x_1, fix any x_2; we have $U_1(x_1) = V_d(x_1, x_2) - U_2(x_2) = kW_d(x_1, x_2) + k_2 - U_2(x_2)$ (proposition 6.4). The latter term is further equal to $k[T_1(x_1) + T_2(x_2)] + k_2 - U_2(x_2) = kT_1(x_1) + k'_2$ where $k'_2 = kT_2(x_2) + k_2 - U_2(x_2)$. Q.E.D.

Definition 6.8

The utilities V_d, U_1 and U_2 which satisfy the property of proposition 6.7 are said to be *conjointly cardinal*. They are defined up to a unique affine transformation (*i.e.* with the same slope $k > 0$).

To sum up, setting aside certain difficulties related to the finite case, difficulties which will come up again in the next section, *we can conclude by stating that separable additive utility functions are cardinal and that the functions V_d, U_1 and U_2 are defined jointly.* This has to be taken into account during their construction. In other words, *we cannot start from any U_1 and U_2 and decide that $U_1 + U_2$ is the decision maker's utility function, even if we know that the decision maker's preorder $\succeq_d$ is additively separable* (example 6.9). This fact is extremely preoccupying, and leaves no margin for error in evaluating the decision maker's utility functions.

Example 6.9

Consider the data of example 4.6. The choice set is $\{A, B, C\}$. With $U_1(A) = 1$, $U_2(A) = 0$, $U_1(C) = 0$, $U_2(C) = 1$ and $U_1(B) = U_2(B) = 0.4$, we saw in 4.6 that the preferences of the decision maker $B \succ C \approx A$ cannot be expressed in the form $U_1 + U_2$. *The decision maker's preorder is nevertheless additively separable.* For if we let $U'_1(A) = 1.5$, $U'_2(A) = 0$, $U'_1(C) = 0$, $U'_2(C) = 1.5$ and $U'_1(B) = U'_2(B) = 1$, it is clear that U_1 and U'_1 (respectively, U_2 and U'_2) represent the same preorder on $\{A, B, C\}$. The decision maker's preorder is represented by $V_d = U'_1 + U'_2$ since $V_d(B) = 2 > V_d(A) = V_d(C) = 1.5$.

The above example shows clearly that the decision maker's preorder may very well be additively separable and be represented by a sum of utilities when we have the 'right utilities' relative to the attributes, but that we have no right to say that any sum of utilities whatever represents the decision maker's preorder. We must therefore, as indicated by proposition 6.7, construct the utilities of the attributes jointly with the decision maker utilities. Before going on to see how we can attempt the precise construction of such an additive utility function (see 6.6), we must ask what properties the decision maker preferences must have for an additively separable utility function V_d actually to exist.

Unfortunately these properties are by no means simple, especially in the finite case, and the reader may wish to skip section 6.5, which is summarized at the beginning of section 6.6.

Aside from cardinality, it is obvious that in many cases, the existence of an additive decision maker utility function enables compensation between criteria (in the sense of definition 5.21).

Example 6.10

Suppose that X is a subset of $|R^2$ and we let $x_i = (x_{i1}, x_{i2})$ and $V_d = x_{i1} + x_{i2}$. Consider $x_1 = (3,0)$, $x_2 = (2,2)$, $x_3 = (3,0)$ and $x_4 = (0,2)$; clearly $[x_1, x_2]$ SC $[x_3, x_4]$ (the same criteria for and against, see section 5.20). However, for the preorder V_d, we have $x_2 \succ x_1$ and $x_3 \succ x_4$, which contradicts 5.21.

Remark 6.11

It must not be thought, however, that additivity necessarily implies compensation. Suppose we have to consider two choice criteria on a set of cars: the price, to be minimized, varying between \$ 20 000 and \$ 40000 in steps of \$ 100, and comfort, to be maximized, represented by a mark between 0 and 5. The utility function $V_d = -$Price + Comfort is non-compensatory and even lexicographic (see section 5.6 and proposition 5.52). For it is the price which decides the choice since a price difference of less than \$ 100 between two vehicles cannot be compensated for by comfort. We thus see that whether a weighted sum is compensatory or not depends on utilities and weights.

On the other hand, it can clearly be seen in example 6.10 that the additivity of the utility function, when the scales of magnitude are comparable (see section 6.11), *allows a loss on one criterion to be compensated by a gain on another.* This allows reasoning of the type: 'I am ready to sacrifice x value units on criterion i to increase criterion j by one unit'. This idea can be locally formalized.

Definition 6.12

Consider an additive utility function $V_d(x)$. The *marginal compensation rate* at the point x of criterion i relative to criterion j is the ratio A_i/A_j where A_j is the small quantity of utility that the decision maker must sacrifice on criterion j in order to gain a small quantity of value A_i on i, such that the complete utility $V_d(x)$ remains unchanged.

In section 6.6 we shall see that under certain conditions this marginal compensation rate can be given mathematical expression.

6.4 Difference additivity models

The additivity of preferences relies on a representation of the preorder $\succcurlyeq_d$ in the form:

$$x \succcurlyeq_d y <==> \sum_j U_j(x) - \sum_j U_j(y) \geq 0 \tag{8}$$

Other more complicated models have been proposed. The first is Tversky's (1969) *additivity of differences*, in which the decision maker preorder is defined by:

$$x \succcurlyeq_d y <==> \sum_j \Phi_j[U_j(x) - U_j(y)] \tag{9}$$

where the real functions Φ_j are strictly increasing and odd $(\Phi_j(-t) = -\Phi_j(-t))$. Condition (9) can be generalized:

$$x \succcurlyeq_d y <==> \sum_j f_j(x_j, y_j) \geq 0 \tag{10a}$$

where the f_j are mappings of $X_j \times X_j$ in $|\mathbb{R}$ satisfying $f_j(x_j,y_j) = - f_j(y_j,x_j)$ or $f_j(x_j,y_j)*f_j(y_j,x_j) \leq 0$ (see Bouyssou, 1986 and Fishburn, 1991).

We can reject the summation in (10a) and propose the model:

$$x \succcurlyeq_d y <==> F\left(f_l(x_1,y_1),\ ...,f_j(x_j,y_j),\ ...\right) \geq 0 \qquad (10b)$$

where F is an increasing (non-decreasing) function of all its arguments (Bouyssou and Pirlot, 1997).

The conditions for relating the existence of the functions Φ_j or f_j to the properties of the preorders are similar to, but more complicated than, those we shall give for additivity in the next section. Here, the reader should see Bouyssou (1986) and Fishburn (1991, 1992b).

The attraction of difference additivity models is that they provide a framework into which can be introduced preferences obtained through an outranking relation (this notion will be introduced in chapter 7), which cannot be done with the simple additivity of formula (8). Note finally that models of the type (9) and (10) which, like additivity, imply cardinality, can serve to define 'x is more preferable to y than t is to z'; see the end of section 2.5. For to measure this difference in the strengths of preferences it is sufficient to write:

$$(x,y) >> (t,z) <==> \sum_j \Phi_j[U_j(x) - U_j(y)] \geq \sum_j \Phi_j[U_j(t) - U_j(z)] \qquad (11)$$

This idea is analyzed by Suppes, Krantz, Luce and Tversky in their book (1989). As we mentioned before in section 2.5, preference strength differences can also be defined algebraically without resorting to the additivity of differences; see Vansnick (1984) and Roy (1985).

We shall not enter any deeper into these models except to set out the theory – itself not simple – on the existence of additively separate preorders. As we mentioned above, the reader may if desired skip the next section.

6.5 The existence of additively separable utility functions

We set the conditions as defined above at the beginning of section 3, of a finite or infinite choice set, $X \subset \pi_j X_j$, $j = 1, 2, ..., n$.

An equivalence relation, it should be remembered, is reflexive, symmetric and transitive, and a *permutation* of a finite set T is a bijective mapping of T onto itself.

Basing our presentation upon that of Fishburn (1970), we shall give a necessary and sufficient condition for a utility to be additively separable.

Definition 6.13 Additivity equivalence relation (E_m)

Consider the elements $x_i = \{x_{i1}, x_{i2}, ..., x_{in}\}$ of X. Let there be m elements $\{x_1, x_2, ..., x_m\}$ of X and m other elements, $\{y_1, y_2, ..., y_m\}$ of X, distinct or not from the previous elements. We say that $\{x_1, x_2, ..., x_n\}$ E_m $\{y_1, y_2, ..., y_m\}$ if and only if there exist n permutations $PER_1, PER_2, ..., PER_n$ such that for all j:

$$(x_{1j}, x_{2j}, ..., x_{mj}) = PER_j (y_{1j}, y_{2j}, ..., y_{mj})$$

In other words, for each space X_j, $j = 1, 2, ..., n$, the elements x_j have overall the same values as those of y_i but not in the same order. As identity is a permutation, the relation (E_m) is reflexive and as the combination of two permutations is another permutation, it is transitive; and by definition it is obviously symmetric.

Examples 6.14

We verify that $\{x_1, x_2\}$ E_2 $\{y_1, y_2\}$ where $y_1 = (x_{11}, x_{22})$ and $y_2 = (x_{21}, x_{12})$. Taking the alternatives, we can verify that $\{U(a_1), U(a_2), U(a_3)\}$ E_3 $\{Ub_1), U(b_2). U(b_3)\}$ where $U(a_1) = (2,1)$, $U(a_2) = (0,2)$, $U(a_3) = (1,0)$, $U(b_1) = (1,2)$, $U(b_2) = (0,1)$ and $U(b_3) = (2,0)$. The reader will note that the numerical values of U_1 and U_2 are the same for a_i and b_i but in different orders, this being the definition of the relation E_m.

Definition 6.15

The preorder $\succcurlyeq$ defined on X is said to satisfy the condition (C_m) if for all families of m elements $\{x_1, x_2, ..., x_m\}$ and $\{y_1, y_2, ..., y_m\}$, whether or not they are distinct, such that: $\{x_1, x_2, ..., x_m\}$ E_m $\{y_1, y_2, ..., y_m\}$ and $\{\forall\, i = 1, 2, ..., m-1 \quad x_i \succcurlyeq y_i \}$ one has $y_m \succcurlyeq x_m$.

Remark 6.16

(C_m) can be written equivalently: $\{x_1, x_2, ..., x_m\}$ E_m $\{y_1, y_2, ..., y_m\}$ and $\{\forall\, i = 1, 2, ..., m \quad x_i \succcurlyeq y_i\}$ imply $\{\forall\, i = 1, 2, ..., m \quad x_i \approx y_i\}$.

Definition 6.17

The preorder $\succcurlyeq$ is said to satisfy the condition (C) if for all $m \geq 2$ it satisfies (C_m).

Theorem 6.18 (Tversky, 1964 and Scott, 1964).

If X is finite, a necessary and sufficient condition for the complete preorder $\succcurlyeq$ to be additively separable is that it satisfy (C).

Proof

The condition is necessary. For simplicity we place ourselves in $X = X_1 \times X_2$. By definition $\{x_1, x_2, ..., x_m\}$ E_m $\{y_1, y_2, ..., y_m\}$ implies:

$$\sum_{i=1}^{m} U_1(x_i) = \sum_{i=1}^{m} U_1(y_i) \text{ and } \sum_{i=1}^{m} U_2(x_i) = \sum_{i=1}^{m} U_2(y_i).$$

Let $U = U_1 + U_2$; then we have $\sum_{i=1}^{m} U(x_i) = \sum_{i=1}^{m} U(y_i)$. Since $x_i \succcurlyeq y_i$ for $i = 1$, 2, ..., $m-1$ is equivalent to $U(x_i) \geq U(y_i)$, the previous equality implies $U(x_m) \leq U(y_m)$, hence $y_m \succcurlyeq x_m$. The sufficiency condition is technically difficult to prove and the reader is referred to Fishburn (1970). Q.E.D.

Condition (C) is obviously very difficult to handle and to verify. In practice, it is not easy to see how to satisfy (C_m) for all families of 2, 3 up to m elements and beyond. Note however that, when X is finite, for every preorder $\succcurlyeq$ there exists m_q such that (C) is equivalent to (C_{mq}). Unfortunately, the m_q values are unbounded (Fishburn, 1970).

Remark 6.19

When X is not finite, condition (C) is not equivalent to a condition (C_{mq}), and is no longer sufficient; it has to be replaced by a stronger condition (Jaffray, 1974).

As we said, the condition (C_m) for $m > 3$ rapidly becomes hard to satisfy. We therefore introduce a condition known as solvability, which is not necessary but which enables the problem to be somewhat simplified. For the sake of simplicity we shall limit ourselves here to the case of two product spaces X_1 and X_2.

Definition 6.20

There is said to be solvability for X and $\succcurlyeq$ if and only if:
(i) for all $x \in X$ and all $y_1 \in X_1$ there exists $y_2 \in X_2$ such that $(x_1, x_2) \approx (y_1, y_2)$.
(ii) for all $x \in X$ and all $z_1 \in X_2$ there exists $z_1 \in X_1$ such that $(x_1, x_2) \approx (z_1, z_2)$.

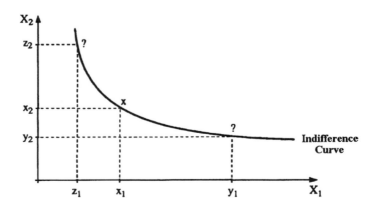

Figure 6.1 Graphic representation of solvability

Remark 6.21

It can be seen in figure 6.1 that the solvability assumption requires that any parallel to an axis meet each indifference class (set of elements mutually indifferent for the decision maker). Thus this condition can only be satisfied, when X is finite, for certain very particular preorders.

Proposition 6.22 (Jaffray, 1982).

If X and $\succeq$ satisfy the assumption of solvability then for $m > 3$, (C_{m-1}) implies (C_m).

Since (C_3) implies (C_2), we deduce that *when (C_3) and the assumption of solvability are satisfied, the condition (C) is satisfied.*
We can also replace (C_3) by a slightly weaker but more intuitive condition (CD).

Definition 6.23

X and $\succeq$ are said to satisfy the condition (CD) if and only if:
$(x_1,x_2) \succeq (y_1,y_2)$ and $(y_1,z_2) \succeq (z_1,x_2)$ imply $(x_1,z_2) \succeq (z_1,y_2)$

Proposition 6.24 (Jaffray, 1982).

If X is finite and with $\succeq$ satisfies the property of solvability and the condition (CD) then (C_m) is satisfied for all $m \geq 2$, and the property of additivity is therefore satisfied.

Remark 6.25

In an infinite space, the so-called *Archimedean conditions* must be added to the above conditions to obtain sufficient conditions. This is the case with the theorem of Luce and Tukey (1964). An Archimedean condition is one which ensures that a sequence of equally spaced (in the sense of utility intervals) elements X may be constructed such that, for a given rank, one element in this sequence will be preferred to any given element.

Debreu's theorem(1960) which also gives a sufficient condition for the existence of an additively separable utility, uses (C_3) and topological conditions. Unlike the theorem of Luce and Tukey, Debreu's theorem does not give constructive conditions. On the other hand, Debreu shows, in every case, the uniqueness up to a strictly positive affine transformation $kU_i + k_i$, $i = 1,2$.

Before going on to construct an additively separable utility function designed to represent $\succeq$, we shall give an alternative interpretation of (C_2). Consider elements of X such that $\{x_1,x_2\}$ $E2$ $\{y_1,y_2\}$ and $x_1 \succeq y_1$; from (C_2) this implies $y_2 \succeq x_2$. If this is expressed in coordinates, two cases must be considered, excluding the trivial case, namely $x_1 = y_1$ and $x_2 = y_2$. We obtain:

1. $x_{11} = y_{11}$ necessarily implies $x_{21} = y_{21}$ and $x_{12} = y_{22}$ and $x_{22} = y_{12}$ otherwise we obtain the trivial case of equality; for the preorders we have $x_1 \succcurlyeq y_1$ implying $y_2 \succcurlyeq x_2$. In other words the difference in choice between x_1 and y_1 stems from the second coordinate since $x_{12} - y_{12} \neq 0$ whereas $x_{11} = y_{11}$. This difference, which is identical to $-(x_{22} - y_{22})$, continues to govern the choice $y_2 \succcurlyeq x_2$ even though the level of the first coordinate is now different from x_{11} since it is equal to $x_{21} = y_{21}$. *Thus for two alternatives at the same level on the first coordinate, the choice will only depend on the second coordinate and this will be so whatever the level on the first.*

2. The second case leads to the same conclusion about the second coordinate. The preferences on the sets X_1 and X_2 are said to be independent. We can formalize this notion.

Definition 6.26

X_1 and X_2 are said to be *preferentially independent* (or *independent in preference*) if:
i) $(x_{11},x_{12}) \succcurlyeq (x_{11},y_{12})$ implies $(z_{11},x_{12})) \succcurlyeq (z_{11},y_{12})$ whatever x_{11} and $z_{11} \in X_1$;
ii) $(x_{11},x_{12}) \succcurlyeq (y_{11},x_{12})$ implies $(x_{11},z_{12}) \succcurlyeq (y_{11},z_{12})$ whatever x_{12} and $z_{12} \in X_2$.

To simplify the next part of the working we shall introduce new notation. For a given vector x, we shall denote by $x_{-i}u_i$ the vector y obtained from x by replacing the *i*th. coordinate of x by u_i. Likewise, if I is a set of indices, $x_{-I}u_I$ is the vector from x in which the coordinates whose indices belong to I have been replaced by the corresponding coordinates of u.

Using this notation, condition i) above is written: $x_{-1}u_I \succcurlyeq y_{-1}u_I$ implies $x_{-1}v_1 \succcurlyeq y_{-1}v_1$; Wakker (1989) also calls this independence relative to coordinate number 1. This can be generalized to any coordinate whatever.

Definition 6.27

The preorder $\succcurlyeq$ on X satisfies *coordinate independence* if and only if for all i, x, y, u_i, v_i:
$x_{-i}u_i \succcurlyeq y_{-i}u_i$ implies $x_{-i}v_i \succcurlyeq y_{-i}v_i$.

It is easy to show (Wakker, 1989) that coordinate independence is equivalent to the property in definition 6.28 below.

Definition 6.28

Overall coordinate independence: for any set of indices I and for all x, y, u_I, v_I
$x_{-I}u_I \succcurlyeq y_{-I}u_I$ implies $x_{-I}v_I \succcurlyeq y_{-I}v_I$

As to the definition of preference independence, this extends to more than two sets.

Definition 6.29

$\{X_1,X_2\}$ is said to be preferentially independent of X_3 if and only if for two alternatives x and y such that $x_3 = y_3$, the decision maker's preferences only depend on values in X_1 and X_2, irrespective of the common level of the third coordinate. Furthermore, if conversely the choice only depends on x_3 and y_3 irrespective of the equal level reached for the first two coordinates $(x_1,x_2) = (y_1,y_2)$, $\{X_1,X_2\}$ and X_3 are said to be *mutually independent*.

Definition 6.30

A set $X = \pi_j\, X_j$ endowed with a preorder $\succcurlyeq$ is mutually independent in preference if for all sub-sets Y of $\{X_1, X_2, ..., X_n\}$, Y is preferentially independent of its complement in $\{X_1, X_2, ..., X_n\}$.

We can see that coordinate independence (definition 6.28) is equivalent to mutual independence in preference (definition 6.30).

The condition for mutual independence in preference is a necessary condition for additivity, but unfortunately it is not sufficient, even in the finite spaces in which we are mainly interested in this book; see Scott and Suppes (1958) and Fishburn (1970).

In the first place, we shall have to assume that each attribute plays a part in the decision maker's choice, *i.e.* that each *coordinate is 'essential'* (Wakker, 1989).

Definition 6.31

A coordinate i is essential if and only if there exist x, u_i and v_i such that $x_{-i}u_i \succ x_{-i}v_i$.

When $n \geq 3$, the coordinate independence is sufficient to bring about additive separability. Surprisingly, when $n = 2$ the situation is a little more complicated and the next theorem requires an extra condition that we now introduce.

Definition 6.32

The hexagon condition (Wakker, 1989)
For all $x_1, y_1, v_1, a_2, b_2, c_2$ $(y_1,a_2) \approx (x_1,b_2)$ and $(v_1,a_2) \approx (y_1,b_2)$ and $(y_1,b_2) \approx (x_1,c_2)$ implies $(v_1,b_2) \approx (y_1,c_2)$.

The hexagon condition results from a particular application of (C_4); this can be represented graphically as shown in figure 6.2.
We must introduce one last condition, also representing part of (C_4), which we shall be using in the next section.

Definition 6.33

(Generalized triple cancellation condition, Wakker, *op.cit.*):
For all x, y, v, w, a_i, b_i, c_i, d_i $y_{-i}b_i \succcurlyeq x_{-i}a_i$ and $v_{-i}a_i \succcurlyeq w_{-i}b_i$ and $x_{-i}c_i \succcurlyeq y_{-i}d_i$
implies $v_{-i}c_i \succcurlyeq w_{-i}d_i$.

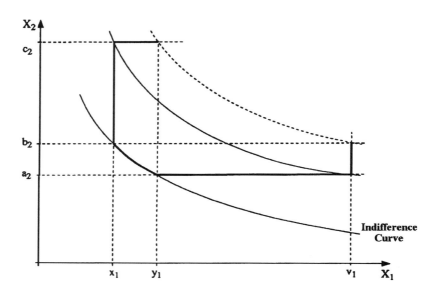

Figure 6.2 Graphic representation of the hexagon condition

Apart from the above conditions which directly concern the preferences, in infinite spaces topological conditions must be introduced on both X and $\succcurlyeq$, which must be continuous (see Fishburn, 1970).

Definition 6.34

The preorder $\succcurlyeq$ is said to be continuous if, whatever y, the sets $\{x \,/\, x \succcurlyeq y\}$ and $\{x \,/\, y \succcurlyeq x\}$ are closed.

Under these conditions, we can state the main theorem of additivity, a theorem which has the advantage of containing all the others.

Theorem 6.35 (Wakker 1989).

If each X_j is an interval in $|R$, if $\succeq$ is a complete and continuous preorder and if each coordinate is essential, then the following properties are equivalent:

i) $\succeq$ is additively separable;

ii) $\succeq$ is coordinate independent ($n \geq 3$) and moreover for $n = 2$ satisfies the hexagon condition;

iii) $\succeq$ satisfies the property of generalized triple cancellation.

In addition, the functions V_d and U_j are jointly cardinal.

This theorem is a generalization of that of Debreu (1960). Given the fact that definitions 6.28 and 6.30 are equivalent, it also contains the classic result of Keeney and Raiffa (1976, p.111). Unfortunately, in view of the first assumption, it does not apply to the case where X is finite. Nevertheless it is on this theorem that the applications are based, including the finite case, as we shall see in the next section. Recently some assumptions, introduced by Gonzales (1996), weaker than those the previous theorem allow to extend theorem 6.35 to products of spaces in which some of them are finite.

In the above theorem, the hexagon condition can be replaced by the Reidemeister condition (Wakker, 1989), also known as the 'corresponding tradeoffs condition' by Keeney and Raiffa (1976).

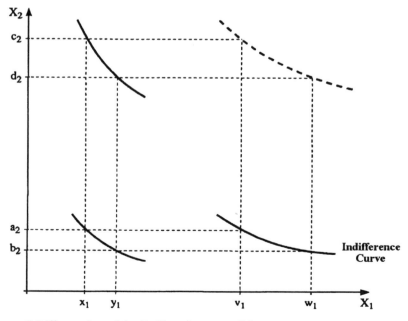

Figure 6.3 Illustration of the Reidemeister condition

Definition 6.36

The Reidemeister condition (named by Krantz *et al.*, 1971).
For all $x_1, y_1, v_1, w_1, a_2, b_2, c_2, d_2$ $(x_1,a_2) \approx (y_1,b_2)$ and $(v_1,a_2) \approx (w_1,b_2)$ and $(x_1,c_2) \approx (y_1,d_2)$ implies $(v_1,c_2) \approx (w_1,d_2)$.

It is easy to verify that, in the case of two coordinates, the triple cancellation condition brings about the Reidemeister condition. Conditions analogous to those in theorem 6.35 are imposed on preference additivity models. Thus Bouyssou and Pirlot (1997) show that a necessary and sufficient condition for condition (10b) to be satisfied is that a cancellation condition comparable to the generalized triple cancellation condition (definition 6.33) be satisfied (*cf.* also Pirlot and Vincke, 1997).

All these conditions have taken us some way away from the finite case, for which there essentially exists only the condition (C), but it is obvious that the notion of preference - or coordinate- independence is an idea which is fundamentally linked to additivity and which deserved mention here.

Finally, note that there is a link between the notions of compensation and coordinate independence (definition 6.27).

Proposition 6.37

If $\succeq$ is a complete preorder then non-compensation implies coordinate independence.

Proof

From definition 5.20, it is clear that $(x_{-i}u_i, y_{-i}u_i)$ MC $(x_{-i}v_i, y_{-i}v_i)$ since the criterion i does not count (equality) and the others have the same values. Suppose $x_{-i}u_i \succeq y_{-i}u_i$; this implies $x_{-i}u_i \succ y_{-i}u_i$ or $x_{-i}u_i \approx y_{-i}u_i$. In the first case the non–compensation implies $x_{-i}v_i \succ y_{-i}v_i$. In the second case $x_{-i}u_i \approx y_{-i}u_i$ implies not($x_{-i}u_i \succ y_{-i}u_i$) and not($y_{-i}u_i \succ x_{-i}u_i$) (proposition 2.15), whence, through non-compensation, not($x_{-i}v_i \succ y_{-i}v_i$) and not($y_{-i}v_i \succ x_{-i}v_i$), i.e. $x_{-i}v_i \approx y_{-i}v_i$. Q.E.D.

The inverse of proposition 6.37 is obviously false because coordinate independence implies the existence of an additive utility function (for $n \geq 3$) which, in certain case (see 6.10, 6.11) is incompatible with non-compensation. When there is non-compensation we may guess that the additive utility, whose existence for $n \geq 3$ we know, is very similar to that of proposition 5.52 giving the lexicographic preorder.

All these considerations have led us some way from practical applications; they are useful, however, for reminding the reader that additivity is by no means an obvious property and above all that it is only true for certain well-chosen functions V_d and U_j which are virtually unique (jointly cardinal), which are perhaps very particular and which have little relation to the decision maker's initial utilities (see example 6.9). Nevertheless, if we pass over these restrictions, assuming that the decision maker is 'naturally' additive, we shall show in the next section that we can derive ways of aggregating preferences.

6.6 Constructing additive utilities

For the benefit of readers who may have skipped the previous section, we begin with a summary of the results needed to use multicriterion methods which are directly based on preference additivity.

a) Under certain conditions, the decision maker's preorder $\succeq$ can be represented by a utility function $V_d = \sum_j U_j$; *this implies that the utilities V_d and U_j are defined jointly and are unique up to an affine transformation (the same slope of the affine function for V_d and U_j).* There is no reason to suppose that any of the U_j representing the n preorders may be added, even when there is existence of an additive decision maker utility function (cf. example 6.9).

b) *When X is finite, the necessary and sufficient condition for the existence of an additive utility function is the condition (C)* (definition 6.17), which is difficult to satisfy. *When X is infinite and $n \geq 3$, the necessary and sufficient condition is that of preference - or coordinate -independence* (definitions 6.28 and 6.30). Roughly speaking, preference independence means that the decision maker's preferences between $x_i = (x_{i1}, x_{i2}, ..., x_{in})$ and $y_i = (y_{i1}, y_{i2}, ..., y_{in})$ depend only on the differences in values of unequal coordinates, independently of the level of equal coordinates. Solvability, Reidemeister or triple cancellation conditions enable more or less satisfiable conditions on the decision maker's preferences to be arrived at.

We have seen that the above results are easier to implement in an infinite set X than in a finite set. There are two ways of going from the discrete to the continuous; the first is to consider probabilities and to say that we are working within uncertainty. In that case all random selections of alternatives are also possible choices, and we thus arrive at the continuous linear utilities of von Neumann-Morgenstern. This is the classical background to the book by Keeney-Raiffa (1976) and the MAUT method (Multi-attribute Utility Theory).

The other way is to consider, for each attribute j, the alternatives a_m and a_M which give minimal and maximal values $U_j(a_m)$ and $U_j(a_M)$ respectively, and to set the problem within the interval $[U_j(a_m), U_j(a_M)]$. This will mean that even if there is no alternative corresponding to an intermediate value within the interval, the values will make sense to the decision maker. This is the position in which we shall place ourselves (see example 6.44).

With the above assumptions about the intervals X_j, the results of section 6.5 tell us the sufficiency conditions to be sure of the existence of an additive utility function. We have seen that the minimal condition is *coordinate independence, or preference independence if the term is preferred.* Looking once again at our familiar model of a choice set $\mathcal{A}$ endowed with n preorders that can be represented by utility functions U_j, we become aware of a rigidity common to many multicriterion methods using such summing procedures as simple weighting: the choice of the decision maker between certain attributes is necessarily independent of the level of satisfaction attained by the other attributes. We develop this idea through two examples.

Examples 6.38

Consider a choice set consisting of cars, with three criteria: price, fuel consumption (miles per gallon) and comfort. The decision matrix is a follows:

	Price	Consumption	Comfort
$a1$	$ 25 000	15	very good
$a2$	$ 25 000	20	good
$a3$	$ 75 000	15	very good
$a4$	$ 75 000	20	good

We can imagine that a decision maker might choose a_2 over a_1, since in a cheap car, fuel consumption is more important than comfort. On the other hand, for a luxury car the choice will most likely be $a_3 \succ a_4$, because at these higher price levels a gallon of gasoline is neither here nor there. The previous preferences are neither coordinate-independent nor mutually preference-independent.

Fishburn (1970) gives a simple and striking example set in time. The first criterion is today's restaurant menu and the second is tomorrow's:

	1st day	2nd day
$a1$	pizza	steak & chips
$a2$	pizza	pizza
$a3$	steak & chips	steak & chips
$a4$	steak & chips	pizza

The decision maker will obviously prefer a_1 to a_2 and a_3 to a_4, so the criteria fail to be independent.

The existence of an additive utility function thus assumes inflexible choice irrespective of the level of satisfaction. The hypothesis of the independence of coordinates may be satisfied, but it is by no means certain that the decision maker's preference $\succeq$ can be expressed in the form $\sum_j U_j$ where U_j values represent criteria; the results in the section above only indicate that there exists V representing the decision maker's preorder and W_j such that $V = \sum_j W_j$. What is the connection between W_j and U_j?

Suppose that coordinate-independence is satisfied and consider $x = (x_1, x_2, ..., x_n)$ and $x' = (x_1, x_2, ..., x'_j ..., x_n)$ which only differ in the jth coordinate. If the decision maker is coherent $x \succcurlyeq x'$ is equivalent to $U_j(x_j) \geq U_j(x'_j)$. But $x \succcurlyeq x'$ is also equivalent to $V(x) \geq V(x')$, which, since all coordinates except the jth are equal, is equivalent to $W_j(x_j) \geq W_j(x'_j)$. Therefore U_j and W_j represent the same preorder and can only differ, if at all, by a strictly increasing transformation.

This does not enable W_j and U_j to be identified. As we have already observed, V and W_j values must be defined jointly up to a unique affine transformation (proposition 6.7). Hence the methods proposed for the practical application of the theoretical framework defined in the previous section are those in which there is *interactive construction of the decision maker's additive utility function, i.e., of V and W_j values together*. Fishburn (1967) listed twenty four different ways of going about this. Here we shall just explain two of them, which can be applied under certainty, as set out by Keeney and Raiffa (1976), the first and most well-known promoters of these methods. The methods we shall show below and the theoretical context described above form the basis of the multicriterion decision methods of the Keeney and Raiffa type; in Europe allusion is often made to the 'American school' to refer to these methods because they have been most widely used in the United States where they occupy a large, perhaps a major part of the multicriterion decision 'market'.

We shall first of all give a theoretical construction based on the property of solvability, limiting ourselves to the case of two criteria for the sake of simplicity.

6.39 The solvability method

Step 1. Start from the worst values x_1^0 for criterion #1 and x_2^0 for criterion #2. Write $U_1(x_1^0) = U_2(x_2^0)$.

Step 2. Take x_1^1 strictly preferred to x_1^0 for criterion #1 and write $U_1(x_1^1) = 1$; this will be the unit of measurement.

Step 3. Ask the decision maker to propose x_2^1 such that (x_1^0, x_2^1) is indifferent to (x_1^1, x_2^0) (solvability). We deduce $U_2(x_2^1) = 1$ since $U_1(x_1^0) + U_2(x_2^1) = U_1(x_1^1) + U_2(x_2^0)$. Likewise, $V(x_1^1, x_2^1) = 2$.

Step 4. Ask the decision maker to give the elements x_1^2 and x_2^2 such that $(x_1^2, x_2^0) \approx (x_1^1, x_2^1) \approx (x_1^0, x_2^2)$; as before we obtain $U_1(x_1^2) = U_2(x_2^2) = 2$.

Step 5. The above procedure is consistent if for the decision maker $(x_1^2, x_2^1) \approx (x_1^1, x_2^2)$, which is none other than the Reidemeister condition. If the decision maker agrees then the construction can continue, otherwise the preorder is not additively separable. The sufficiency condition has thus been tested by the interactive procedure (see figure 6.4).

Following steps. x_1^k and x_2^k are constructed from one value to the next in such a way that $u_1(x_1^k) = U_2(x_2^k) = k$. At each step, we check that the Reidemeister condition is satisfied and thus obtain a net $X_1 \times X_2$ with the values of U_j at each point of intersection. Between points on the net U_j can be estimated by interpolation or by the following method.

The solvability method tells us that the decision maker's preferences are defined by a utility function $V = U_1 + U_2$ where the U_1 values are those which have been constructed. However, for practical reasons it is often convenient to evaluate each criterion on a scale from 0 to 1. When the utility function $V = \sum_j W_j$ is additively separable, there is nothing to tell us that the scales are thereby normalized. In the solvability method we have seen that, though the lower end of the scale can be set to 0, the upper end is not fixed. In the next method, on the other hand, we shall fix the upper and lower points, but we shall seek the aggregate utility function in the form $V = \sum_j w_j U_j$ where w_j values are positive real numbers whose sum is 1 (weights).

Clearly it is sufficient to write $w_j U_j = W_j$ to get back to the usual additive form, and the two formulations are essentially equivalent. In way we arrive at the most common formulation of the problem as seen through MAUT.

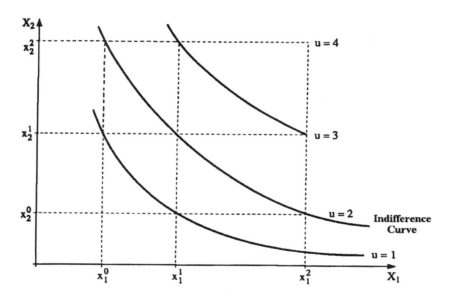

Figure 6.4 The solvability method

6.40 The classical MAUT problem formulation

We seek $V = w_1 U_1 + w_2 U_2$ representing the decision maker's preferences $\succeq$ on $X = X_1 \times X_2$ with, in addition, $0 \leq U_j \leq 1$ and $w_1 > 0, w_2 > 0, w_1 + w_2 = 1$.

Definition 6.41

Given two values x_1 and x_2 belonging to X_1, x_m is said to be *the point equidistant (in preferences) between x_1 and x_2* if and only if there exist y_1 and y_2 such that:
$(x_1,y_1) \approx (x_m,y_2)$ and $(x_m,y_1) \approx (x_2,y_2)$
For $n > 2$ if there exist x and y such that:
$x_{-1}x_1 \approx y_{-1}x_m$ and $x_{-1}x_m \approx y_{-1}x_2$

In other words, if the decision maker is willing to give up $(y_1 - y_2)$ to go from x_1 to x_m, he will be willing to go from x_m to x_2 at the same cost $(y_1 - y_2)$; see figure 6.5.

Proposition 6.42

If the Reidemeister condition (in the case of two criteria) or the coordinate-independence condition is satisfied, then the definition of the equidistant point is independent of the choice of y_1 and y_2 (respectively, x and y).

Proof

The case $n = 2$. Let x_m be such that (x_1,y_1) and $(x_m,y_1) \approx (x_2,y_2)$ and let y'_1 and y'_2 be such that $(x_1,y'_1) \approx (x_m, y'_2)$. The Reidemeister condition implies $(x_m,y'_1) \approx (x_2,y'_2)$.

The case $n = 3$. In the context we are concerned with, coordinate independence is equivalent to generalized triple cancellation (theorem 6.35). Let x_m be the equidistant point such that $x_{-1}x_1 \approx y_{-1}x_m$ and $x_{-1}x_m \approx y_{-1}x_2$ and in addition satisfying $x'_{-1}x_1 \approx y'_{-1}x_m$. The generalized triple cancellation condition when applied twice over implies $x'_{-1}x_m \approx y'_{-1}x_2$. Q.E.D.

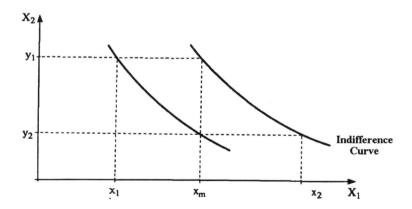

Figure 6.5 The equidistant point

Now we have proved the proposition 6.42, we can go on to describe the *equidistant point method* for two criteria.

6.43 The equidistant point method

Step 1. Determine the worst value x_0 and the best value x_1 relative to the first criterion. Write $U_1(x_1) = 1$ and $U_1(x_0) = 0$.

Step 2. Ask the decision maker for the point equidistant in preference between x_0 and x_1; call this $x_{0.5}$ and let $U_1(x_{0.5}) = 0.5$.

Step 3. Ask the decision maker for the point equidistant between x_0 and $x_{0.5}$; call this $x_{0.25}$ and let $U_1(x_{0.25}) = 0.25$. Do the same for $x_{0.5}$ and x_1 to obtain $x_{0.75}$.

Step 4. Verify that $x_{0.5}$ is the point equidistant between $x_{0.25}$ and $x_{0.75}$. This allows the consistency of the decision maker's replies to be checked but is not a sufficient condition for additivity. In this respect the method is less reliable than the previous one.

Following steps. When a sufficient number of points have been found by the same method, plot the function U_1 by interpolation from these points. Then do the same for U_2.

Determining the weights w_j.

Choose two indifferent points $(x_1,y_1) \approx (x_2,y_2)$; this implies $w_1U_1(x_1) + w_2U_2(y_1) = w_1U_1(x_2) + w_2U_2(y_2)$; as the values $U_1(x_1)$, $U_2(y_1)$, $U_1(x_2)$ and $U_2(y_2)$ are known, this enables w_1 and w_2 to be determined by the above equation in conjunction with the equation $w_1 + w_2 = 1$.

Generalization

The solvability method is intrinsically suited to two criteria. To generalize it, it must be used on the two free criteria left after blocking out the level of the $n - 2$ other criteria; this procedure is legitimate through preference independence. But it will, at best, only allow a section of the additive utility function V to be constructed. We have to begin again for the other levels, and the difficulty and combinatorial complexity of the procedure can be well imagined: it is virtually inoperable. The equidistant point method, on the other hand, is easily generalized because each function U_j is constructed separately. However, the latter method does not allow the sufficiency condition for additivity to be verified.

Example 6.44

We end with a three-criterion example. Suppose we have to choose cars in terms of their price (in thousands of dollars), their comfort and their fuel consumption in miles per gallon. The decision matrix is shown below.

Although we are in the discrete case, we can fictitiously consider that we are reasoning under $X = [20,70] \times [4,10] \times [14,26]$ such that the topological assumptions of theorem 6.35 are satisfied (see the commentary at the beginning of the section). This is equivalent to saying that any triplet of X represents a possible choice. The first thing to do is to cursorily question the decision maker to make sure that coordinate-independence is satisfied. We now proceed to the dialogue stage. One protagonist is the analyst (A) and the other the decision maker (D).

	MIN Price (k$)	MAX Comfort	MAX consumption
a_1	70	10	14
a_2	60	9	20
a_3	50	9	18
a_4	45	8	22
a_5	40	9	16
a_6	40	7	24
a_7	30	6	20
a_8	30	7	22
a_9	20	5	24
a_{10}	20	4	26

A. You consider that (50,9,20) is preferable to (60,9,20); would you also consider that (50,5,24) is preferable to (60,5,24)?

D. Naturally

A. You consider that (60,10,22) is preferable to (60,8,22); would you also consider that (20,10,26) is preferable to (20,8,26)?

D. Yes.

A. You consider that (70,9,18) is preferable to (70,7,22); would you also consider that (20,9,18) is preferable to (20,7,22)?

D. (It is not as easy to answer as before; see the remark at the beginning of the section. Assume the decision maker's answer is 'yes' to avoid invalidating coordinate independence.)

We continue the questioning until we are convinced that it is reasonable to assume coordinate independence, and we can then go on to determine the utilities proper by using the equidistant point method.

a) Determining U_1.

We start off with $U_1(70) = 0$ and $U_1(20) = 1$.

A. What is the lowest level of fuel consumption which would make you indifferent between (20,7,20) and (45,8,?).

D. I would be indifferent between (20,7,20) and (45,8,26).
(This serves to focus thoughts on the more difficult question to follow).

A. Where does the price have to lie so that $(20,7,20) \approx (?,8,26)$ and $(?,7,20) \approx (70,8,26)$? (Here we recognize the question relating to the equidistant point).

D. 50 (We note a slight inconsistency compared to the decision maker's previous response, and this may be the opportunity to refine the responses: the decision maker could now say that $(20,7,20) \approx (50,8.5,26)$ and $(50,7,20) \approx (70,8.5,26)$, which is consistent with the previous response.).

A. (Now that the decision maker has understood the system, we can ask the questions directly.)
Where does the price lie such that $(20,7,20) \approx (?,8.5,26)$ and $(?,7,20) \approx (50,8.5,26)$?

D. 40.

A. Where does the price lie such that $(50,7,20) \approx (?,8.5,26)$ and $(?,7,20) \approx (70,8.5,26)$?

D. 60.

With these and possibly a few more intermediate points we can construct the curve U_1 (see figure 6.6).

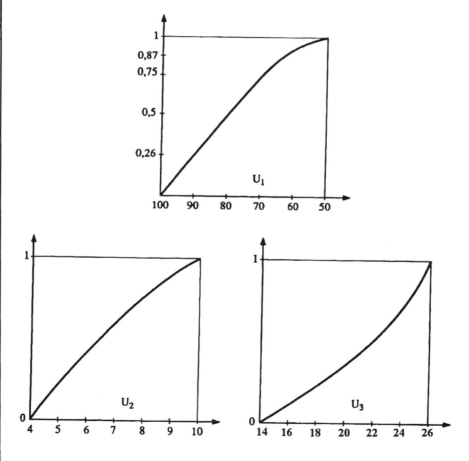

Figure 6.6 Utility functions for example 6.44

b) constructing U_2

U_2 is constructed by the same process. We firstly have $U_2(10) = 1$ and $U_2(4) = 0$. We launch the question/response procedure straight away.

A. Where does the price lie such that $(50,4,20) \approx (60,?,26)$ and $(50,?,20) \approx (60,10,26)$?

D. 6.5.

A. Where does the price lie such that $(50,4,20) \approx (60,?,26)$ and $(50,?,20) \approx (60,6.5,26)$?

D. 5.

A. Where does the price lie such that $(50,6.5,20) \approx (60,?,26)$ and $(50,?,20) \approx (60,10,26)$?

D. 8.

As before, from these responses and other intermediate points if necessary, we can construct the curve U_2 (see figure 6.6).

U_3 *is constructed in the same way* and we give the resulting curve directly in figure 6.6. It now remains to determine the points w_j.

c) Determining the weights

We can note that $V(20,4,14) = w_1U_1(20) + w_2U_2(4) + w_3U_3(14) = w_1$, and likewise $V(70,10,14) = w_2$ and $V(70,4,26) = w_3$. To determine $V(20,4,14)$ we can ask the decision maker what value of comfort $(70,?,14)$ is indifferent to $(20,4,14)$. If the response is 9, this will lead to the equation:

$w_1 = w_2U_2(9)$ or $w_1 = 0.9w_2$, since $U_2(9) = 0.9$.

Similarly, we ask the question: for what consumption is $(70,4,?)$ indifferent to $(20,4,14)$? If the response is 26, this will lead to:

$w_1 = w_3U_3(26)$ or $w_1 = w_3$.

Finally since we have $w_1 + w_2 + w_3 = 1$, then $w_1 = w_3 = 0.32$ and $w_2 = 0.36$, which completes the determination of V. We can then complete the decision matrix by reading off values from the curves U_j.

This analysis gives the ranking: $a_6 \succ a_8 \succ a_{10} \succ a_9 \succ a_4 \succ a_5 \succ a_3 \succ a_7 \succ a_2 \succ a_1$. This is quite consistent with a decision maker who is fairly sensitive to consumption as indicated by U_3.

	price	U_1	comfort	U_2	consump-tion	U_3	V
a_1	70	0	10	1	14	0	0.36
a_2	60	0.25	9	0.9	20	0.35	0.516
a_3	50	0.5	9	0.9	18	0.25	0.564
a_4	45	0.62	8	0.75	22	0.5	0.628
a_5	40	0.75	9	0.9	16	0.1	0.596
a_6	40	0.75	7	0.58	24	0.75	0.688
a_7	30	0.9	6	0.4	20	0.35	0.544
a_8	30	0.9	7	0.58	22	0.5	0.656
a_9	20	1	5	0.2	24	0.75	0.632
a_{10}	20	1	4	0	26	1	0.64

The method of determining weights that we have just used in example 6.44 is actually the compensation method from paragraph 4.7. This being the case, we now take another look at the notion of compensation rates in the continuous case (definition 6.12). In our example, we have just seen that the functions U_j are continuous by construction and, we may assume, differentiable and hence V will have the same properties.

Proposition 6.45

If the functions V and U_j are derivable, then the marginal compensation rate at the point x of criterion i relative to criterion j is equal to $U'_j(x_j)/U'_i(x_i)$.

Proof

At the point x, consider the equation of the equi-satisfaction curve $V(x) = \sum_j U_j(x_j) =$ constant. To keep the same satisfaction, if we increase criterion i by δi, then we must decrease criterion j by δj. The satisfaction will not have changed if:

$$U_i(x_i + \delta_i) + U_j(x_j - \delta_j) = 0 \tag{12}$$

Equation (12) can also be written: $\delta_i[U_i(x_i + \delta_i)/\delta_i] - \delta_j[U_j(x_j - \delta_j)/ (-\delta_j)] = 0.$

When the δ values tend to zero, we obtain: $\delta_i/\delta_j = U'_j(x_j)/U'_i(x_i)$. Q.E.D.

Example 6.46

We now illustrate proposition 6.45 through example 6.44. At the point (40,7,24) corresponding to a_6, we have approximately $U'_1(x_1) = 0.026$, $U'_2(x_2) = 0.17$ and $U'_3(x_3) = 0.13$. By using the above proposition, this means that 'locally' the decision maker considers that a 1 supplementary m.p.g. in fuel consumption is 'worth' 5 k\$ while one unit of comfort worth 6,5 k\$. This type of tradeoff, especially when expressed as here in monetary form, is very useful in making local decisions. Thus, if the decision maker finds a car with performance figures (45,7,26), he will immediately know that he will prefer this to a_6, since he will have obtained 2 supplementary m.p.g. in fuel consumption for 5 k\$ instead of the 10 k\$ that he was willing to pay.

At this point the reader will have noticed that methods for determining additive utilities are long and tedious, and that the questions that must be addressed to the decision maker are not always easy to answer. In practice, the whole of the interactive part is generally computerized and the U_j functions are constructed automatically from the decision maker's responses. Nevertheless much is demanded of the decision maker, and his ability to respond consistently to subtle questions may well be in doubt. Some experiences shown that a direct evaluation of the weights generally does not give the same values as using one of the previous MAUT methods (Olson *et al.* , 1995). There are weaknesses on the theoretical side, too, for we know that even if we assume that decision maker utility is additive, U_j and V are defined conjointly and must be precisely tuned (see example 6.9). Using the method of construction described, we can never be completely certain that we have obtained the right functions, even though the method is partially consistent.

On the other hand, one of the method's advantages is that it is easily extended to utilities in the domain of the uncertain. We have decided not to include this aspect of decision here, but it is nevertheless of great interest and a considerable part of Keeney and Raiffa (1976) is devoted to it. Its other advantage is that it provides a lot of information, assuming that the latter is considered reliable, in particular through the tradeoff ratios.

Finally note that in spite of everything that has been said and proved above we may continue to think, against all the evidence, that the cardinal utilities corresponding to each criterion, once they have been constructed by, *e.g.*, the equidistant point method or one of the methods described in section 2.4.4, are the 'right utilities', and to deduce from this, without further verification, that the decision maker utility is of the form $V(x) = \sum_j w_j U_j(x)$. The weights must then be determined, and this can be done by one of the methods in chapter 4; we have just seen that the compensation method of section 4.7 was used in the equidistant point method. An initial interpretation of the results of this chapter could be that they form a theoretical justification for the use of the methods of chapter 4, provided the decision maker possesses an additively separable preorder, which would be subsequently verified by the decision maker certifying that his choices are consistent with those determined by the weighted sum. This is obviously an arguable point of view since the decision maker is necessarily influenced by what the expert proposes; to be scientifically rigorous and honest, it would be better to check first that the decision maker preorder is additive, but as we have seen this is unfortunately difficult—hence the tendency to adopt the initial interpretation as a 'less bad' course.

6.7 The UTA method

The UTA method we describe next does not have entirely the same objectives as the MAUT method we have just seen, but, like the latter, it is based on the idea that the decision maker preferences are additive. From this starting point, and using partial information, it attempts to reconstruct the said preferences.

6.7.1 Theoretical principles

In the UTA method (**UT**ilité **A**dditive) a computerized mathematical procedure is used to evaluate the utility functions associated with each criterion within a context of global comparisons that the decision maker makes in a sub-set of alternatives, called the *reference set*. Here the term global, or holistic, refers to the 'subjective' comparison of two alternatives while simultaneously bearing in mind the whole set of attributes—as we saw in chapter 4. *Each alternative is considered as a whole and there is no separation—even mental—between criteria.* Once the utility functions associated with each criterion have been evaluated, a decision maker's additive utility function enabling all the alternatives in the choice set to be evaluated is constructed. In the UTA method it is presupposed that the utilities U_j associated with each criterion are piece-wise linear. The process is then one of determining the parameters of these functions to be consistent with the decision maker's preferences stated during global comparisons of the alternatives in the reference set. This matching, often called ordinal regression, leads finally to solving a linear program.
A certain number of post-optimal analyses are also required in this method. The basic UTA method, described in Jacquet-Lagrèze and Siskos (1982), has seen numerous applications and benefited from subsequent improvements (the PREFCALC and MINORA software packages—see below and chapter 10). Now we shall go a little further into the details.

We shall denote by $U_j(a)$ the utility of an alternative a, belonging to a choice set $\mathcal{A}$, relative to criterion j ($j = 1, 2, ..., n$). We shall assume, after any necessary translation, that for any alternative we have $U_j(a) \geq 0$. UTA assumes the decision maker utility V can be expressed in the form:

$$V(a) = \sum_j U_j(a) \qquad (13)$$

As we saw for MAUT (at the end of 6.39), equation (13) is equivalent to:

$$V(a) = \sum_j w_j V_j(a) \qquad (14)$$

where values of w_j can be interpreted as weights and where the functions V_j are normalized. For the moment we shall set aside the interpretation in terms of weights and come back to it later.

We define the *ideal point* in $\mathcal{A}$ as the fictional alternative whose coordinates separately achieve the best value for each criterion $a^M = (a^M_1, a^M_2, ..., a^M_n)$, and the *anti-ideal* point as the alternative whose coordinates separately achieve the worst value for each criterion $a^m = (a^m_1, a^m_2, ..., a^m_n)$. We shall take up these ideas again in more detail in section 8.3; here, an intuitive notion will be sufficient. We normalize the utilities by setting:

$$V(a^M) = 1 \Longrightarrow \sum_j U_j(a^M{}_j) = 1 \tag{15}$$

and $V(a^m) = 0 \Longrightarrow U_j(a^m{}_j) = 0$ for all j $\tag{16}$

If we denote by A_r the sub-set of $\mathcal{A}$ serving as reference set for questioning the decision maker, then for each pair $(a,b) \in A_r$, the decision maker will state his global preference or indifference and this then allows us to say, using a preorder (see 2.5) with indifference threshold s:

the decision maker states $a \succ b \Longrightarrow V(a) > V(b) \Longrightarrow [V(a) - V(b)] > s$ $\tag{17}$
the decision maker states $a \approx b \Longrightarrow V(a) \approx V(b) \Longrightarrow |V(a) - V(b)| \leq s$ $\tag{18}$

For each alternative in A_r the calculated utility function $V'(a)$ differs from the actual value $V(a)$ by an error $\sigma(a)$:

$$V'(a) = V(a) + \sigma(a)$$

Expressions (17) and (18) can thus be re-written:

$$a \succ b \Longrightarrow \sum_j [U_j(a_j) - U_j(b_j)] + \sigma(a) - \sigma(b) > s \tag{19}$$

$$a \approx b \Longrightarrow \sum_j |U_j(a_j) - U_j(b_j)| + \sigma(a) - \sigma(b) \leq s \tag{20}$$

Assuming there is transitivity the decision maker need only perform $(m_r - 1)$ comparisons of alternatives where m_r is the number of alternatives in A_r.

As we have already said, the functions $U_j(x)$ are supposed to be piece-wise linear. To define the pieces we choose n_j points in the interval $[a^m{}_j, a^M{}_j]$ where the function U_j is defined, giving:

$$x_h = a^m{}_j + [(h - 1)/(n_j - 1)][a^M{}_j - a^m{}_j]$$

with h varying from 1 to n_j and the parameter n_j defined by the decision maker. Finally we have to define the values $U_j(x_h)$, the intermediate values $U_j(x)$ being defined by linear interpolation.

Each function U_j is monotone increasing or decreasing according to whether the criterion is to be maximized or minimized; we thus have:

if U_j is increasing, $U_j(x_{h+1}) - U_j(x_h) \geq 0$ $h = 1, 2, ..., n_j$
if U_j is decreasing, $U_j(x_{h+1}) - U_j(x_h) \leq 0$ $h = 1, 2, ..., n_j$ $\tag{21}$

With the constraints (15), (16) and (19) through (21), and the objective function:

$$\text{minimize } z = \sum_{a \in Ar} | \sigma(a) | = \sum_{a \in Ar} [\sigma^+(a) - \sigma^-(a)] \tag{22}$$

we obtain a linear program with the following dimensions:

variables	Number
$U_j(x_h)$	$\sum_j n_j$
$\sigma^+(a), \sigma^-(a)$	$2m_r$

constraints	number
(15)	1
(16)	n
(19) and (20)	$m_r - 1$
(21)	$\sum_j (n_j - 1)$

plus the constraints associated with non-negativity. The constraints mostly take the form of inequalities so there will in general be an infinite number of feasible points from which will be chosen the solution(s) which maximize z (defined in (22)). We shall call the value of this optimum z^M. Practical experience with the method has shown that feasible solutions are also obtained using other objective functions, for example to achieve a finer match, as defined through Kendall or Spearman correlation coefficients, between the preorder calculated by the method and the preorder given by the decision maker in the reference set.

This is why the UTA method does not stop at the previously found optimal solution, but, using very powerful post-optimal analysis processes, continues to explore the convex envelope of solutions of the type $z \leq z^M + kz^M$, i.e. the set of solutions which do not exceed z^M by more than a fraction k of z^M. At the end of the process, UTA outputs a value $U_j(x_h)$, this being the mean of the values obtained by post-optimal analysis. From the $U_j(x)$ values obtained $V(a)$ is evaluated for all the alternatives in $\mathcal{A}$, belonging or not to A_r.

The above operations can all be performed easily using the software PREFCALC, written by the authors of the UTA method. The current version (3.0) can handle ten-criteria problems with 200 alternatives, including around 30 alternatives in the reference set and five linear interpolation points in the definition of U_j values (see chapter 10 for further details). Since its appearance in 1983, many copies of PREFCALC have been sold and a new improved version has been launched (see UTA PLUS, Chapter 10).

6.7.2 An example of an application

In the work of Jacquet-Lagrèze and Siskos referred to earlier (1982) there is an interesting application of UTA to automobile choice with six criteria and ten cars in the reference set. To show the possibilities of the method we shall apply it to our usual candidate selection problem, although it should be noted that the small number of alternatives – six – would not normally justify the use of the method.

Example 6.47 Candidate selection by the UTA method.

The table below reproduces the data once again, with values of the coordinates of the ideal and anti-ideal points for the criteria to be maximized or minimized. The bottom lines in the table show the actual and possible extreme values for the criteria; the latter are not necessarily attained by the alternatives in the table, as for example in the case of the two criteria on the right where we have taken the extreme values on the measuring scale as ideal and anti-ideal values.

Alternatives	STU MAX $C1$	EXP MAX $C2$	AGE MIN $C3$	INT MAX $C4$	TES MAX $C5$
Albert	6	5	28	5	5
Blanche	4	2	25	10	9
Donald	5	7	35	9	6
Emily	6	1	27	6	7
Georgia	6	8	30	7	9
Helen	5	6	26	4	8
Best	6	8	25	10	9
Worst	4	1	35	4	5
Ideal point	6	8	25	10	10
Anti-ideal point	4	1	35	0	0

Figure 6.7 Problem data and ideal and anti-ideal points

Once the data have been entered, the PREFCALC software immediately calculates the ideal and anti-ideal values. It then asks us to set the number n_j of interpolation points corresponding to the $(n_j - 1)$ linear segments that we wish to consider for each criterion j; we shall take $n_j = 3$, giving two linear segments in this example. PREFCALC also suggests the number of reference alternatives we should be considering in A_r, including among these the ideal and anti-ideal points if desired. In our example PREFCALC suggests seven alternatives; we actually have only eight, including the ideal and anti-ideal. We shall therefore just take half the initial alternatives together with the ideal and anti-ideal points making a total of five alternatives. This will lead to greater uncertainty in the determination of $U_j(x_h)$ values but will still allow us to go on with the demonstration.

Lastly, PREFCALC asks us to make global preference comparisons in the set A_r. We choose to place the alternatives of Georgia, Albert and Emily in the set A_r because they have widely differing experience, the criterion which we consider to be the most important here (though there is in fact no need to formalize how much

important, it is simply additional knowledge which can inform the choice of *Ar.*)
From here we can now tell PREFCALC that our preferences in *Ar* are as follows:

Alternatives	Preference order
Ideal	1
Georgia	2
Albert	3
Emily	4
Anti-ideal	5

With these data, PREFCALC solves the linear problem and provides us with the
evaluated functions $U_{j,}$, or V_j to be exact (see (14)), which are shown graphically in
figure 6.8.

Unless otherwise stated, PREFCALC assumes that the reference alternatives are
equidistant in utility V over the interval [0,1] and of course decreasing in the order
indicated by the decision maker (see the bottom right-hand frame of figure 6.8).

These values serve to determine the weights w_j in expression (14), V_j taking
values within [0,1]. Values of w_j appear in the upper right-hand corners of the
frames in figure 6.8.

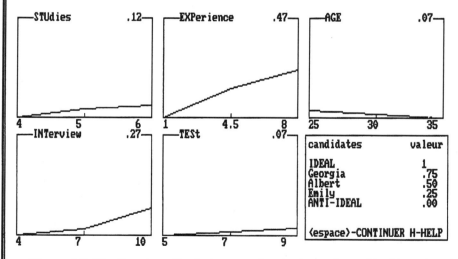

Figure 6.8 The functions V_j relative to each criterion estimated by PREFCALC

Finally, using expression (14), PREFCALC estimates the value $V(a)$ of all the
alternatives in $\mathcal{A}$. These are shown in the table following figure 6.9.

Alternative	$V(a)$
Ideal	1
Georgia	0.75
Helen	0.68
Donald	0.63
Albert	0.50
Blanche	0.43
Emily	0.25
Anti-ideal	0.00

Figure 6.9 Values $V(a)$ estimated by PREFCALC

Despite the limitation referred to above concerning the number of alternatives in the reference set, the results differ little from those we obtained by the simple weighting method of chapter 4: Georgia $\succ$ Donald $\succ$ Helen $\succ$ Blanche $\succ$ Albert $\succ$ Emily. There is some similarity between the two processes, though UTA is more powerful and general in the choice of $V(a)$ which is piece-wise linear and not linear as for the weighted sum method.

PREFCALC is an interactive method: calculations can be repeated after changing the assumptions made in the choice of ideal and anti-ideal points, in the number of interpolation points n_j and in the choice of reference alternatives over which the decision maker's preferences can be varied. It is also possible to give values of the weights w_j directly instead of calculating them. All these features facilitate the work of the software user.

6.7.3 Features and extensions.

The UTA method has been used an appreciable number of times in fields as diverse as purchasing choice (Jacquet-Lagrèze and Shakun, 1984), freeway route planning (Siskos and Assimakopoulos, 1989), strategic choice in small and medium businesses (Richard, 1983) and wholesale distribution network management (Moscarola and Siskos. 1983).

The advantages of the method include the useful hypothesis of non-linearity of $V(a)$, assuming of course that it is additive. Note too that the method requires only a preorder (non cardinal) of the alternatives in the reference set. Lastly, there is the post-optimal sensitivity analysis to the method's credit. On the down side is the fact that the method relies on a rather complex calculation process, though this is largely handled by PREFCALC.

But despite all we have said in favor of the method, global comparisons within the reference set are a source of difficulty which should not be underestimated. If no 'natural' reference set exists the results can be considerably affected. The central role of these comparisons in the UTA method brings it close to the alternative-comparison methods that we shall be looking at in section 8.2, the best known of which is by Zionts (1981). The UTA method however differs in at least two essential ways from comparison methods which prevent its inclusion with the latter. The first is that the function $V(a)$ is not directly derived from the comparisons. The second is that the comparisons we require the decision maker to make in comparison methods are sequential and follow each other until a satisfactory set of weightings is achieved and a 'best' alternative is identified, whereas in UTA the decision maker gives his preferences once and for all at the start.

We shall conclude this chapter by citing the variants and extensions of the UTA method. Jacquet-Lagrèze *et al.* (1987) proposed the use of the UTA method as an intermediate stage in solving a linear multicriterion program. The UTA method is used to adjust a utility function V to the set of non-dominated extremal alternatives which are solutions to the multicriterion program. The function V is then used to evaluate the feasible solutions to the initial program.

Stewart (1987) proposed introducing some aspects of UTA (non-linear utility function, minimization of decision maker inconsistency) into Zionts' progressive information aggregation procedure (1981) and into those of Koksalan *et al.* (1984) and Korhonen *et al.* (1984), both derived from Zionts (see section 8.2). The aim was to use the UTA method in the selection problem to eliminate the alternatives one by one.

But perhaps the most pertinent variant up to now has been the MINORA (Multicriteria Interactive Ordinal Regression Analysis) method of Siskos (1986), and this is for two main reasons. The first is the improvement in the objective function z when we take into account the fact that the error $\sigma(a)$ can take any sign in expressions (19) and (20); this is in fact how we have considered it. This possibility, which did not feature in the 1982 paper, has actually been proposed by Siskos and Yannacopoulos (1985). The second reason is the incorporation of MINORA into an ambitious DSS with multiple interaction opportunities where the decision maker can step in to achieve a better tuning of the weights w_j to the data of the problem. In the cited paper by Siskos (1986) and in those of Siskos and Zopounidis (1987) and Hadzinakos *et al.* (1991) the reader will find several useful applications of MINORA: evaluating sales outlets in a network of stores, evaluating investments for a venture-capital firm and developing a policy to cope with earthquakes in Greece.

7 OUTRANKING METHODS

7.1 Introduction

In this chapter we describe and analyze a wide range of multicriterion decision methods gathered under the heading of outranking methods. They all involve the fertile concept of the outranking relation, which emerged during the sixties through the influence of French workers. Among the latter the name of B. Roy stands out as the initiator of the famous ELECTRE method, which is the prototype outranking method. Roy has become the leader of a whole generation of workers dedicated to promoting multicriteria decision and increasing knowledge in this domain.

Since then a large number of variants and closely related methods have been devised and applied by both students and professionals, mainly in Europe; it is for this reason that this branch is frequently referred to as the 'European school of multicriteria decision', in contrast to the American school which is much more attached to the methods of multi-attribute additivity described in the last chapter. Adverse criticism of outranking methods compared to multiattribute utility has always come from the States side of the Atlantic, this criticism being aimed mainly at their so-called lack of theoretical foundations and at their pragmatism. The argument goes back a long way and has involved many authors, but more recently new angles have emerged which we feel to be very interesting and on which we shall be commenting later (sections 7.2.2 and 7.2.3). For the moment we shall simply note that the theoretical framework of outranking methods has changed during the nineties giving them stronger foundations.

This chapter is organized as follows. In the first section (7.2) we define the basic elements underlying the idea of outranking. Section 7.3 is devoted to the first of these methods: ELECTRE and its variants. In section 7.4 we describe in detail the highly accomplished PROMETHEE method and give an example for the reader to appreciate its possibilities. The last section contains a wide range of other outranking methods, no less important than the others from various points of view but outside the scope of this book for all but a summary of the main features.

7.2 Outranking relations

7.2.1 The intuitive ideas

The methods we shall be looking at in this chapter are all based on a pairwise comparison of alternatives. What we do is to compare alternatives systematically, criterion by criterion, as in the method of Condorcet (chapter 5). This is unlike the UTA method (see section 6.7) and unlike the methods of comparison of alternatives that we shall be looking at in section 8.2, in which alternatives are compared globally and not criterion by criterion.

After having made the comparisons we construct a *concordance coefficient* c_{ik} associated with each pair of alternatives (a_i, a_k). There are numerous ways of calculating this coefficient, but to give a clear idea of the process we shall give a simple yet widely used method which we shall formalize more precisely in the next section. Here, c_{ik} is defined as the sum of the weights of the criteria for which the alternative a_i is better than or equivalent to a_k. On calculating c_{ik} for all the pairs of alternatives in $\mathcal{A}$ we obtain an $m \times m$ matrix called the *concordance matrix* $C = (c_{ik})$. This matrix is then used in various ways depending on the precise method. In some, which we can call concordance methods, the matrix is practically the only information used to produce the aggregate preorder (see 7.5). In other methods, further notions are introduced.

As we have said, outranking methods use the concept of concordance; this is completed by the notion of *discordance*, reflecting in concrete terms what would seem to have been Condorcet's original idea: when an alternative a_i is at least as good as an alternative a_k for a *majority of criteria, and there exists no criterion for which a_i is substantially less good* than a_k, we can safely say that a_i outranks a_k. Using modern terminology we write $a_i \, S \, a_k$.

An *outranking relation* such as the one we have just referred to defines a preorder, which may be partial, in the hitherto undifferentiated set of non-dominated alternatives for the product   preorder . The price which has to be paid in constructing this relation is to accept a degree of subjectivity in choosing a procedure from the various possibilities which exist for making sense of the above italicized features of Condorcet's idea. From this come the various variants that we shall be meeting together with some accusations of 'recipe' sometimes attached to the proposed ideas.

We shall illustrate these ideas with a very simple example.

Example 7.1

Consider a choice set with four alternatives, and consider three criteria to be maximized, as shown in the decision matrix below.

Suppose all the criteria are equally important, *i.e.* all the weights w_j are equal to 1. The estimations all belong to the interval $[0,100]$.

		Criteria		
		1	2	3
Alternatives	a	90	10	100
	b	100	0	100
	c	90	100	90
	d	50	50	100

Comparing alternatives *a* and *b*, we can state that *a* *outranks* *b* (abbreviated to a *S b*), this being a perfect illustration of Condorcet's idea. The reader will agree that the difference of 10 in criterion #1 is not an 'outstanding' inferiority sufficient to cancel the superiority or equality of a over b for the other two criteria.

For the same reason, and since the situation is perfectly symmetrical, we must also admit that *b S a* and likewise *c S a*. However, we could refuse *c S b*, although this decision could be reversed if we change the assumption of equality of weights. Note too that we can refuse to accept *b S d* and *d S b*; it follows that alternatives *b* and *d* are not ranked by the outranking relation and remain *non-comparable*.

Finally, note that though we have *c S a* and *a S b*, we do not have *c S b*: an outranking relation can very well lead to *intransitivity*.

The above example shows that outranking relations depend largely on the values given to the weightings w_j and the indifference thresholds s_j on which will depend the points at which discordance occurs. We shall now try to give more precise definitions of the concepts introduced so far.

7.2.2 The basic concepts

Recent work by Roy (1990) has united and clarified his own and others' efforts (including Bouyssou, Jacquet-Lagrèze, Roubens, Vansnick and Vincke) with the aim of giving a solid theoretical basis to outranking methods. Here we shall outline this work, simplifying a little where necessary.

We shall be using the concept of pseudo-criterion introduced in section 2.5, as this will allow us to express the decision maker's preferences in the most general way. For each criterion j, $j = 1, 2, ..., n$, we define the indifference threshold q_j and the preference threshold p_j and we assume that $p_j > q_j$. The threshold q and p correspond respectively to thresholds s_1 and s_2 in definition 2.41, section 2.5. These thresholds will be used to compare the evaluations $U_j(a)$ and $U_j(b)$ of the alternatives a and b relative to criterion j.

We define the preference relation $\succ_j$ relative to criterion j as follows:

$$a \succ_j b \Longleftrightarrow U_j(a) > U_j(b) + p_j \tag{1}$$

The indifference relation $\approx_j$ is defined by:

$$a \approx_j b \Longleftrightarrow U_j(b) - q_j \leq U_j(a) \leq U_j(b) + q_j \tag{2}$$

The criterion j is thus represented by a pseudo-order (see 2.41). We now define the *outranking relation S_j associated with criterion j*, which covers all cases where a is not strictly worse than b:

$$a \, S_j \, b \iff U_j(a) \geq U_j(b) - q_j \tag{3}$$

For each pair of alternatives (a, b), we define the set of criteria $C(a, b)$ such that a is not strictly worse than b by:

$$C(a, b) = \{j \, / \, U_j(a) \geq U_j(b) - q_j \} \tag{4}$$

We shall call this sub-set the *concordance set*. Likewise we define the *discordance set* between a and b as the sub-set of criteria for which b is indisputably preferred to a.

$$D(b, a) = \{j \, / \, U_j(b) > U_j(a) + p_j\} \tag{5}$$

These two sets more or less correspond to Condorcet's idea of 'at least as good' for $C(a, b)$ and 'manifestly inferior' for $D(b, a)$. We can state the obvious proposition:

Proposition 7.2

For any pair of alternatives (a, b) we have $C(a, b) \cap D(b, a) = \varnothing$.

Proof

Inequalities (4) and (5) cannot simultaneously be satisfied since $q_j < p_j$. Q.E.D.

It may happen that criteria do not belong to any of the above sub-sets; this is the case if we have neither $a \succ_j b$ nor $b \succ_j a$; from the definitions this will occur if:

$$U_j(b) - p_j \leq U_j(a) < U_j(b) - q_j \tag{6}$$

If (a, b) satisfies (6) b is said to be *weakly preferred* to a (see paragraph 2.5), and this is denoted $b \succ^w_j a$, where the 'w' for weak is written as superscript to leave room for the criterion j as subscript.

In the figure 7.1 we show the values $U_j(b)$ together with the values $U_j(b) \pm q_j$ and $U_j(b) \pm p_j$. These values define zones, and if $U_j(a)$ lies in a given one of these zones, the corresponding relation indicated in VISION 1 will hold.

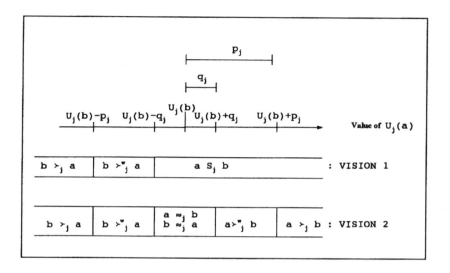

Figure 7.1 The concept of weak preference

VISION 2 offers another interesting point of view. The zone $a\ S_j\ b$ is divided up according to the three possibilities; we see clearly how weak preference defines an intermediate zone between strict preference and indifference and is interpreted as a hesitation between indifference and strict preference.

Coming back to our sub-sets of criteria, we can define a sub-set C_w for *weak preference* of b over a:

$$C_w(b, a) = \{j\ /\ U_j(b) - p_j \leq U_j(a) < U_j(b) - q_j\} \tag{7}$$

this being the set of criteria such that $b \succ^w_j a$. Again, we can state an obvious result from the definitions.

Proposition 7.3

The three sets $C(a,b)$, $C_w(b,a)$ and $D(b,a)$ form a partition of the set of criteria.

What we aim to do is to construct an overall outranking relation which aggregates the pseudo-orders relative to each criterion, based on the 'importance' of the above three sets. The way this 'importance' is measured varies according to the outranking method. We shall call the values of 'importance' associated with each set $f_c(a,b)$, $f_{cw}(b,a)$ and $f_d(b,a)$; in the next section we shall see what values are adopted by ELECTRE I. We can now define the following coefficients.

Definition 7.4

The *concordance coefficient* $c_{a,b} = f_c(a,b) + f_{cw}(b,a)$ (8)
The *discordance coefficient* $d_{a,b} = f_d(b,a)$
(the reader should note carefully the inversion in the order of a and b: the discordance of a relative to b is measured by how much b is preferred to a, *i.e.* $f_d(b,a)$)

From these coefficients we can construct the *concordance matrix* ($m \times m$) $C = (c_{ik})$ containing the coefficients $c_{a,b}$ for all the pairs of alternatives (a_i, a_k), and likewise the *discordance matrix* ($m \times m$) $D = (d_{ik})$ formed from the coefficients $d_{a,b}$. Outranking methods in general are based on these matrices.

7.3 The ELECTRE method

7.3.1. The various versions

The ELECTRE method (from the French **EL**imination **Et** **C**hoix **T**raduisant la **RE**alité) – or rather the various versions of the method – are well known in Europe as evidenced by the abundance of literature on the subject and applications which use it. Since the advent of the original version, ELECTRE I (Roy, 1968a), other versions have succeeded it; these are summarized in figure 7.2 below.

Note that, since 1978, all versions of ELECTRE incorporate the notion of pseudo-criterion mentioned in the section above. Version II and especially version III use the notion of fuzzy set.

ELECTRE version	First reference	Type of criterion	Weights required	Fuzzy	Type of problem
I	Roy (1968a)	simple	yes	no	selection
II	Roy and Bertier (1973)	simple	yes	a little	ranking
III	Roy (1978)	pseudo	yes	yes	ranking
IV	Roy & Hugonnard (1982)	pseudo	no	no	ranking
IS	Roy and Skalka (1985)	pseudo	yes	no	selection

Figure 7.2 The various versions of ELECTRE

The weights featuring in table 7.2 should be understood as a measure of the importance the decision maker attaches to the criteria rather than as any formal substitution ratio, whether marginal or not, since the values associated with each criterion are not aggregated into a global value such as the sum in the simple weighting method. This applies to all ELECTRE methods, which are not as strongly compensatory as the standard weighting method; we shall come back to this point. However, the use of 'weights' to calculate the concordance and discordance coefficients does involve compensation between criteria. ELECTRE IV is the only method which does not use weights, but functions through a set of outranking relations, embedded in each other, which are constructed progressively. The MELCHIOR method of Leclerq (1984) and the TACTIC method of Vansnick (1986), which we shall be referring to in 7.5.2, are also based on these ideas.

For the problem formulation, we have reintroduced the terms used in 2.6.2. Versions I and IS of ELECTRE yield a global outranking relation defined over the alternatives, the relation being represented by an oriented graph. The relation is dependent on preference thresholds defined by the decision maker. From this outranking relation, the information required to make a selection can be extracted. The other versions of ELECTRE give an aggregate preorder of the alternatives. To illustrate better the principles involved, we shall describe the first version of ELECTRE in more detail.

7.3.2 ELECTRE I

After describing ELECTRE I by giving the relevant propositions and proofs, we shall apply it to an example.

In ELECTRE I, there are only criteria, not pseudo-criteria; consequently $C(a,b) = \{j \mid a \succcurlyeq_j b\}$ and $D(b,a) = \{j \mid b \succ_j a\}$. If we consider two alternatives $a_i = (a_{i1}, a_{i2}, ..., a_{in})$ and $a_k = (a_{k1}, a_{k2}, ..., a_{kn})$ where, with the usual notation, $a_{ij} = U_j(a_i)$, then assuming that the criteria are maximized, we also obtain $C(a_i, a_k) = \{j \mid a_{ij} \geq a_{kj}\}$ and $D(a_k, a_i) = \{j \mid a_{ij} < a_{kj}\}$. From this the *concordance coefficient* of definition 7.4 is defined as follows:

$$c(a_i, a_k) = c_{ik} = \sum_{j \in C(a_i, a_k)} w_j \qquad (9)$$

where values of $w_j > 0$ represent the weights expressing the importance of the criteria, once normalized such that $\sum_j w_j = 1$.

Proposition 7.5

The condition $0 \leq c_{ik} \leq 1$ always holds. Furthermore, $c_{ik} = 1$ if and only if a_i dominates a_k for the product preorder and likewise $c_{ik} = 0$ if and only if a_k dominates a_i for the product preorder.

Proof

The inequality is obvious since $\sum_j w_j = 1$. Note that $c_{ik} = 1$ is equivalent to $a_i \preccurlyeq_j a_k$ for all j, and this is indeed the definition of dominance for the product preorder (definition 3.1.) Q.E.D.

The *discordance coefficient* is defined a little differently:

$$d(a_k, a_i) = d_{ik} = (1/\delta)\max_{j \in D(ak, ai)}(a_{kj} - a_{ij}) \tag{10}$$

where $\delta = \max_j \max_{i,k}(a_{kj} - a_{ij})$.

In other words, d_{ik} represents the maximum difference in utility between a_i and a_k for criteria where a_k strictly dominates a_i; this difference is divided by the maximum possible intra-criterion difference for all the alternatives and all the criteria.

Proposition 7.6

The inequalities $0 \leq d_{ik} \leq 1$ are always true. Furthermore, $d_{ik} = 0$ if and only if a_i dominates a_k for the product preorder .

Proof

The first statement is obvious. The second results from the fact that $d_{ik} = 0$ is equivalent to $D(a_k, a_i) = \varnothing$. Q.E.D.

Clearly it is necessary to normalize the criteria for d_{ik} to be more or less comparable, nevertheless their values are sensitive to the normalization procedure adopted.

On the other hand, the *concordance coefficients* c_{ik} are purely ordinal in the sense that their values only depend on the preorder $\succcurlyeq_j$ and not on U_j. Of course they do depend on the weights, which can be problematical as we shall see in section 7.3.3. However, *concordance is clearly non-compensatory*, in other words the outranking relation defined by $a_i S\, a_k \Longleftrightarrow c_{ik} \geq c_{ki}$ is, as the reader will see immediately, non-compensatory in the sense of definition 5.21; numerous results on this can be found in Bouyssou and Vansnick (1986).

If now we set the values of two parameters $0 < t_c < 1$ and $0 < t_d < 1$, which we shall call respectively *concordance and discordance thresholds*, and which represent the minimum value that we wish c_{ik} to take and the maximum discordance that we are prepared to accept for d_{ik}, we can finaly define an outranking relation.

Definition 7.7

a_i is said to *outrank* a_k, (denoted $a_i\, S\, a_k$,) if and only if $c_{ik} \geq t_c$ and $d_{ik} \leq t_d$.

Proposition 7.8

The relation S is reflexive. When it is transitive, the preorder defined by S extends the product preorder .

Proof

The first part of the proposition results from the fact that $c_{ii} = 1$ and $d_{ii} = 0$. As for second part, note that according to propositions 7.6 and 7.5, if a_i dominates a_k for the product preorder, then clearly $a_i \, S \, a_k$. In other words, the relation S is a finer one than the product preorder, ranking more pairs of alternatives, and when this expression has a sense (see below), its Pareto set is included in that of the product relation (proposition 3.25.) Q.E.D.

The relation S is represented graphically as a so-called *outranking graph* using the conventions of figure 2.6, *i.e.* an arc from a_i to a_k exists when $a_i \, S \, a_k$.

The main problem with S is that it is not necessarily transitive. This can be seen from example 7.1 where although we obtained $c \, S \, a$, $a \, S \, b$, we did not have $c \, S \, b$ (see also examples 7.12 and 7.13). In cases where S is intransitive, S is not a preorder and the notion of Pareto set does not exist. For a way in to the selection problem enabling a recommendation to be given to the decision maker, Roy (1968a) makes use of a well known graph notion, that of the kernel (see Berge, 1970).

Definition 7.9

A *kernel* associated with an outranking relation S is any sub-set N of alternatives in $\mathcal{A}$ such that:
i) for any alternative $a_k \in \mathcal{A} - N$, there exists an alternative $a_i \in N$ such that $a_i \, S \, a_k$, the so-called *stability property* of graphs;
ii) whatever the alternatives a_i and a_k belonging to N, we have neither $a_i \, S \, a_k$ nor $a_k \, S \, a_i$, the so-called *absorption property* of graphs.

Figure 7.3 shows how easy it is to find graphs without kernels and with several kernels.

The first statement in definition 7.9 is that the kernel N is complete, in a slightly weaker sense than in definition 3.9; the second states that N is in a certain sense minimal. Definition 7.9 is a generalization of the Pareto set (see proposition 3.11) as proposition 7.10 shows.

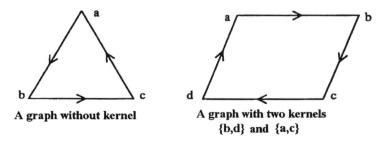

A graph without kernel A graph with two kernels
{b,d} and {a,c}

Figure 7.3 Various kernels in a graph.

Proposition 7.10

If S is transitive and if the Pareto set (or set of efficient points, see 3.8) of S relative to $\mathcal{A}$ is minimal complete, then the kernel is unique and concurrent with the set of efficient points.

Proof

Let N be a kernel and O the set of efficient points. Since O is minimal there exist no indifferent pairs within it. Clearly $O \subset N$ because any element x of O will be such that in $\mathcal{A}$ there does not exist any y such that $y \succ_s x$ where $\succ_s$ is the asymmetric part of S. If x does not belong to N, there exists $z \in N$ such that $z\,S\,x$. Then either z will belong to O, which is absurd because of minimality, or z will not belong to O and there exists $y \in O$ such that $y \succ_s z$. By transitivity, $y\,S\,x$, which is once again absurd since they both belong to O.

Let x belong to O and not to N; from the definition of a kernel, i), there exists an alternative $z \in N$ such that $z\,S\,x$, which is absurd since $x \in O$. Therefore the only possible kernel is O. Q.E.D.

Given the above proposition together with propositions 3.11 and 3.13, *in the case where $\mathcal{A}$ is finite and provided S is transitive, then the kernel is the Pareto set.* We can therefore say that the notion of the Pareto set is generalized by that of the kernel, as stated above.

Proposition 7.11

All acyclic graphs (definition 5.31) admit a single kernel.

The proof of this proposition, which goes back to games theory (von Neumann and Morgenstern, 1944), is to be found in Berge (1970, theorem 4, page 299).

Also known in graph theory are algorithms for finding the kernel. In sub-section 7.3.3 we shall be giving a few more recent commentaries on the subject, but first let us have an actual application example of ELECTRE 1.

Example 7.12

Candidate selection (ELECTRE I method)

We take the usual normalized data (figure 4.2, chapter 4), reproduced here for convenience (figure 7.4).

We calculate the concordance and discordance coefficients according to formulas (9) and (10). Thus for example:

$$c_{23} = c_{BD} = 0.10 + 0.20 + 0.20 = 0.50$$
$$d_{23} = d_{BD} = \text{Max}[(0.156 - 0.125), (0.241 - 0.069)]/[0.276 - 0.034)$$
$$= 0.172/0.242 = 0.71$$

Alternatives	Criteria				
	C_1	C_2	C_3	C_4	C_5
Albert	0.188	0.172	0.168	0.122	0.114
Blanche	0.125	0.069	0.188	0.244	0.205
Donald	0.156	0.241	0.134	0.220	0.136
Emily	0.188	0.034	0.174	0.146	0.159
Georgia	0.188	0.276	0.156	0.171	0.205
Helen	0.156	0.207	0.180	0.098	0.182
Weights	0.25	0.25	0.10	0.20	0.20

Figure 7.4 The data for the candidate selection example.

We perform the same calculation for all the pairs and finally obtain the concordance matrix $C = (c_{ik})$ and discordance matrix $D = (d_{ik})$.

$C =$

	A	B	D	E	G	H
A		0.50	0.35	0.50	0.35	0.45
B	0.50		0.50	0.75	0.50	0.50
D	0.65	0.50		0.45	0.20	0.70
E	0.75	0.25	0.55		0.35	0.45
G	0.90	0.70	0.80	0.90		0.90
H	0.55	0.50	0.55	0.55	0.10	

$D =$

	A	B	D	E	G	H
A		0.51	0.40	0.19	0.43	0.28
B	0.43		0.71	0.26	0.86	0.57
D	0.14	0.28		0.16	0.28	0.19
E	0.57	0.40	0.86		1.00	0.71
G	0.05	0.30	0.20	0.07		0.10
H	0.13	0.61	0.51	0.20	0.30	

By setting the thresholds at $t_c = 0.6$ and $t_d = 0.4$ we can determine the relation S. Thus for example $G\,S\,D$ since $c_{GD} = 0.80 > t_c$ and $d_{GD} = 0.20 < t_d$. For clarity we use here the matrix representation of S defined in 5.24 (the values of the main diagonal are irrelevant here). We obtain the following matrix S:

$S =$

	A	B	D	E	G	H
A	0.5	0	0	0	0	0
B	0	0.5	0	1	0	0
D	1	0	0.5	0	0	1
E	0	0	0	0.5	0	0
G	1	1	1	1	0.5	1
H	0	0	0	0	0	0.5

The graph associated with the relation S is shown in figure 7.5 (the reflexivity arcs have been omitted).

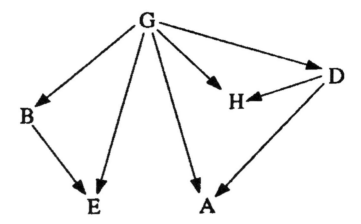

Figure 7.5 Graph of the outranking relation

In this case, S is seen to be transitive, the kernel is therefore unique and since Georgia outranks everybody else, it is equal to $\{G\}$, the Pareto set of the relation S. As we know, if the values of the thresholds are changed, other results will be obtained, and these are shown in figure (table) 7.6.

t_c	t_d	Number of arcs in the graph	Alternatives in the kernel
0.5	0.5	13	G
0.6	0.4	8	G
0.7	0.3	6	G,B
0.8	0.2	3	G,B,D
0.9	0.1	3	G,B,D
0.9	0.08	2	G,B,D,H
0.9	0.05	1	G,B,D,H,E
0.95	0.05	1	G,B,D,H,E
1	0	0	G,B,D,H,E,A

Figure 7.6 Sensitivity of the kernel to variations in the thresholds.

In the example above the kernel is always unique. As expected, when threshold requirements are tightened (t_c increasing, t_d decreasing), there are fewer and fewer pairs in the relation and the kernel grows; this same phenomenon occurs with the

Pareto set (see section 3.2). In the limit, $t_c = 1$ and $t_d = 0$ and we find the product preorder (proposition 7.8), which in our case is void. Note finally that the fact that G appears in all the kernels does not constitute a definitive reason for choosing G if a single candidate is to be selected, for an alternative very different from the others and not in the relation S could also share the same property. Nevertheless and added to the former observation about G, as G is alone in some kernels, these two facts altogether induce to think on G as the best choice.

Given that ELECTRE is designed to deal with the selection problem, no more than what we have just mentioned should be deduced from table 7.6. In particular, the temptation should be resisted to deduce from it the ranking $G \succ B \succ D \succ H \succ E \succ A$, which is the order of appearance in the kernel. As we shall see next, this is not what kernels are for.

7.3.3 Discussion of the method and extensions

The example above brings out various properties which deserve further comment. Firstly, the outranking relation is of course partial and there are certain alternatives which are not in the relation, such as E and A in figure 7.5. This incomparability, which is natural in view of the definitions, turns up quite generally in outranking methods, and especially in ELECTRE I.

As we have already noted, intransitivities can also arise. In our example, there are none for $t_c = 0.6$ and $t_d = 0.4$; but on taking $t_c = 0.55$ and $t_d = 0.45$, we obtain among other relations $D\,S\,H$, $H\,S\,E$, although D and E are not comparable.

We can even obtain a circuit, as we shall see in the next example.

Example 7.13 Existence of a circuit in an outranking relation

Consider an example with three criteria to be maximized, having three identical weights, with the following decision matrix:

		Criteria		
		1	2	3
Alternatives	*a*	70	50	50
	b	60	70	40
	c	50	60	60
	d	55	55	55
	e	100	20	80

We can see that the three alternatives a, b and c have values which identify them as a circular permutation of the Condorcet triplet type (example 5.17); d is quite average and e rather different from the other alternatives.

The concordance and discordance matrices are as follows:

		a	b	c	d	e
	a	1	2/3	1/3	1/3	1/3
	b	1/3	1	2/3	2/3	1/3
C =	c	2/3	1/3	1	2/3	1/3
	d	2/3	1/3	1/3	1	1/3
	e	2/3	2/3	2/3	2/3	1

		a	b	c	d	g
	a	0	0.4	0.2	0.1	0.6
	b	0.2	0	0.4	0.3	0.8
D =	c	0.4	0.2	0	0.1	1
	d	0.3	0.3	0.1	0	0.9
	e	0.6	1	0.8	0.7	0

If we set the concordance to $t_c = 2/3$ and the discordance to $t_d = 0.5$, the graph of the relation will be as shown in the left graph of figure 7.7.

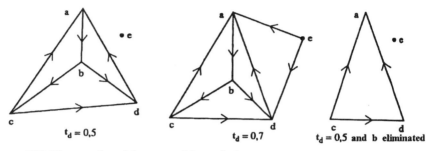

Figure 7.7 The graphs of the outranking relations.

For $t_d = 0.5$, three circuits exist: $a \to b \to c \to a$, $a \to b \to d \to a$ and $a \to b \to c \to d \to a$. There is no kernel. If we change the discordance by taking $t_d = 0.7$ (figure 7.7, middle graph), circuits still exist but $\{b,e\}$ is a kernel. With $t_d = 0..5$ or $t_d = 0.7$, if we eliminate alternative b, and hence the circuits (figure 7.7, rightmost graph), $\{e,c\}$ becomes the kernel, which is a good illustration of the fact pointed out in example 7.12, that an alternative (in this case e) which is not in S because of strong discordance, for example, may be in the kernel when it exists.

As the existence of a circuit can result in the non-existence of the kernel or of its non-uniqueness, Roy (1968a), in the ELECTRE method, recommends eliminating circuits before seeking the kernel. From a selection point of view, we can consider that if a circuit $a \to b \to c \to a$ exists, then the decision maker is incapable of choosing among its elements and we consider the circuit $\{a,b,c\}$ as a single fictional alternative $s = \{a,b,c\}$ taking the place of $\{a,b,c\}$ in the graph. As the relation S is extended to s through the rules below, the new graph without a circuit has a kernel (proposition 7.11).

Definition 7.14

Extension rules:
If s takes the place of a circuit, we say that:
  $a\,S\,s$ if there exists $x \in s$ such that $a\,S\,x$
  $s\,S\,a$ if there exists $x \in s$ such that $x\,S\,a$

In example 7.13 with $t_d = 0.7$, we let $s= \{a,b,c,d\}$; applying rules 7.14 implies $e\ S$ s, and the new kernel is $\{e\}$. Thus the recommended strategy with ELECTRE I is to prefer the alternative e which appears superior, assuming strong discordance, to a, b, c and d which are indifferent.

However inconvenient the above difficulties of incomparability and intransitivity may appear, we should remember that they do reflect the great wealth of results which can be obtained by these methods, this being due to their tolerance to possibilities that other methods proscribe (weak preferences, quasi- and pseudo-criteria). But this very wealth is also a weakness, at least from a standardization point of view, since it leaves a wide margin of assessment to the decision maker. The methods can lead to substantially different results for the same problem and they involve considerably complex calculations.

One of the most complete studies of this problems of the outranking relation is that of Vanderpooten (1990a), in which prescriptive rules of interpretation are drawn up (see also the recent books by Schärlig, 1996 and Maystre et Bollinger, 1999). Thus in the case of ELECTRE I, he recommends systematically selecting the kernel if the graph has no circuit. If there are circuits, he suggests two procedures for eliminating them to give a graph with no circuit. The first method cited is that of Roy (1968a) described above. Since this procedure involves a large loss of information, especially if the graph is highly connex (*i.e.* has many arcs), Vanderpooten suggests another procedure: break the circuit where the decision maker considers this is most justified (the weakest preference) by eliminating an arc $a_k\ S\ a_i$. This gives a relation with no circuit. Thus in example 7.13, with $t_d = 0.5$, if we eliminate the arc $a \rightarrow b$, the kernel is $\{a,b,e\}$, which is a selection, large in number but reasonable.

Other solutions have also been proposed, such as for example the *quasi-kernel* of Hansen *et al.* (1976); these are interesting but apparently little-used ideas, probably because further theoretical difficulties arise. Thus, though all graphs with or without a circuit possess a quasi-kernel (Chvatal and Lovasz, 1974), there is no proof that it is unique (Crama and Hansen, 1983).

Bearing in mind that the graph of the outranking relation depends on the concordance and discordance thresholds t_c and t_d chosen by the decision maker, as we saw in example 7.12, analysis of the results becomes still more complicated. Vetschera (1986, 1988) studied this important question in depth and produced a software package, IDEAS, specially designed to analyse this dependence.

The discordance coefficients are also affected by the choice of utilities and consequently by normalization. To overcome this weakness, variants of ELECTRE I have been designed; one such is that of Vincke (1989), based on the definition of a discordance set made up of pairs of values of criteria that the decision maker considers too discordant and which will result in elimination of S; in this way we avoid normalization prior to the measurement of discordance. The TACTIC method, which we describe below (7.5.2), is also based on this idea.

Another noteworthy aspect of ELECTRE I is that it is impossible, by definition, for one alternative that is dominated by another to appear in the kernel, even if this alternative outranks all the others. This is the phenomenon that Schärlig (1985) called the 'second brilliant', and is the reason why successive appearances of alternatives in the kernel when strengthening the thresholds should not be considered as a ranking method.

One way of avoiding any wrong interpretations is to avoid the notion of kernel in displaying the results! In work along these lines, Vidal and Yehia Alcoutlabi (1990) propose making use of the matrix of the relation S by means of a quadratic optimization. This is equivalent to maximizing or minimizing an objective function to find a relation close to S, but having the right properties (this is the idea expressed in section 5.5). Following the same lines, Cook *et al.* (1988) propose minimizing the number of arcs to be inverted to arrive at a transitive relation S, an idea already put forward by Slater (see remark 5.46 b). Likewise, Oral and Kettani (1990), using a weak notion of kernel, transform ELECTRE I into a non-linear integer optimization model which is solved by a special algorithm due to Glover (1975).

Roy and various others have designed and produced other versions of ELECTRE (II, III and IV) which lead directly to ranking of all the alternatives (the γ problem). The idea of ELECTRE II is to consider the alternatives belonging to a circuit as indifferent, following the basic principle of ELECTRE I. Next, alternatives ranked first are those which are not outranked and are removed from the choice set; the alternatives which are ranked second now become those which are not outranked and so on. In this way we obtain a complete preorder which can of course include many ties. We shall not give details of the algorithms for these methods here, but the interested reader can refer to the reviews in Crama and Hansen (1983), Vincke (1989) and Roy (1990), the book by Roy and Bouyssou (1993, 1998) and the source references in table 7.2. These articles deal also with the use by ELECTRE IS, III and IV of generalized coefficients of concordance and discordance (definition 7.4) based on the notion of pseudo-criteria. Perny (1998) gives a generalized comprehensive view of ELECTRE-like methods in which the concordance and discordance relationships are modeled as pairwise fuzzy functions on $\mathcal{A} \times \mathcal{A}$. It is obviously a great asset to be able thereby to introduce more highly tuned decision maker preferences than the usual preorder. Lastly, Yu Wei (1992) deals with the problem of distribution into pre-defined sub-groups (the classification problem), this results in a new version of ELECTRE, namely ELECTRE TRI (*tri* = sorting in English), see Mousseau *et al.* (1999). This problem is also addressed by Perny (1998) in a fuzzy setting.

The idea of taking any outranking relation and 'extending' it, in a sense to be defined, in a preorder relation is analyzed by Vincke (1991 and 1992). Vincke lists twenty two possible procedures, including those of ELECTRE II above, and others using the notion of flow, which we shall be defining in section 7.16. Vincke studies these various procedures from the point of view of the properties that can reasonably be expected from the 'extension'.

To end on a practical note, the ELECTRE methods are up to now the ones which have had the largest number of software packages devoted to them. Many were developed in the LAMSADE laboratory under B.Roy. There do exist others, such as

IDEAS referred to above (Vetschera, 1988) and ELECCALC by Kiss *et al.* (1992) which have some very interesting features. This ELECCALC, which is based on ELECTRE II, enables interactive calculation of the coefficients of concordance and discordance from global comparisons made by the decision maker from a sub-set of the choice set (the reference set method, as in UTA, chapter 6.7). More ample information on these software packages will be given in chapter 10.

The wide range of problems that the ELECTRE method has been applied to includes the choice of subway stations in Paris (Huggonard and Roy, 1983), the choice of chemical pollution monitoring systems (Siskos *et al.*, 1986) and the selection of computer equipment (Fichefet, 1985). Numerous references are to be found in Crama and Hansen (1983) and Roy (1990), Maystre and Bollinger (1999) and Vincke (1999b).

7.4 The PROMETHEE method

The PROMETHEE method (Preference **R**anking **O**rganization **METH**od for **E**nrichment **E**valuations) is one of the most recent outranking methods; the first published reference to it is in Brans *et al.* (1984), where previous applications are in fact cited. The most complete and didactic references are in Brans and Vincke (1985) and Brans *et al.* (1986). It is being used more and more frequently, especially for problems of location: hydroelectric stations (Mladineo *et al.*, 1987), stores in a competitive environment (Karkazis, 1989), garbage disposal sites (Briggs *et al.*, 1990, Vuk *et al.*, 1991); applications also include financial assessment (Mareschal and Brans, 1991, Hens *et al.*, 1992), etc. The main feature claimed for this method is that it is perfectly intelligible for the decision maker, and the present authors agree that it is indeed one of the most intuitive of multicriterion decision methods.

PROMETHEE makes abundant use of the notion of pseudo-criterion (see 2.5 and section 7.2.2 above). Consider any criterion j to be maximized and two alternatives $a_i = (a_{i1}, a_{i2}, ..., a_{in})$ and $a_k = (a_{k1}, a_{k2}, ..., a_{kn})$ where as usual $a_{ij} = U_j(a_i)$. We assume that $a_{ij} \geq a_{kj}$; to find out how the two alternatives are situated in relation to criterion j, we can define $d_{ik} = a_{ij} - a_{kj}$ and consider the usual preference function relative to j:

$$S_j (a_{i,} a_k) = S_j (d_{ik}) = \begin{cases} 0 & \text{if } d_{ik} = 0 \text{ (indifference)} \\ \\ 1 & \text{if } d_{ik} > 0 \text{ (strict preference)} \end{cases}$$

Here, $S_j (d_{ik})$ is a function of $[0,1]$ in $\{0,1\}$.

In 2.5 we saw an initial generalization of this idea under the name of *quasi-criterion*: we simply consider a threshold $q > 0$ which we call the *indifference threshold* and write:

$$S_j \ (a_{i,} \ a_k) = S_j \ (d_{ik}) = \begin{cases} 0 & \text{if } d_{ik} \leq q \quad \text{(indifference)} \\ \\ 1 & \text{if } d_{ik} > q \quad \text{(strict preference)} \end{cases}$$

To avoid jumping from 0 to 1 we introduce two thresholds, q (indifference) and p (strict preference) and the *linear-preference pseudo-criterion*:

$$S_j \ (a_{i,} \ a_k) = S_j \ (d_{ik}) = \begin{cases} 0 & \text{if } d_{ik} \leq q \quad \text{(indifference)} \\ (d_{ik} - q)/(p - q) & \text{if } q < d_{ik} \leq p \\ 1 & \text{if } d_{ik} > p \quad \text{(strict preference)} \end{cases}$$

With this definition, $S_j \ (d_{ik})$ is a function of [0,1] in [0,1] and the jump is eliminated; the function is shown graphically in figure 7.8. We shall see from this figure two other types of function $S_j \ (d_{ik})$ which can be used.

Once the decision maker has defined the type of pseudo-criterion he wishes to use for each criterion j, the *preference index*, c_{ik} can be calculated, assuming weights $w_j > 0$ normalized to unity ($\sum_j w_j = 1$) have been given:

$$c_{ik} = \sum_j w_j S_j (a_i, a_k) = \sum_j w_j S_j (d_{ik}) \tag{11}$$

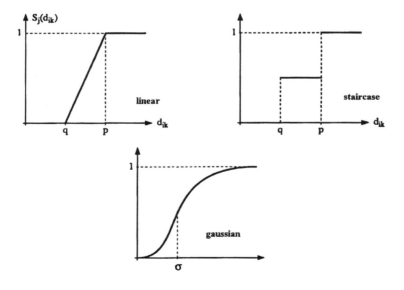

Figure 7.8 Various types of 'pseudo-criteria'

Proposition 7.15

The preference index c_{ik} is null if and only if a_i is indifferent to a_k for all criteria; the preference index c_{ik} is equal to 1 if and only if a_i is strictly preferred to a_k for all criteria.

Proof

The proposition is self-evident. Q.E.D.

The information obtained can easily and usefully be displayed in the form of a graph by introducing the input, output and net flow at each node (alternative).

Definition 7.16

The *output flow* at i $\quad \phi_i^+ = \sum_k c_{ik}$

The *input flow* at i $\quad \phi_i^- = \sum_k c_{ki}$

The *net flow* at i $\quad \phi_i = \phi_i^+ - \phi_i^-$

The PROMETHEE method bases its conclusions on information about these flows.

Definition 7.17

In the method PROMETHEE I:
a_i outranks a_k if and only if $\quad \phi_i^+ > \phi_k^+$ and $\phi_i^- < \phi_k^-$

$\qquad\qquad$ or $\quad \phi_i^+ > \phi_k^+$ and $\phi_i^- = \phi_k^-$

$\qquad\qquad$ or $\quad \phi_i^+ = \phi_k^+$ and $\phi_i^- < \phi_k^-$

In all other cases a_i does not outrank a_k.

A consequence of definition 7.17 is that the outranking relation is a partial preorder . If we wish to obtain a complete preorder, net flows can be used with the PROMETHEE II method with the following definition.

Definition 7.18

In the method PROMETHEE II: $\quad a_i$ outranks a_k if and only if $\phi_i \geq \phi_k$

We can illustrate the above results through a simple example.

Example 7.19 Candidate selection (PROMETHEE method)

We make no apology for once again taking our favorite example, and we repeat the original data below.

Alternatives	Criteria				
	1 MAX	2 MAX	3 MIN	4 MAX	5 MAX
Albert	6	5	28	5	5
Blanche	4	2	25	10	9
Donald	5	7	35	9	6
Emily	6	1	27	6	7
Georgia	6	8	30	7	9
Helen	5	6	26	4	8
Normalized weights	0.25	0.25	0.1	0.2	0.2

Figure 7.9 Data for candidate selection.

To illustrate the method, suppose we consider criterion #3 (age) as a linear 'pseudo-criterion' with thresholds $p = 5$, $q = 0$ (type 3 pseudo-criterion in PROMETHEE terminology); the other criteria remain 'normal' ones.

We now calculate the preference indices c_{ik} and the flows according to formula (11) and definition 7.16. The results are shown in figure 7.10.

c_{ik}	A1	A2	A3	A4	A5	A6	ϕ^+
A1 Albert	0	0.50	0.35	0.25	0.04	0.45	1.59
A2 Blanche	0.46	0	0.50	0.69	0.30	0.42	2.37
A3 Donald	0.65	0.50	0	0.45	0.20	0.45	2.25
A4 Emily	0.42	0.25	0.55	0	0.06	0.45	1.73
A5 Georgia	0.65	0.50	0.80	0.65	0	0.90	3.50
A6 Helen	0.49	0.50	0.30	0.47	0.08	0	1.84
ϕ^-	2.67	2.25	2.50	2.51	0.68	2.67	

Figure 7.10 Preference indices and flows for example 7.19.

With the above data, PROMETHEE will conclude that:

> A2 outranks A1, A3, A4 and A6;
> A3 outranks A1, A4 and A6;
> A4 outranks A1;
> A5 outranks A1, A2, A3, A4 and A6;
> A6 outranks A1.

This result is illustrated in the graph in figure 7.11.

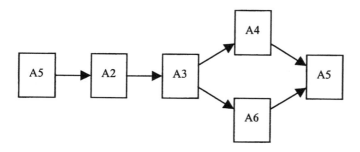

Figure 7.11 Graph of outranking from PROMETHEE I

Finally, if we calculate the net flow $\phi_i = \phi_i^+ - \phi_i^-$ in order to apply PROMETHEE II and obtain a complete preorder , we obtain the ranking $A5 \succ A2 \succ A3 \succ A4 \succ A6 \succ A1$ (figure 7.12.)

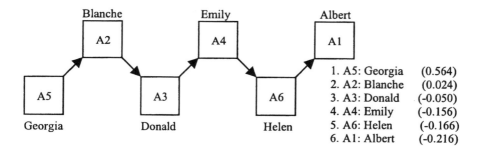

1. A5: Georgia   (0.564)
2. A2: Blanche   (0.024)
3. A3: Donald   (-0.050)
4. A4: Emily   (-0.156)
5. A6: Helen   (-0.166)
6. A1: Albert   (-0.216)

Figure 7.12 Total preorder from PROMETHEE II

For comparison, the simple weighting method of chapter 4 gave the result $A5 \succ A3 \succ A6 \succ A2 \succ A1 \succ A4$ for the same example (it has to be said that the data are slightly different here since the age criterion has become a linear pseudo-criterion). As we see, the two methods give the same winner, Georgia, but differ substantially in the ranking of Blanche and Emily. Given the difference in basic assumptions and philosophy in the two methods, these differences are hardly surprising and can serve as a lesson that in multicriterion analysis there is no 'natural' optimum! Note that if we had considered age as a 'normal' criterion instead of a linear pseudo-criterion the final ranking from PROMETHEE II becomes $A5 \succ A2 \succ A3 \succ A6 \succ A4 \succ A1$, differing from that previously obtained only in the inversion of the ranking of Emily and Helen. Thus we find, rather surprisingly, the same result as from the non-orthodox use of ELECTRE I introduced at the end of section 7.3.2, and this deserves closer attention. A recent study by Salminen *et al.*, (1998) shows that PROMETHEE gives results that are closer to the linear sum than ELECTRE III; clearly PROMETHEE has some of the flavor of a weighted sum.

We end with a few additional pieces of information about PROMETHEE. First, there is the software package PROMCALC (**PROM**ethee **CALC**ulations), produced by the original authors. This is quite a well designed program which we ourselves used for the calculations in example 7.19 and for the accompanying figures. We shall give a fuller description in chapter 10.

A certain amount of subjectivity is involved in the PROMETHEE method, especially in setting the thresholds of the pseudo-criteria, rather similarly to what happens in ELECTRE I with concordance and discordance thresholds; however, according to Brans *et al.* (1986), PROMETHEE has the advantage over ELECTRE of robust results able to survive threshold modifications. Note that these thresholds do not enter at the same point in the two procedures: in ELECTRE I, they are directly involved in the outranking relation, with the result that they have a strong influence on the relation, whereas in PROMETHEE they are involved in the preparatory phase of criterion definition. PROMETHEE has been given an axiomatic definition making it an aggregation method satisfying the conditions of neutrality (the aggregate preorder does not take account of the names of the alternatives), monotone (the social preorder reacts the right way when the number of people preferring one alternative to another increases) and various other more complicated conditions; see Bouyssou and Perny (1992).

Other versions of PROMETHEE are able to treat more complicated decision-making situations, especially in probabilized uncertainty (Mareschal, 1986; d'Avignon and Vincke, 1988). Versions III, IV and V of PROMETHEE are designed to treat various aspects of decision; PROMETHEE V (Brans and Mareschal (1992)), for example, deals with the problem of multicriterion selection from a sub-set of alternatives with constraints on the types of alternative to select.

Mareschal (1988) investigates the sensitivity of results to changes in weightings (this function is included in PROMCALC). In this way, the value of criteria can be assessed and any which are weak or null can be eliminated. An example of this type of approach is to be found in Fernández (1991) together with interesting extensions and theoretical considerations on the use of PROMETHEE for structuring more complex decision maker preferences.

Lastly, Diakoulaki and Koumoutsos (1991) have developed an extension of PROMETHEE using the concepts of the ideal and the non-ideal (see 8.3 for a definition of these terms), which, according to the authors, produces a more precise and robust preorder than PROMETHEE itself. Finally, Gomes and Lima (1992) have produced the method TODIM, based on a notion rather similar to that of net flow as in PROMETHEE.

7.5 Other methods

7.5.1 Concordance methods

This is the name given to methods based mainly on concordance coefficients c_{ik} which we have already met; these are obtained from pairwise comparisons of alternatives. These methods rarely use the concept of discordance. The concordance coefficients are then used to evaluate the alternatives a_i through a simple function, e.g. of the type $V(a_i) = \sum_k c_{ik}$ leading to a ranking of the alternatives.

A large and representative collection of methods of this type has been assembled by what we might call the Dutch School of multicriteria analysis; it has been almost entirely Dutch authors who have introduced, perfected and applied these methods to diverse problems such as regional planning, location of public services, etc. These methods are dubbed 'qualitative' by the authors, *i.e* the necessary initial information is mainly ordinal, both for weights and for utilities associated with each criterion. The ordinal information is then either used directly or converted into numerical values in various ways depending on the method. The recent work of Janssen *et al.* (1990) gives a good overview of the situation; more detailed information is to be found in Nijkamp and Voogd (1985) and Rietveld (1980).

Among these methods, EVAMIX by Voogd (1983) deserves mention, as it is based on coefficients of concordance and discordance, computed from ordinal and cardinal data, which are eventually aggregated. There is also the 'Expected Value' method originally proposed by Schlager (1968) which, though not actually a concordance method, uses data cardinalization (weights and utilities) carried out by a statistical procedure which is explained in Rietveld (1984) to achieve ranking by simple weighting.

Of all the Dutch methods, one of the most well known and representative is REGIME by Hinloopen *et al.* (1983). Briefly, the starting point is an $m \times n$ matrix $E = (e_{ij})$ based on ordinal evaluations. e_{ij} values are such that $e_{ij} > e_{kj}$ means that for criterion j, alternative i is preferred to alternative k (ranks could be used; see figure 5.1, chapter 5). The relative importance of the criteria is given by an ordinal weight vector $w = (w_1, w_2, \dots w_n)$ such that $w_j > w_h$ indicates that criterion j is more important than criterion h.

To compare a given pair of alternatives (a_i, a_k) in relation to a criterion j, we construct the 'REGIME signs matrix' $R = (r_{j,ik})$ with dimension $n \times m(m-1)$ where $r_{j,ik} = +1$ if $e_{ij} > e_{kj}$, $r_{j,ik} = -1$ if $e_{ij} < e_{kj}$ and $r_{j,ik} = 0$ if $e_{ij} = e_{kj}$. In fact, since $r_{j,ik} = -r_{j,ki}$, there will only be $n \times m(m-1)/2$ comparisons to make.

A probabilistic vector w^* is associated with the weight vector w. It is simplest to consider a uniform distribution which is compatible with the weights expressed; to do this we can carry out n draws from a uniform distribution over [0,1], rank the drawn values and match them with criteria in the order of the weights given by the decision maker. With this vector the weighted sum is calculated:

$$v_{ik} = \sum_j w^*_j \, r_{j,ik}$$

which can be interpreted by saying that $v_{ik} > 0$ means that alternative a_i is globally preferred to alternative a_k. On calculating the sum $v_i = \sum_k v_{ik}$ we end up with a ranking of the alternatives.

The critical feature of the method is the assumption of uniform distribution used in constructing w^*, but the authors make the point that this is the most reasonable assumption to make in the absence of any initial information and that other types of distribution would be easily incorporated if information were available. It should be remembered too that, as for any pairwise comparison method, the complexity of the calculations increases as $m!$, and the authors recommend limiting the number of alternatives to twenty. The REGIME method is regularly used for actual problems (see Janssen *et al.*, 1990) and is one of the methods included in the SIAD DEFINITE (see chapter 9).

7.5.2 Other outranking methods

Four more methods are worth at least a brief mention; though they are less widely used than ELECTRE and PROMETHEE, they are often cited.

The ORESTE method (**O**rganization, **R**angement **E**t **S**ynthèse de données rela**T**ionn**E**lles by Roubens (1982) requires only an ordinal evaluation of the criteria and weights, *i.e.* preorders, with ties allowed, which are quantified by the method of average rank (section 5.2.1.); this operation is carried out before actually applying ORESTE, which is a two-phase outranking method. In the first phase a complete aggregated preorder is constructed in several successive steps. The first step is to carry out a projection of a 'position matrix', *i.e.* a matrix whose columns represent the alternatives ranked according to each criterion, and where the columns are in turn ranked in decreasing order of importance of criteria; the projection is preformed on to a straight line of arbitrary origin by means of a distance. Several distances can be considered according to the type of information held by the decision maker, and they must fulfil certain coherence conditions. The matrix of distances thus obtained is then transformed into an 'overall ranking matrix', *i.e.* the coefficients are ranked from 1 to $n \times m$. Finally the overall ranking matrix is aggregated by Borda's method. In the second phase, a partial preorder is constructed; this is aggregated by means of an outranking relation using three parameters: concordance, discordance and indifference. Roubens (1982) showed that the second method satisfies the axiom of unanimity or the Pareto axiom (see section 5.4), but not that of indifference of irrelevant alternatives which he considers less fundamental. This author then applies his method to the classic problem of choice of computer. Pastjin and Leysen (1989) analyzed and developed the ORESTE method in detail, clarifying and interpreting the meaning of the parameters used and the results obtained. Our presentation here is based on this work.

The TACTIC method (**T**raitement des **A**ctions **C**ompte **T**enu de l'**I**mportance des **C**ritères) was devised by Vansnick (1986). In this method, for each criterion j the decision maker possesses basically two preorders: an 'ordinary' preorder $\succeq_j$ with asymmetric part $\succ_j$, expressing the preferences, and a pairwise 'veto' relation $\succ^V_j$ where $a_i \succ^V_j a_k$ means that the decision maker thinks that the difference between a_i and a_k for criterion j is great enough for 'a_k to be indisputably unacceptable compared with a_i'; this is the expression of a veto against a_k. As in ELECTRE, we let $C(a,b) = \{j \mid a \succ_j b\}$ and $c(a_i, a_k) = c_{ik} = \sum_{j \in C(ai,ak)} w_j$ (formula 9). Choosing a number $0 < \alpha \leq 1$, we can define an outranking relation by:

$$a_i \, S \, a_k \iff 1) \; c_{ik} \geq \alpha \, c_{ki} \quad \text{and} \quad 2) \; \{j \mid a_k \succ^V_j a_i\} = \varnothing$$

In other words, the outranking is based on traditional concordance and on the 'non veto'. In TACTIC, the cardinal character of discordance found in ELECTRE I is eliminated while conserving the working principle of that method, enabling this type of method to be seen more axiomatically as a special case of the non-compensatory methods, based entirely on the concordance index (see Bouyssou and Vansnick, 1986).

The MAPPAC method (Multicriterion Analysis of Preferences by means of Pairwise Alternatives and Criterion comparisons) was first published in Matarazzo (1986), and then in later works (Matarazzo, 1990 and 1991). Its most original feature is to be based on pairwise comparisons of pairs of alternatives (a_i, a_k), considering only pairs of criteria (C_h, C_j) each time. In this way, a 'fundamental preference index' $r_{hj}(a_i, a_k)$ can be calculated after normalizing the utilities by dividing by the maximum deviation (Max$_i$ a_{ij} – Min$_i$ a_{ij}) and weighting the criteria.

Once this has been done, the r_{hj} indices are aggregated into global indices $r(a_i, a_k)$ which are used to construct more or less complicated outranking relations with indifference threshold, veto threshold, etc. The PRAGMA method (Preference RAnking Global frequencies in Multicriterion Analysis), devised by the same author, (Matarazzo, 1988), is based on a 'profile' of each alternative, once normalized and weighted as in MAPPAC. These 'profiles' are actually a special case – for two criteria – of the 'value paths' of Schilling *et al.* (1983). The profile comes from visualizing the alternatives in the form of a polygonal line joining the various values according to different criteria (figure 7.13). Using these data, a 'global frequency matrix' is obtained which, the author states, contains valid information for the decision maker, complementing the results of the MAPPAC method.

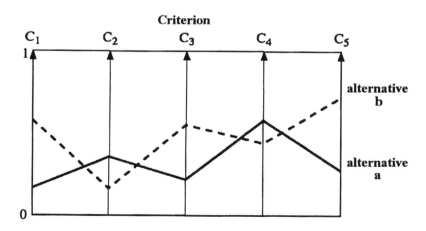

Figure 7.13 The value paths of Schilling *et al.* (profiles).

Lastly, the MELCHIOR method (Méthode d'ELimination et de Choix Incluant les relation d'ORdre) of Leclercq (1984) is a semi-ordinal outranking method, since it only requires a preorder R_p, which may be partial, to define the importance of the criteria (the weighting), and pseudo-orders for the criteria. The central concept is of 'masking' criteria as follows. Let a_i and a_k be two alternatives, let X be the set of criteria j such that $a_i \succ_j a_k$ and let Y be the set of criteria j such that $a_k \succ_j a_i$. We can construct a bipartite graph (X, Y, U) in which the vertices are the criteria separated into two sub-sets X and Y and in which the arcs are the preference relations between criteria. In other words, $(j,h) \in U$ if and only if $j\ R_p\ h$. Then X is said to mask Y when U contains a complete matching between Y and a sub-set of X (for an account of this notion of matching in bipartite graphs, see Berge, 1970 or Wilson, 1972). The notion of 'matching' is central to MELCHIOR, with variants being more or less linked to the choice of parameters: the preference relation R_p between criteria, the various types of mask, etc. This is an interesting method from a theoretical point of view, firstly because it contains ELECTRE IV as a special case, and secondly because it introduces into multicriterion methods the extremely fertile notion of bipartite graphs, with the associated number theorems such as that of Hall giving a simple necessary and sufficient condition for the existence of matching. The practical possibilities of the method, on the other hand, look rather weak given the sheer complexity of the author's numerical example and the relatively few workers who have followed it up.

8 OTHER MULTICRITERION DECISION METHODS

8.1 Introduction

In the previous chapters, we gave fairly detailed descriptions of the discrete multicriterion decision methods that we think are the most widely distributed, analyzed and utilized, with both the present and the next few years in mind. We gave these methods pride of place because we feel that they best illustrate the basic concepts of multicriterion decision theory.

There do however exist a large number of other methods relying on a range of methodologies, described in the literature and more or less widely used to solve multicriterion problems. Several of these methods make use of various multicriterion decision ideas that we met in previous chapters, but do not easily fit into the methodological classification of those chapters. We think it well worth including these methods in this additional chapter, and we shall be describing the ones with special interest in more depth. Note that many of these methods are being continually updated by their authors and it is likely that the relative importance given here to some of them has to be increased in a near future.

In section 8.2 we shall be discussing a fairly large set of methods that can be gathered under the title *alternative comparison methods*, these being based on a certain number of global or holistic comparisons between pairs of alternatives; the term 'holistic' is used here in contrast to comparisons criterion by criterion (see chapter 7). We shall devote some space to Zionts' method (1981) which is probably the earliest.

Section 8.3 is devoted to a whole set of procedures, all of them based on *ranking alternatives according to how far they are from an ideal (or anti-ideal) alternative*. One of the difficulties here is to give a precise meaning to the intuitive concept 'how far'.

Section 8.4 is on *permutation methods* with the example of QUALIFEX devised by Paelinck (1976). Section 8.5 covers a very diverse set of methods which are described in detail in the literature but which have seen little use up to now.

8.2 Alternative comparison methods

As we said above, this term covers a set of methods making use not only of the basic information in the decision matrix but also of holistic comparisons of certain alternatives. Since these comparisons are made by the decision maker the methods are interactive with progressive input of information. Here we should remind the reader of the difference between the global comparison of alternatives (saying whether one alternative is or is not preferred to another) and the criterion by criterion comparison used in the outranking methods described in the previous chapter.

One feature of alternative-comparison methods is that the comparisons enable the criterion weight vector to be constructed indirectly instead of the latter being among the initial input of data in the multicriterion problem. It is actually possible that this key idea originated in the literature on marketing, because there the problem is tackled in the following way: determine the importance for the consumer (for us, the decision maker) of certain features (criteria) in a range of competing products (the alternatives) by comparing pairs of products. The LINMAP method of Srinivasan and Shocker (1973a and b), one of the first in the field, is still a widely used standard method. We shall be describing it in section 8.3 as it is based on the notion of distance from an ideal alternative.

In 1977, after the success of LINMAP and his own earlier work in the field, S. Zionts and his team at the New York State University at Buffalo developed a new method based on global pairwise comparisons of alternatives. From this stemmed numerous variants, each more or less an improvement, which we shall be describing after the basic method.

8.2.1 Zionts' basic method

Although earlier publications exist, we consider that Zionts (1981) was the first widely known work describing the method in detail. As we have said, the method is based on earlier work, especially the continuous multicriterion decision method of Zionts and Wallenius (1976); it might even be said that it is largely an adaptation of the continuous method to the discrete case.

One important feature common to all these alternative-comparing methods is that they assume the existence of a perfectly defined decision maker utility function to be found interactively (this is the first of the games described in section 9.3.1). This multi-attribute utility function is generally assumed to be linear, *i.e.* additive (in the sense of chapter 6), while the utilities relating to each criterion are also assumed to be linear; this is the case in the method of Zionts to which we shall now turn our attention. But unlike the methods described in chapter 6, the present multiattribute utility function remains implicit, *i.e.* no analytical expression for it is sought.

Another important idea used in Zionts' method is that of the convex-efficient alternative which we now introduce. For the basic concepts involved in convexity used in this chapter the reader may refer back to sections 3.2 and 4.2 and to the various textbooks, especially Rockafeller (1970).

Definition 8.1

Given a set X of alternatives a_i, an alternative a is *convex-efficient relative to X* if it is not strictly dominated by any strict convex combination of alternatives in X. In other words, there exists no $0 < \lambda_i < 1$, $\sum_i \lambda_i = 1$ nor any alternatives a_i such that a is strictly dominated by $\sum_i \lambda_i a_i$.

We illustrate the idea of a convex-efficient alternative with the following example.

Example 8.2

Convex-efficient alternatives.
Consider a set of three alternatives and two criteria to be maximized:

		Criteria	
		C_1	C_2
Alternatives	A	2	8
	B	4	4
	C	8	2

We can see that the three alternatives are efficient (they are non-dominated or Pareto optimal). On the other hand, B is not convex-efficient relative to $\{A, B, C\}$, for if the alternatives are represented graphically in criterion-space (figure 8.1), we see that B is efficient for the product preorder (*cf.* proposition 3.24). B, however, is strictly dominated by, among others, $\frac{1}{2}A + \frac{1}{2}C$, since $U(\frac{1}{2}A + \frac{1}{2}C) = (5,5)$.

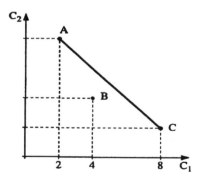

Figure 8.1 Example of a non convex-efficient alternative.

The graphical representation clearly shows how B is 'below' the line $[A,C]$, representing the set of convex combinations of A and C, *i.e.* the points:

$$\lambda_1 A + \lambda_2 C \text{ where } 0 \leq \lambda_i \leq 1 \text{ and } \lambda_1 + \lambda_2 = 1$$

The convex-efficient alternatives are actually efficient alternatives from the convex envelope of $\mathcal{A}$ (definition 4.7). Proposition 4.8 will therefore appear in a slightly different form.

Proposition 8.3

Suppose that the utility U of a decision maker be represented by a weighted sum of criteria (an additive utility); then an alternative will be efficient if and only if it is convex-efficient.

Proof

The condition is necessary. We now show that if an alternative a is not convex-efficient then it is not optimal. Let $(\sum_i \lambda_i a_i) \succ a$; then $U(\sum_i \lambda_i a_i) > U(a)$, and through linearity this means $\sum_i \lambda_i U(a_i) > U(a)$. If each $U(a_i)$ were such that $U(a_i) \leq U(a)$ we would have $\sum_i \lambda_i U(a_i) \leq \sum_i \lambda_i U(a) = (\sum_i \lambda_i)U(a) = U(a)$; consequently there necessarily exists a_i such that $U(a_i) > U(a)$ and a is therefore not optimal.

The condition is sufficient. We show that if a does not maximize $U(x)$ then it is not convex-efficient. There exists $y \in \mathcal{A}$ such that $U(y) > U(a)$. Let $0 \leq \lambda < 1$; consider $z = \lambda a + (1-\lambda)y$; we have $U(z) = \lambda U(a) + (1-\lambda)U(y)$, which implies $U(z) > U(a)$, e.g. $\frac{1}{2}U(y) + \frac{1}{2}U(a) > \frac{1}{2}U(a) + \frac{1}{2}U(a) = U(a)$, hence a is not convex-efficient. Q.E.D.

Example 8.4

This is a numerical illustration of example 8.2.
Suppose $U(x) = w_1 U_1(x) + w_2 U_2(x)$ where $0 < w_i < 1$ and $w_1 + w_2 = 1$. We will then have $U(A) = 2w_1 + 8w_2 = 2w_1 + 8(1-w_1) = 8-6w_1$. Likewise, we obtain $U(B) = 4$ and $U(C) = 6w_1 + 2$.
A will be preferred to B whenever: $8-6w_1 > 4 \implies w_1 < 2/3$.
C will be preferred to B whenever: $6w_1 + 2 > 4 \implies w_1 > 1/3$
As can be seen, whatever the value of w_1, either A or C is better than B, and both are for $1/3 \leq w_1 \leq 2/3$. No weight w_1, and therefore no weight w_2, can be found such that B is efficient (proposition 8.3).

Applying proposition 8.3, we see that, for the decision maker who is seeking the best alternative – assuming linear utility – nothing is lost if a non convex-efficient alternative is removed. This alternative may of course be the 'brilliant second.' However, this difficulty is dealt with in Zionts' method which we now explain. We give a rather simplified description, omitting some technical details which, though necessary for the method to work, need not be dwelt on in understanding the principle. We give the method in the form of an algorithm.

Algorithm of the Zionts method

Step 1: Initial reference alternative.
- Choose an arbitrary weight vector w or an initial estimation given by the decision maker, or weights equal to $1/n$ (where n is as usual the number of criteria).
- Using these weights, calculate the best (for the time being) alternative by simple weighting; in what follows we shall call this the 'reference alternative'.

Step 2: Compare the reference alternative with its neighbors.
- Amongst all the convex-efficient combinations in the choice set, find the alternatives neighboring the reference alternative, which we shall call 'neighbors' in what follows. This search is carried out using linear programs. These programs include constraints brought about by the preferences expressed by the decision maker in steps 3 and 5.
- Ask the decision maker to compare the reference alternative with all its neighbors; the result will be a list of neighboring alternatives strictly preferred to the reference alternative; we shall call these the 'preferred list'.
- If the preferred list is void, go to step 7.

Step 3: Calculate the weights
- Take an alternative from the preferred list and eliminate it from the list. The preference thus expressed, in terms of the function U, constitutes an extra constraint in addition to those already similarly expressed in the previous steps, for a linear program whose variables will be the weights being sought; this program is a variant of that in step 2. If the program has a solution, which will then be a weight vector compatible with all the preferences expressed up to now, go to step 4.
- If no solution exists, eliminate the constraint from the decision maker's last response and re-run the program.
- If there is still no solution after eliminating all the constraints, go to step 6.

Step 4: Obtaining the maximal solution.
- Look for the optimal alternative using the weights obtained in the previous step. This will be referred to as the 'optimal solution' in what follows.

Step 5: Comparison with the reference alternative.
- Ask the decision maker to compare the optimal solution with the reference alternative. According to his stated preference, generate a new constraint for the linear programs.
- If it is the optimal solution which is preferred then this becomes the reference alternative. Go to step 2.
- Otherwise go to step 6.

Step 6: *A neighboring alternative becomes a reference alternative.*
- If the list of preferences is void, go to step 7.
- Otherwise make an arbitrary choice of any one of the alternatives remaining in the preferred list to be the new reference alternative, and go to step 2.

Step 7: *End.*
- Using the hitherto calculated weights, work out the final ranking of the alternatives. END.

The above algorithm always ends since the number of convex-efficient vertices is finite and, as can easily be proved, no convex-efficient alternative can return, once eliminated from the preferred list. According to Zionts (1981), the entire process can be looked at as 'an expression of the preferences (of the decision maker) through progressive restriction of the set of possible weights'.

Practical applications of the method have shown that it can be used effectively in problems with up to 200 alternatives, for the number of comparisons required from the decision maker remains acceptable, and does not exceed about a hundred in the worst case. However the number of criteria must remain within six or seven, beyond which it is no longer possible to rely seriously on global comparisons owing to the cognitive limitations of the brain (see the start of section 4.6).

It is fairly clear that, for the use of weights to be legitimate, Zionts' method assumes cardinal utilities for each criterion (see chapter 6). If the utilities were purely ordinal, they would have to be 'cardinalized' but then the change of utility function would lead to serious complications, if only to justify the recourse to convex-efficient alternatives, a fact recognized by Zionts himself (see for example section 6.9). This among other weaknesses has triggered an energetic research effort to perfect the method, and yielded results that we shall now summarize briefly.

8.2.2 Variants on the basic method

The number of extensions to Zionts' method is very large and we shall attempt to give a brief overview from the abundant and expanding literature.

1. Korhonen, Wallenius and Zionts (1984) generalize the basic method to a decision maker's utility function (implicit) which is quasi-concave, strictly increasing. It is not without interest in this connection to note that, since the decision maker's utility function remains implicit, there obviously cannot be any *a priori* verification that it is endowed with any property such as linearity or quasi-concavity. It is at most a question-begging function justifying the elimination of a large number of alternatives which are supposed to be dominated if the decision maker's utility has the right property. Neither does *a posteriori* verification have any sense since the algorithm presupposes the form of the function

Korhonen *et al.* (1984) also introduce the concept of domination by a convex cone to eliminate alternatives which will not by any stretch of the imagination be chosen by the decision maker, thereby avoiding wasteful comparisons. This concept is implicitly present in Kornbluth (1978), and primarily in the thesis of Koksalan (1984), both having been under the supervision of Zionts. Koksalan's thesis led to improvements in the basic method, published by Koksalan, Karwan and Zionts (1984), Koksalan (1989), Taner and Koksalan (1991), Koksalan and Taner (1992) and Salminen (1992). Malakooti (1988) also proposed an improvement to Zionts' method; using the gradient he eliminates 'dominated' alternatives, not in a convex cone as in Korhonen *et al.* (1984) but in a half-space. This method thereby reduces the number of comparisons required from the decision maker, but on the other hand requires information about the gradient.

2. The treatment of ordinal or mixed (ordinal and cardinal) utilities is dealt with by Koksalan, Karwan and Zionts (1986) following on from previous work by Karwan *et al.* (1985) on continuous multicriterion decision restricted to integer solutions.

3. Korhonen (1988), using Wierzbicki's (1980) concept of 'reference direction' in continuous multicriterion decision, together with the 'visual method' of Korhonen and Laakso (1986), enables the decision maker to compare alternatives shown to him in a highly graphical form. He also proposes, where the decision maker utility is quasi-concave, using the above-mentioned method by Korhonen, Wallenius and Zionts (1984) in such a way that the decision maker can eliminate all the alternatives until only one remains. He produced the software package VIMDA (Visual Interactive Method for Discrete Alternatives) which handles ten-criteria choice for up to 500 alternatives (see chapter 10).

4. Khairullah and Zionts (1987) propose a method for comparing alternatives, based on Khairullah's doctoral thesis (1992), which differs from the basic method by making use of Bayes' probability theorem applied to the decision maker's preferences; this being founded on an old psychometric tradition (Thurstone, Bradley, Terry, David etc.) of probabilistic models for pairwise comparisons. A recent review of these ideas is to be found in David (1988).

Basically, the relative preference for each alternative is measured by the probability p_i of each alternative a_i being chosen by the decision maker from among the remaining alternatives (choice without restoration) in a series of global pairwise comparisons. Determining these probabilities would require a total of $m(m - 1)/2$ comparisons for each alternative. Obviously a only part of these comparisons is actually carried out. Once the comparisons have been made, an optimization problem, with equality restraints, can be formulated mathematically in which a complete   preorder is sought which minimizes the inconsistency of the final preorder with respect to the pairewise comparisons given by the decision maker (an idea similar to those studied in section 5.5). Before carrying out this optimization, the main problem, limiting the number of comparisons as a function of probabilities, can be tackled in various ways. The method appears advantageous in comparison with LINMAP and Zionts' method, if we can rely on an analysis carried out by the authors using their PRODALT software.

5. Badran (1988) gives a rather original method which he claims overcomes certain difficulties such as controlling independence, which is encountered when using additive methods (see chapter 6). He does however requires the values of utilities associated with each criterion j to be ranked according to a partition S_j of the set of values. From this he generates a sequence of alternatives (those in the choice set plus any fictional alternatives) such that, in this sequence, two successive alternatives differ only in the value of a single criterion, j for example, and for this criterion, by one division of the partition S_j. The decision maker is then asked to make comparisons between successive alternatives in the sequence on an arbitrary scale of values which is given once for all preferences (very strong preference, strong preference, medium preference etc.); through these comparisons the decision maker's utility function can be constructed approximately. Clearly, one great weakness of the method is the number and difficulty of the questions asked of the decision maker; another weakness is the arbitrariness in the partition of values for each criterion. Nevertheless, comparing alternatives which differ only by a single criterion is easier than comparing alternatives in general.

8.3 Methods involving distance from an ideal alternative

8.3.1 The concept of the ideal alternative

The methods we have just looked at feature the 'ideal' alternative which the decision maker would, if he could, choose without hesitation. Unfortunately, there is no such thing as the ideal, as we all know. This mythical 'ideal' does not therefore figure among the possible choices and the decision maker will try to get as close as possible to the 'ideal' by choosing the alternative which is 'closest to the ideal', in a way we shall be defining more precisely below.

The concept of the ideal alternative plays a large part in various scientific fields, especially in the psychometrics literature on choice, where a notion of the absolute ideal is used (Coombs, 1958). The notion of ideal point in the field of multicriterion decision seems to have been introduced by Geoffrion in an internal report at Stanford in 1965. Roy (1968a) mentions it and Benayoun *et al.* (1971) use the concept in the STEM multicriterion decision method. It was Zeleny (1973) however, who made it the central feature in his 'compromise solution', the latter being the 'closest to the ideal'. According to Zeleny (1982, p.154), the concept of ideal and compromise solution 'represents more than a mere practical technique; it constitutes an assumption about the underlying rationality of the decision process in man'. Since his initial work, the notion of ideal point has spread to other domains such as group decision with the 'utopian point' of Yu (1973).

The minimization of distance was probably first introduced into the field of continuous multicriterion decision by Charnes and Cooper (1961) in their work on 'goal programming'. This idea fits naturally into linear programming. In goal programming, it is not the distance from an ideal which is minimized, but the distance from the decision maker's goal, and the latter can be realistic rather than ideal. In fact if it is too realistic the closest alternative may turn out not to be Pareto optimal (*cf.* figure 9.10).

As far as we know, the first use of the notion of the ideal in discrete multicriterion decision was LINMAP by Srinivasan and Shocker (1973a, 1973b). We shall go into this in more detail in section 8.3.3. We should also mention 'discrepancy analysis' by Nijkamp (1979) in which alternatives are ranked according to their correlation coefficient with an ideal alternative. Some 'multivariant methods' which we shall be looking at in 8.5 also use an ideal point in the first steps. However it is in the method TOPSIS (section 8.3.2) that the concept of the ideal comes to the fore in multicriterion decision, being the basis of a highly specific working method.

Before describing the methods, we shall give formal definitions of the basic notions. As usual, we shall have alternatives a_i, $i = 1, 2, ..., m$ and a decision matrix (a_{ij}) where $a_{ij} = U_j(a_i)$, $j = 1, 2, ..., n$. By changing the utilities (see 2.7) we can assume, without limiting generality, that all the criteria are to be maximized and that $a_{ij} \geq 0$.

Definition 8.5

An *ideal point* (in $|R^n$) is the point a^M defined by $a^M = (a^M_1, a^M_2, ..., a^M_n)$ where $a^M_j = \text{Max}_i \, a_{ij}$. The alternative a^M is called the *ideal alternative*.

According to definition 8.5, the ideal point is obtained by maximizing each criterion separately. Clearly the ideal alternative a^M is not in the choice set $\mathcal{A}$, otherwise the problem of multicriterion choice would be solved, since it would suffice to choose a^M which, by definition, dominates all the alternatives in $\mathcal{A}$. The alternative a^M is a 'fictional alternative', also known as the *'fantom maximum'* as it does not exist as a possible choice; it is sometimes known as the *'zenith'*. From an interactive point of view, Roy (1974) introduced the apt term *'focal point'* to describe the ideal alternative the decision makers tries to attain.

Definition 8.6

The *anti-ideal point* (in $|R^n$) is the point a^m defined by $a^m = (a^m_1, a^m_2, ..., a^m_n)$ where $a^m_j = \text{Min}_i \, a_{ij}$. The alternative a^m is called the *anti-ideal alternative*.

Certain authors refer to the anti-ideal point as the *'fantom minimum'* or the *'nadir'*.

Axiom 8.7 The so-called axiom of rationality of choice (Zeleny, 1982).

It is rational to choose an alternative nearest to ideal or furthest from anti-ideal.

Whether an alternative is near or far, it is assumed that a distance has been defined. There are many possible distances in $|R^n$ which are not equivalent from a multicriterion point of view, but the commonest is the Minkowski distance introduced in section 2.7, which we repeat here for the reader's convenience.

The Minkowski metric m_p between two points $x = (x_1, x_2, ..., x_n)$ and $y = (y_1, y_2, ..., y_n)$ in $|R^n$ is defined by:

$$m_p = [\sum_j |x_i - y_i|^p]^{(1/p)} \text{ for } p \geq 1$$

The most commonly used values for p are $p = 1$, $p = 2$ and $p = \infty$.

If we take $p = 1$ (block distance), we obtain: $m_1 = \sum_j |x_i - y_j|$.

For $p = 2$ (Euclidean distance) we obtain: $m_2 = [\sum_j (x_i - y_j)^2]^{1/2}$.

For $p \to \infty$, since the limit of $= [\sum_j |x_i - y_i|^p]^{(1/p)}$ is equal to $\text{Max}_j |x_i - y_j|$ (the Tchebycheff distance), we obtain: $m_\infty = \text{Max}_j |x_j - y_j|$.

Using the triangular inequality, it can be proved that the distance is a decreasing function of p, i.e. $m_1 \geq m_2 \geq m_\infty$, and an intuitive justification of this property can be found in Cohon (1978, p. 181 et seq.).

We can take any one of the above distances to measure the distance from the ideal or anti-ideal. We can also apply weighting differently to each coordinate j to arrive at the central concept of weighted distance in multicriterion analysis.

Definition 8.8

The *weighted distance* between two alternatives:
$a_1 = (a_{11}, a_{12}, ..., a_{1n})$ and $a_2 = (a_{21}, a_{22}, ..., a_{2n})$
is defined as the sum: $d_p = [\sum_j w_j^p |a_{1j} - a_{2j}|^p]^{1/p}$ for $p \geq 1$.

We have already seen how these various notions of distance underlie the normalization of utilities (see 2.7). The same applies to the normalization of weights, though we have always given prominence to division by the sum, which is related to m_1.

Before putting the above notions to use in a multicriterion choice method, we must choose:

a) the type of distance used for normalizing utilities and weights;
b) the choice of one or other of the alternatives in axiom 8.7: proximity to the ideal or remoteness from the anti-ideal;
c) the type of distance used to measure this proximity or remoteness.

This flexibility in setting up the method is actually a weakness because for a given problem the results generally differ according to the arbitrary choices made among the above possibilities. The method TOPSIS takes this difficulty into account, as we shall now see.

8.3.2 The method TOPSIS.

TOPSIS (Technique for Order Preference by Similarity to Ideal Solution) by Hwang and Yoon (1981) solves dilemma b) above by considering both the distance from the ideal and from the anti-ideal. To show that it is indeed a dilemma which leads to differing results, consider figure 8.2. Here we show five alternatives A, B, C, D and E with a choice of two criteria, so that we remain within $|R^2$. We also show the ideal and anti-ideal points. It is immediately obvious that if we use the usual Euclidean distance d_2 with equal weights, point C is closest to the ideal, and D is furthest.

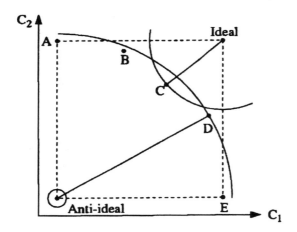

Figure 8.2 Illustration of the notions of distance to the ideal and the anti-ideal.

TOPSIS solves the dilemma in the choice between the ideal and the anti-ideal by making use of an idea that Dasarathy (1976) applied to data analysis. For each alternative $a_i = (a_{i1}, a_{i2}, ..., a_{in})$, we calculate the weighted distances $d^M_p(a_i)$ and $d^m_p(a_i)$ to the ideal and anti-ideal respectively, according to the chosen metric p:

$$d^M_p(a_i) = [\sum_j w_j^p \mid a^M_j - a_{ij} \mid^p]^{1/p} \tag{1}$$

$$d^m_p(a_i) = [\sum_j w_j^p \mid a^m_j - a_{ij} \mid^p]^{1/p} \tag{2}$$

From (1) and (2) the *similarity ratio* can be calculated:

$$D_p(a_i) = d^m_p(a_i)/(d^M_p(a_i) + d^m_p(a_i)) \tag{3}$$

and this varies from $D_p(a^m) = 0$ for the anti-ideal point to $D_p(a^M) = 1$ for the ideal point. Lastly, $D_p(a_i)$ is used for the final ranking of the alternatives.

Articles on the use of distance from the ideal as a method have appeared in the literature. Batanovic (1989) describes the choice of an urban traffic control system in Belgrade in which the distance d_p is calculated in an original way by minimizing the sum of the differences between the preorders of each criterion and the aggregate preorder from d_p. In Yoon and Hwang (1985), TOPSIS is used for the location of factories. We shall take our usual example of candidate selection.

Example 8.9 Candidate selection.

The data of the problem are those of figure 4.2 normalized, reproduced here in Figure 8.3. The normalization method used for the weights and the utilities is division by the sum (procedure #3, section 2.7 in association with the metric m_1). The ideal and anti-ideal points are normalized in the same way.

We shall calculate the distances and the similarity ratio using formulas (1) and (3). For the sake of uniformity, we shall still use the metric m_1, though this is not obligatory. The results of the calculations are given in table 8.4. The last but one column shows the ranking according to $D_1(a_i)$, and, for comparison, the last column shows the ranking obtained through simple weighting (chapter 4). In this table the distances d^M_1 and d^m_1 have been multiplied by 10 000 for clarity, but this does not of course affect the final result.

Alternatives	Criteria				
	C_1	C_2	C_3	C_4	C_5
Albert	0.188	0.172	0.168	0.122	0.114
Blanche	0.125	0.069	0.188	0.244	0.205
Donald	0.156	0.241	0.134	0.220	0.136
Emily	0.188	0.034	0.174	0.146	0.159
Georgia	0.188	0.276	0.156	0.171	0.205
Helen	0.156	0.207	0.180	0.098	0.182
Weights	0.25	0.25	0.10	0.20	0.20
Ideal	0.188	0.276	0.188	0.244	0.205
Anti-ideal	0.125	0.034	0.134	0.098	0.114

Figure 8.3 Normalized data for candidate selection

Alternatives	$d^M_1(a_i)$	$d^m_1(a_i)$	$D_1(a_i)$	Rankings	
				TOPSIS	Simple weighting
Albert	7060	5845	0.453	5	5
Blanche	6750	6155	0.477	4	4
Donald	4075	8830	0.684	2	2
Emily	9070	3835	0.297	6	6
Georgia	1780	11125	0.862	1	1
Helen	5985	6920	0.536	3	3

Figure 8.4 Application of TOPSIS to candidate selection

As can be seen from table 8.4, the results from TOPSIS are exactly the same as those from simple weighting. Is this just a coincidence? The following proposition shows that it is not.

Proposition 8.10

Using the distance $d^M{}_1$, simple weighting and the method of minimization of the distance from the ideal produce the same ranking

Proof

We have $d^M{}_1(a_i) = \sum_j w_j \ |a^M{}_j - a_{ij}| = \sum_j w_j \ (a^M{}_j - a_{ij}) = k - \sum_j w_j a_{ij}$ where k is a constant. Therefore minimizing $d^M{}_1(a_i)$ is equivalent to maximizing $\sum_j w_j a_{ij}$, which is simple weighting. Q.E.D.

Proposition 8.10 shows that TOPSIS and simple weighting will give the same ranking whenever TOPSIS is equivalent to minimizing the distance to the ideal for $d^M{}_1$, which is the case in our example as can be seen in table 8.4. Also evident from the table is the fact that maximizing the distance to the anti-ideal produces the same ranking again. This too is not due to chance as proposition 8.11 shows.

Proposition 8.11

$d^M{}_1(a_i) + d^m{}_1(a_i) =$ constant is always true. Therefore, in this metric, TOPSIS, maximizing the distance to the anti-ideal and minimizing the distance to the ideal will produce the same ranking.

Proof

We have $d^M{}_1(a_i) = \sum_j w_j \ |a^M{}_j - a_{ij}| = \sum_j w_j \ (a^M{}_j - a_{ij})$ and $d^m{}_1(a_i) = \sum_j w_j \ |a^m{}_j - a_{ij}| = \sum_j w_j (a_{ij} - a^m{}_j)$, therefore $d^M{}_1(a_i) + d^m{}_1(a_i) = \sum_j w_j a^M{}_j - \sum_j w_j a^m{}_j$ does not depend on a_i. Let the constant be k; it follows that $D_1(a_i) = d^m{}_1(a_i)/k$, hence TOPSIS and maximizing the distance to the anti-ideal give the same result. Now $d^M{}_1(a_i) = k - d^m{}_1(a_i)$ and this requires the minimizing of $d^M{}_1(a_i)$ be equivalent to the maximizing of $d^m{}_1(a_i)$. Q.E.D.

What we have just stated for the metric m_1 cannot be generalized. When other metrics are used for calculating distances and/or normalizing, the various results are in general distinct. Table 8.5 shows the final rankings obtained in our example 8.9 of candidate selection. Substantial changes are evident, especially in the distance from the ideal.

Alternatives	$d^M{}_1$	$d^M{}_2$	$d^M{}_\infty$	D_1	D_2	D_∞
Albert	5	4	3	5	4	4
Blanche	4	5	5	4	5	5
Donald	2	2	1	2	2	2
Emily	6	6	6	6	6	6
Georgia	1	1	2	1	1	1
Helen	3	3	4	3	3	3

Figure 8.5 Various rankings in terms of the distances used

To end on a more technical note, we can define a *'compromise set'* as the set of alternatives ranked first either with TOPSIS or by minimizing the distance from the ideal, and varying the metric used to define the distance, including the metrics used for normalizing the utilities and those of the weights. In our example the compromise will simply be equal to {Georgia, Donald}. The decision maker then has to choose the final alternative from the compromise set, basing his choice on hunch and instinct.

Yoon (1987), not content with the consensus compromise solution, has tried to measure the 'credibility' of each metric. Without going into details, suffice it to say that he obtains a table of values of credibility μ_p associated with each metric m_p which depend on the dimensions of the problem (criteria and alternatives). The distance he finally recommends using is a weighted average of the three standard distances:

$$d = \mu_1 d_1 + \mu_2 d_2 + \mu_\infty d_\infty$$

According to Yoon's results, the distance d_1 is the most credible and d_∞ the least, the gap becoming wider and wider as the dimensions of the problem increase. Given propositions 8.10 and 8.11, this means that TOPSIS as modified by Yoon approaches simple weighting more and more closely as the number of alternatives, and above all the number of criteria, increase. Finally, TOPSIS method is extended to continuous multicriterion decision making (Lai *et al.*,1994), while another extension to fuzzy distances to ideal and anti-ideal points is in Liang (1999).

8.3.3 LINMAP and variants

In chapter 4 we mentioned the method LINMAP (**LIN**ear programming technique for **M**ultidimensional **A**nalysis of **P**reference) for evaluating weights. Here we are concerned with it as a multicriterion decision method which is both a method of comparing alternatives (analyzed in section 8.2 where LINMAP was mentioned) and one of the distance methods that we are studying in this chapter.

In Srinivasan and Shocker (1973a) is to be found an initial LINMAP method using the Euclidean distance d_2 to measure distance from the ideal. Starting from the decision matrix and the decision maker's preferences for global comparisons of alternatives, LINMAP produces the weights and the ideal point a^M. The decision maker's preferences form a set $\mathcal{P}$ of pairs of alternatives (a_i, a_k) such that $a_i \succ a_k$.

Once the weights and a^M are known, the final ranking is given by the distance from the ideal.

Without going into all the details of the above-mentioned reference, LINMAP operates by minimizing the discrepancy between the values in the decision matrix and the decision maker's preferences expressed in the set $\mathcal{P}$. This discrepancy is measured by the sum, for the criteria concerned, of the differences in distance to an unknown ideal point (with a weighted distance d_2 with unknown weights). We finally end up with a linear optimization program (since the quadratic terms cancel out), with the solution of the weight vector w and the 'ideal' vector a^M. The second version of LINMAP has an analogous procedure using a distance d_1.

$\mathcal{P}$ does not necessarily have to contain all the $m(m-1)/2$ pairs, but the more comparisons there are the greater the accuracy of the weights. Practically speaking, if $m < n$ (fewer alternatives than criteria) the method is of low accuracy and reliability; nor is the method reliable if more than seven criteria are used since, as we have already said several times, the global comparisons must then be treated with caution. Finally, note that the method can accept inconsistency (intransitivities) in the decision maker's responses provided there are not too many of them.

These restrictions have not affected the popularity of LINMAP in the field of marketing and its authors have produced several versions of a software package (CONJOINT LINMAP) in widespread use; this package enables the possibilities of the method to be extensively explored (see chapter 10).

Pekelman and Sen (1974) produced an improvement to LINMAP which remained fairly close to the original method and in particular to the resulting linear program model. This improvement is designed for situations where there is only a small number of alternatives and it requires fewer comparisons than in LINMAP to be made by the decision maker. The sacrifice is that this method only enables the weights to be calculated, the ideal point being obtained in some other way. The reader may well feel that the price is not too high given that it would seem easy to calculate a^M from the decision matrix A (see section 8.3.1). Note, however, that in certain cases (see AIM, section 8.3.4), it is necessary to use a level of aspiration (see section 3.4) which is considerably more subjective in character than the ideal point. In any case the ideal point will be hard to locate if the A matrix transforms are carried out as they are in some multidimensional methods (see 8.5.2).

After analyzing the above situation, Horsky and Rao (1984) produced a method called LINPAC producing better results than LINMAP by asking the decision maker for more precise information about each comparison, a strategy which can reduce the number of comparisons to be made.

8.3.4 Other distance methods

Before we leave distance-based procedures, here are two recent methods with an algorithm built around the idea of distance from the ideal; in fact the originality of the methods lies in their interactivity, and so we shall be referring to them again in chapter 9. Interactivity also features in the *evolving focal point method* of Roy (1974) and this too will be discussed in the next chapter.

The AIM (Aspiration-level Interactive Model) of Lotfi *et al.* (1992) is, according to the authors, eclectic from a theoretical point of view; they have striven above all for user-friendliness. The basic process consists of seeking an 'optimal' alternative which minimizes the distance (which can be chosen to be Euclidean or Tchebycheff) from an 'aspiration level', fixed by the decision maker for each criterion. The AIM package containing the method (see chapter 10 for details) enables progressive and interactive modification of these aspiration levels. With this software, the decision maker has a screen display of information (see figure 9.13) including, for each criterion, the best and worse values, the current aspiration level and the closest alternative to this aspiration level for one of the two possible distances. Also displayed is the percentage of alternatives above the current aspiration level, criterion by criterion, and for the whole of the criteria. In a later step, an analysis of outranking of the ELECTRE type enables the decision maker to choose between the 'optimal' alternative at a given stage, and the nearby alternatives. According to Lotfi *et al.* (1992), the method stands up well beside others such as the AHP method of Saaty (1980). In Yoon *et al.* (1991), the method is extended to continuous multicriterion decision. Apart from the promising experimentation of the authors, we do not know of any other applications.

The method TRIPLE C (Circular Criteria Comparison model) of Angehrn (1989) is also based on the idea of interactive exploration; the final ranking is arrived at through minimization of the distance from the ideal using the distance d_1 and prior normalization by division by the length of the scale (procedure #2, section 2.7). Its general 'decision maker-centered' philosophy goes further than this, the main idea being user 'learning' through the concept of interactive visualization which, according to the author, '...facilitates and stimulates the progressive structuring of the problem, exploration of the space of alternatives and explanation/communication of the results of the decision process'. We shall come back to these ideas in chapter 9, and in chapter 10 we shall describe the features of the Macintosh software package which, like AIM, forms the substance of the method.

8.4 Permutation methods

The original idea of using this type of method in multicriterion analysis is due to Jacquet-Lagrèze (1969); the idea was then taken up by Guigou (1973). Pérez (1991) points out that the basic permutation method is formally identical to the procedure proposed by Kemeny and Snell (1962, chapter 2) for the aggregation of individual opinions. The methods did not however really take shape until the arrival of the method QUALIFLEX of Paelinck (1976) and workers at the Rotterdam Institute of Economy who used it in numerous applications and promoted it. Recently the same team launched the software QUALIFLEX for PC compatible microcomputers (see chapter 10).

None of these variants demand a great deal of information from the decision maker: ordinal utilities and cardinal weights are all that is required. If it were not for the weights, these methods would be ordinal in the same way as those studied in chapter 5. There also exist versions which accept ordinal weights at the expense of results becoming more complicated to interpret (see section 10.5).

Numerous applications of the method have been described by the Dutch school, who appear at the present time to be the only users. Thus, Paelinck (1976) describes a case of regional planning and another (1977) of airport siting. Ancot and Paelinck (1982) apply the method to four cases of urban resource management (water, garbage disposal, public transport and public services).

The basic idea behind the method is to compare each permutation in the order of alternatives against the information contained in the decision matrix. For each permutation an index called the '*value index*' is calculated; this is equal to the difference between a measure of the concordance of pairs (in the sense of chapter 7) and the pre-order induced by the permutation, minus a measure of the discordance for the same pairs (refer to the calculation in example 8.12 below). The permutation which gets the best value index is deemed to be the best. As we have just said, calculation of the concordance coefficients presupposes knowledge of a cardinal weight vector w; variants of the method accept ordinal weights. This will all become clear with an example.

Example 8.12 Market investment.

An individual wishes to purchase a certain number of shares, and has the choice between shares A, B and C. He considers three criteria:
C_1 profitability in annual percentage
C_2 solvency (qualitative)
C_3 liquidity (qualitative)

Thus we have two qualitative criteria (necessarily ordinal) and one quantitative criterion. The performance table is shown in figure 8.6, which also shows the weight vector.

		Criteria		
		Return C_1	Solvency C_2	Liquidity C_3
Alternatives	A	11%	++	+
	B	10%	+++	+
	C	10%	+	++
Weights	w	0.5	0.3	0.2

Figure 8.6 Share investment data

First, the $m!$ permutations of the m alternatives in the choice set are made. Here there are $3! = 6$ possible permutations which, with the corresponding rankings, are:

Permutation	Ranking	Permutation	Ranking
P_1	$A \succ B \succ C$	P_4	$C \succ B \succ A$
P_2	$B \succ A \succ C$	P_5	$C \succ A \succ B$
P_3	$B \succ C \succ A$	P_6	$A \succ C \succ B$

For each permutation P_k we obtain an $m \times m$ matrix $M^k = (m_{ij}^k)$, in which the rows and columns are ranked in the order associated with the permutation and where the coefficients m_{ij}^k express the concordance of the pair of alternatives (a_i, a_j). Each value of m_{ij}^k is obtained from the sum of the weights of the criteria for which a_i is equal to or greater than a_j.

Thus in our example, if we consider the permutation P_2, we obtain the matrix M^2:

$$M^2 = \quad \begin{array}{c|c|c|c} & B & A & C \\ \hline B & - & 0.5 & 0.8 \\ \hline A & 0.7 & - & 0.8 \\ \hline C & 0.7 & 0.2 & - \end{array}$$

For example, calculation of m_{12}^2 gives 0.5 since B is equal to or better than A for criteria C_1 and C_3, leading to $m_{12}^2 = w_2 + w_3 = 0.3 + 0.2$.

In this way, the coefficients m_{ij}^k of the upper triangle of the matrix $(i < j)$ indicate concordance of the decision maker's evaluations with the ranking represented by P_k. Likewise, in the lower triangle the values of m_{ij}^k $(i > j)$ indicate discordance between the decision maker's evaluations and the ranking induced by P_k. The value index I_k which we mentioned above, associated with each P_k, is equal to the sum of the concordances minus the discordances:

$$I_k = \sum\nolimits_{i < j} m_{ij}^k - \sum\nolimits_{i > j} m_{ij}^k$$

In our example we will thus have:

$$I_2 = 0.5 + 0.8 + 0.8 - 0.7 - 0.7 - 0.2 = 0.5$$

For all permutations we obtain:

$$I_1 = 0.9 \quad I_2 = 0.5 \quad I_3 = -0.7 \quad I_4 = -0.9 \quad I_5 = -0.5 \quad I_6 = 0.7$$

Permutation P_1 gives the best index, and so the final aggregation from the method will be:

$$A \succ B \succ C$$

The reader may wonder what should be done if the decision maker gives only one preorder for the weights (ordinal weight structure). The following shows what happens if the only information is $w_1 \geq w_2 \geq w_3$.

The fact of the weights being ordinal does not prevent us from making their sum equal to 1, and so our information on w can be summarized as:

$$w_1 + w_2 + w_3 = 1 \tag{4}$$
$$w_1 \geq w_2 \tag{5}$$
$$w_2 \geq w_3 \tag{6}$$

The geometrical representation in $\mathbb{R}^3$ of the set of vectors $w = (w_1, w_2, w_3)$ satisfying constraints (4) to (6) is shown by the hatched area in figure 8.7.

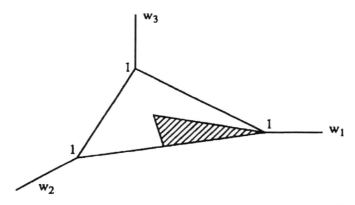

Figure 8.7 The set of (feasible) weights satisfying the constraints in $\mathbb{R}^3$

The hatched area is represented in the plane $w_1 + w_2 + w_3 = 1$ in figure 8.8.

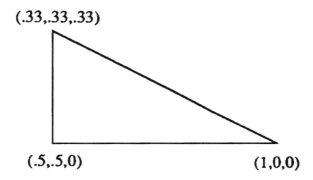

Figure 8.8 The set of feasible weights in the plane $w_1 + w_2 + w_3 = 1$

If we have no further information on the weights we have to consider all the weight vectors belonging to the triangle in figure 8.8, and for each one perform the same calculation as in example 8.12. A very useful theorem obtained by Paelinck (1976) enables the calculation to be shortened by only considering extremal points (vertices) of the weight polyhedron, *i.e.*, in our example, the vertices of the triangle. Paelinck's theorem was later proved more simply by Claessens, Lootsma and Voogd (1991), and the algorithm proposed by Paelinck (op. cit. p.69) is as follows: 'Evaluate each permutation for all the extremal points and choose the best in each case. If one of them is the best for two extremal points, then it is best for the line joining the two points. If two different permutations are optimal at two extremal points, there exists an intermediate point on the line joining them where both permutations are optimal, but in any case the permutations common to the two extremal points are the best. Likewise, inside the polyhedron it is the permutations common to the extremal points which dominate.' Our next example shows this put into practice.

Example 8.13 Market investment (continued).

Suppose our (ordinal) weights preference structure is: $w_1 > w_2 > w_3$.

For each permutation P_k we can easily calculate how many times each weight w_j is involved in addition or subtraction in calculating I_k. Thus, for permutation P_2 the matrix M^2 will be:

$$M^2 = \begin{array}{c|c|c|c} & B & A & C \\ \hline B & - & w_2 + w_3 & w_1 + w_2 \\ \hline A & w_1 + w_3 & - & w_1 + w_2 \\ \hline C & w_1 + w_3 & w_3 & - \end{array}$$

and hence:

$$I_2 = (w_2 + w_3) + (w_1 + w_2) + (w_1 + w_2) - (w_1 + w_3) - (w_1 + w_3) - (w_3) = 3w_2 - 2w_3 .$$

From these calculations, we can construct the following table which shows, below each permutation P_k, the number of times each weight appears, with its sign, in the calculated index I_k.

	W	P_1	P_2	P_3	P_4	P_5	P_6
+++	w_1	2	0	-2	-2	0	2
++	w_2	1	3	1	-1	-3	-1
+	w_3	-2	-2	0	2	2	0

In the above table, the column P_2 shows that w_1 features in I_2 with coefficient 0, w_2 with coefficient 3 and w_3 with coefficient -2. For each value of the vector w, respecting the weights preference structure, is then easy to calculate the index I_k.

Thus, for $w_1 = 0.5$, $w_2 = 0.3$ and $w_3 = 0.2$ we have $I_2 = (3 \times 0.3) - (2 \times 0.2) = 0.5$. As we said, it is actually sufficient to do the calculation for the vertices of the polyhedron of admissible weights. Having done this, we can then select the optimal permutation. The set of optimal permutations is shown in the following table.

w_1	w_2	w_3	Optimal permutations
1	0	0	P_1 and P_6
0.5	0.5	0	P_1 and P_2
0.33	0.33	0.33	P_1, P_2 and P_6

Figure 8.9 summarizes the results in graphical form. In the figure, we see that permutation P_1, *i.e.* $A \succ B \succ C$, appears as the optimal permutation inside the triangle of our feasible region ($w_1 > w_2 > w_3$). If we had accepted equal weights, such as $w_2 = w_3$, the hypotenuse of the triangle in figure 8.9 would also have been optimal and permutation P_6 ($A \succ C \succ B$) would have been a candidate. Note that in both cases it is A which is preferred.

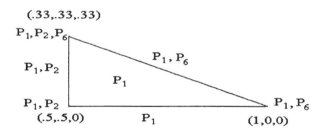

Figure 8.9 The optimal permutations in the feasible region of weights

We mentioned earlier that all the above calculations can be performed and interpreted conveniently using the software package QUALIFLEX. This program is actually quite a lot richer, enabling, for example, random and/or systematic weight structures to be generated (see chapter 10).

It is interesting to note that Paelinck (1983) formalized the basic algorithm of the permutation method as a special case of a QAP (Quadratic Assignment Problem); a fairly recent exposé of this problem is to be found in Finke *et al.* (1987). Such a method is very elegant and attractive; it is this algorithm which is used in the QUALIFLEX package.

We conclude this section with a quick assessment of the permutation methods. One of their advantages is their flexibility in handling the decision maker's information, which may be ordinal or cardinal. There are two main drawbacks; the first is that the complexity of its algorithm makes it a blackbox undermining the desirable decision maker's comprehension and trust face to the obtained results. The second drawback is that the results can be difficult to interpret, especially when the weights are ordinal, even when the software has facilities for testing the sensitivity of weights, as does QUALIFLEX package. We should bear in mind too that, underlying the calculation of value indices, is the assumption of additivity with all the restrictions that this carries (see chapter 6). A permutation method can in fact be interpreted as a method of exhaustive concordance, since we are concerned with the concordance of every possible ranking of alternatives, characterized by a permutation, and not simply in the concordance of pairs. The relatively rare use of the methods thus belies their interest, and they are likely to become more widely used in future.

8.5 Miscellaneous methods

8.5.1 Introduction

We end this chapter and our survey of multicriterion aggregation methods with a brief look at a few methods which, for various reasons, (of minor interest, too recent, too specific, no or few applications, etc.) have not seen such wide use as the methods we have discussed up to here.

First, we present those methods which can be qualified as 'multidimensional' as they are all based on methods borrowed from multidimensional statistical analysis of data, the aim being to reduce the dimension of the data array (each attribute corresponding to a dimension). Finally, we briefly mention various methods which are hard to classify under a single heading.

8.5.2 Multidimensional methods

We begin with the methods based on MDS (MultiDimension Scaling), a technique which is well described by Kruskal and Wish (1978). The *'structural map'* of Rivett (1977) appears to be the first attempt to use this method in the field of multicriterion decision. Starting from an alternatives dissimilarity matrix, which may for example be derived from a concordance matrix (see section 7.5), and using MDS, a two-dimension 'map' may be derived which interrelates alternatives according to the extent of their dissimilarities. To put this in more visual terms, MDS performs the operation which is the inverse of reading the distances between towns on a road map: starting with the distances, the towns are positioned as accurately as possible to reflect those distances. In terms of multicriterion decision, if we have been careful to place the ideal alternative on our map, the alternative chosen will be the one which is 'closest' to the ideal.

A second method suggested by Stewart (1981) is '*correspondence analysis*'. Starting from a decision maker's statement of a pair of alternatives between which he is indifferent, a series of graphs is constructed in two dimensions; from these a more limited sub-set of alternatives can be determined by means of a simple graphic procedure. Cheung (1991, 1992) has recently given the method a mathematical structure (coefficients of similarity and χ^2 distances). Thus formalized the method appears as a variant of the MDS line.

Stewart (1981) also proposed another method, using *factorial analysis* to reduce the number n of criteria in the initial problem to a sub-set of hypercriteria which 'explain' most of the decision matrix. Briefly, in criterion space, each alternative is seen as a point $|R^n$ and the set of alternatives as a cluster of points; factorial analysis seeks the projection onto a space of the smallest possible dimension such that the shape of the cluster is conserved as much as possible (minimal loss of information as measured by *percentage of 'explanation'*); the axes on which the projection is made are the '*main components*'. In our case, if we limit ourselves to only two main components or '*hypercriterion*', the graphical representation enables us easily to choose (see section 3.2) a sub-set of non-dominated alternatives. The method has been tested by, among others, Massam and Askew (1982), who do not give it a very favorable assessment.

Our last reference to a multidimensional method is that of its 'hidden' use in the software GAIA accompanying the PROMCALC package based on PROMETHEE (see section 7.4). GAIA uses analysis into main components, as described above, to produce a graphic representation of the concordance or discordance between the criteria and the relative positions of alternatives (see sections 9.4 and 10.3). Vetschera (1992) analyzed and generalized this type of method.

8.5.3 Other methods

A simple method which has been around for a long time (see Hwang and Yoon, 1981) consists of considering, on a normalized decision matrix, the minimal value $Min_j\ a_{ij}$ for an alternative a_i on the set of criteria, and to rank the alternatives in according to the value of these minima. The alternative ranked in first position achieves $Max_i Min_j\ a_{ij}$, hence the name *Maximin method*. If we apply the maximin method to the normalized matrix of table 4.2, we obtain the ranking: Georgia (Min = 0.156), Donald (Min = 0.134), Albert (Min = 0.114), Helen (Min = 0.098), Blanche (Min = 0.069), Emily (Min = 0.034). Of course, while the Min is ordinal, Maximin will be sensitive to the scales chosen to measure the utilities.

In fact, with normalized utilities $a_{ij} = U_j(a_i)$ belongs to the interval [0,1]. Each U_j can be interpreted as a 'fuzzy set' on $\mathcal{A}$ (*cf.* Hwang and Yoon, 1981). The value a_{ij} then corresponds to the 'proximity' of alternative a_i to an ideal alternative for criterion j. The closer a_{ij} is to 1, the more ideal a_i is considered to be for criterion j (see Dubois and Koning, 1991). Here, the operators for composing fuzzy sets (Dubois and Prade, 1980 and Bouchon-Meunier 1993) may be considered as aggregation operators. Thus $Min_j\ a_{ij}$ is the intersection of the 'fuzzy sets' U_j.

Likewise, Max$_j$ a_{ij} is the union of 'fuzzy sets' U_j. There does in fact exist an instance of the *Maximax* method of choice which consists of choosing the alternative a_i which achieves Max$_i$ Max$_j$ a_{ij} (*cf.* Hwang and Yoon, 1981). This method of choice is obviously hard to justify. Among the numerous possible fuzzy operators, Dubois and Koning (1991) look for those which satisfy certain properties of the Arrow theory axioms type (see section 5.4). In terms of decision, this results in a reasonable interpretation of the fuzzy operators used.

Now we will consider more recent methods which are more sophisticated than Maximin. Korhonen is a very active Finnish worker who, alone or in collaboration, has devised a large number of new multicriterion methods, generally eclectic in their principles and therefore hard to classify. Leaving aside their continuous multicriterion methods (Korhonen and Laakso, 1986), when we look at Korhonen's propositions in the field of discrete multicriterion analysis we can identify five different directions. Two of them were mentioned in section 8.2: generalization of Zionts' method (Korhonen et al, 1984) and the VIMDA method (Korhonen, 1988). It should however be noted that these methods borrow features of other methods than that of Zionts, and VIMDA in particular could also be considered as a variant of the distance from the ideal type of method (section 8.3).

One of the sources of inspiration for two of Korhonen's methods to be described below is the model proposed by Bowman and Colantoni (1973) and their successors (see section 5.5). In the context of voting, this method enables to obtain an aggregated preorder therefore transitive. The idea, we recall, is to look for a transitive relation as close as possible (optimization) to a given pairwise relation resulting, for example, from a simple majority vote. Mathematically, this gives a linear integer program (see 5.5).

Thus, Korhonen and Soismaa (1981) propose a two-step method. In the first step, we seek an aggregation, not necessarily transitive, which minimizes the gap between the relations obtained by considering the preferences alternative pair by alternative pair, for each criterion, while at the same time being compatible with the decision maker's global comparisons. The result is a continuous linear optimization problem with several objectives, and this is handled by the multiobjective method of Zionts and Wallenius (1976). During the second phase, starting with the weights obtained in the first phase, transitivity is forced following the methodology of Bowman and Colantoni. In this way, by asking for comparisons of pairs of alternatives according to the procedure in the method of Zionts and Wallenius, the decision maker does not have to set any weights *a priori*.

The HIRMU method (Hierarchical Interactive Method for Ranking alternatives with MUltiple criteria) of Korhonen (1986) is also based on the Bowman and Colantoni model, extended to problems with hierarchically structured criteria. The method utilizes a visual interactive procedure similar to that described in Korhonen and Laakso (1986), which leads to a rank correlation matrix of the Spearman type (see Kendall, 1970, p.8 *et seq.*) between criteria. This matrix is then used in an optimization model of the Bowman and Colantoni type. An application of the method to the choice of computer systems is described in Tanner (1991).

One of the last ideas from Korhonen is that of dynamic processing of the choice set, *i.e.* the set of alternatives is not fixed once and for all, but can change through the introduction of new alternatives. In Korhonen *et al.* (1986), Korhonen and Wallenius (1986) and Salminen *et al.* (1989) such a method is presented and analyzed; essentially it involves an active procedure designed to determine the analytical form of the decision maker's utility function. This determination is based on the decision maker's responses in comparing pairs of alternatives. At each iteration can be calculated the probability of finding better alternatives outside the choice set being considered at that iteration. With the help of this information, the decision maker can elect to stop if he is satisfied with the best alternative so far obtained, or to continue over one more iteration, enlarging the choice set and at the same time bearing in mind the cost of the process.

The *linear assignment* method was introduced by Blin (1976) and subsequently improved by Bernardo and Blin (1977) and Blin and Dodson (1978). This method only requires one preorder for each criterion and a scale of cardinal weights; it could actually be regarded as an advanced weighting method for ordinal utilities. At the start we consider the $m \times m$ matrix M, defined by $M = (m_{ik})$ such that m_{ik} is equal to the sum of the weights w_j of the criteria j for which the alternative a_i is ranked at k. If there are ties $(a_{ij} = a_{kj})$ for one criterion j, that criterion is doubled up into j_1 and j_2, each with a weight $w_j/2$; then for j_1 we place a_i in the rank above a_k, whilst for j_2 we do the reverse. If there are more than two ties, say p ties, we divide the criteria as much as necessary $(p!)$ so that all possible tie rankings are represented.

If, as is usually the case, the weights are normalized such that their sum is equal to 1, the matrix M is doubly stochastic, *i.e.* the sum of the coefficients in each column and each row is equal to $\sum_j w_j = 1$ as the reader can readily verify. Doubly stochastic matrices have interesting properties which can be exploited. Each coefficient m_{ik} measures the contribution of a_i in an aggregated ranking, assuming that a_i is at the aggregated ranking k. The greater m_{ik}, the more valid is the assumption that a_i is at an aggregated ranking k. The global problem thus becomes:

$$\text{maximize } \sum_i m_{i\sigma(i)} \text{ on the set of permutations } \sigma \in \text{Per}\{1, 2, ..., m\}.$$

Given the doubly stochastic nature of M, the outcome is a particularly simple linear program.

The great drawback of the linear assignment method is the weakness of its theoretical properties, as demonstrated by Pérez (1991, 1994) and Pérez and Barba-Romero (1995), see section 12.2; this probably accounts for why it has been little used despite its apparently attractive features. Pérez looked for consistency properties in certain types of multicriterion methods. By 'consistency properties' we mean properties of the Arrow theorem axioms type (section 5.4), such as the axioms of 'unanimity' and 'irrelevant alternatives', together with other weaker ones that have not been mentioned (see Villar, 1988 or Pérez, 1991). What Pérez showed was that linear assignment satisfies none of the usual coherence properties. When certain modifications are introduced, the situation is slightly improved, but in that case Blin's method is transformed into a little-known method of voting proposed by Cook and Seiford (1978).

Lastly there is the MOOTBA method (**MultiObjective Optimization Theory Based Approach**) of White and Sage (1980), which may be described as an interactive linear weighting method; the decision maker is required to propose a partial preference structure for the alternatives, in graphic form. From this and the weights given by the decision maker, the method sorts the criteria into hierarchical levels using simple weighting. This hierarchical structure enables alternatives to be eliminated interactively until the decision maker is satisfied with the selection and stops.

9 COMPUTERS, ARTIFICIAL INTELLIGENCE, INTERACTIVITY AND MULTICRITERION DECISION

Multicriterion decision has been linked to computer methodology almost since the beginning; the first interactive methods, such as those of Benayoun and Tergny (1969), and outranking methods of the ELECTRE type (Roy, 1968) were designed for computer use, the former with linear programming (the simplex algorithm) and the latter with its rather lengthy calculations of outranking matrices. From 1970 through 1980, computers were treated merely as a way of performing the required calculations rapidly. As we saw in section 1.2, it was not until the beginning of the eighties that those concerned with multicriterion decision began thinking about ways and means designed more specifically for computers.

Computer methodology partners multicriterion decision in three distinct ways: in the performing of complicated calculations, in the use of 'artificial intelligence' and, in the general context of decision aid, interactivity. In the following sections, we shall be looking further into these points, and especially the last two.

9.1 The complexity of calculations

In chapter 3 (figure 3.7) we gave the limits of complexity of the calculations involved in searching for a Pareto optimum. We saw that the complexity of these calculations increases rapidly with the number of alternatives. This complexity, more or less a function of $O(m^2 \times n)$ (where m is the number of alternatives and n the number of criteria) occurs in all methods based on pairwise comparisons of alternatives, such as Condorcet voting and outranking methods of the ELECTRE type (where in addition there is the search for circuits in the graph, which is exponential). Within reason (of the order of 10 through 100 alternatives and 10 criteria), computers can handle this easily, but 'hand' calculation is obviously out of the question. A good practical idea of complexity can be had from the minimum configurations required by some applications for microcomputers (*cf.* Chapter 10). In some permutation methods (section 8.4), the complexity is also exponential limiting the number of alternatives and criteria to deal with. In outranking methods, complexity is considerably increased through the handling of graphs in which each alternative is a vertex, *e.g.* in the elimination of circuits and seeking a node; this is why related applications impose a limited number of alternatives.

The standard algorithms of operational research are utilized by some multicriterion methods, especially in continuous multicriterion decision. This is particularly true in the linear programming found in a supporting role in methods such as UTA (Jacquet-Lagrèze and Siskos, 1982), the methods of Zionts and Wallenius (1976, 1983), Korhonen and Laasko (1986) and many others. Linear integer programming is also found in the method of Bowman and Colantoni (see section 5.5). None of these methods can be used without a computer.

In sum, the complexity of calculation in multicriterion methods is comparable with operational research algorithms to within a mutiplying factor n. Here, n is the number of criteria which is at most around twenty, and so the difference is not very great.

9.2 Artificial intelligence (AI) and multicriterion decision

At the start of the AI adventure, Newell and Simon (1972) claimed two main ideas in their seminal book: that computers are able to handle symbolic as well as numerical computing, and that the general framework of human reasoning is heuristic. We will return to the second idea in the third section of this chapter. Let us start with symbolic representation.

The environment in which multicriterion decision analysis developed was mainly of the OR type, and this is more amenable to numerical than symbolic representation. This is probably why, though there has been a great deal of work on aggregation using quantified alternatives (*i.e.* alternatives represented as numerical vectors), there are very few papers on the more qualitative forms of reasoning such as the design of the alternatives and/or criteria themselves.

Four main families of AI can be distinguished: rule-based, object-based, neural network and logic-based. We begin with the use of AI in multicriterion decision management. (MCDM)

9.2.1 Aggregation and rules

One of the first ideas which comes to mind is the analogy between elementary aggregation and the classical rules of logic concerning propositions and predicates, as used in expert systems. This idea first appeared in Bohanec *et al.* (1983). Let us illustrate it with a simple example. Suppose that in choosing an automobile, braking and acceleration are considered important, noise and capacity are to be taken into consideration and price is a non-negligible criterion.

The decision maker will express his preferences in the following way. Consider automobiles A and B.

If BRAKING(A) = GOOD and if ACCELERATION(A) = SATISFACTORY
Then TECHNICAL SPECIFICATIONS(A) = GOOD

If NOISE(A) = MEDIUM and if CAPACITY(A) = HIGH
Then COMFORT(A) = MEDIUM

If TECHNICAL SPECIFICATIONS(A) = GOOD and if COMFORT(A) = MEDIUM and if PRICE(A) = MEDIUM Then CHOICE (A) = ACCEPTABLE

For vehicle B the rules will be as follows:

If BRAKING(B) = MEDIUM and if ACCELERATION(B) = MEDIUM
Then TECHNICAL SPECIFICATIONS(B) = MEDIUM

If NOISE(B) = HIGH and if CAPACITY(B) = MEDIUM
Then COMFORT(B) = MEDIOCRE

If TECHNICAL SPECIFICATIONS(B) = MEDIUM and if COMFORT(B) =
MEDIOCRE and if PRICE(B) = MEDIUM Then CHOICE (B) = RESERVED

What is the connection with multicriterion analysis? Just this: we can interpret each of the terms, Braking, Acceleration, Noise, Capacity and Price as criteria of choice for the decision maker; these are the attributes that will influence the decision. Within the choice set, the decision maker will define a preorder for each of the attributes. This preorder is expressed through ordered symbolic values of the type {VERY BAD, POOR, MEDIOCRE, MEDIUM, GOOD, EXCELLENT}. What we can also observe in our example is the grouping of low-level criteria into more aggregated criteria. Thus noise and capacity are aggregated into comfort. The dependencies can be shown as a tree diagram (Figure 9.1)

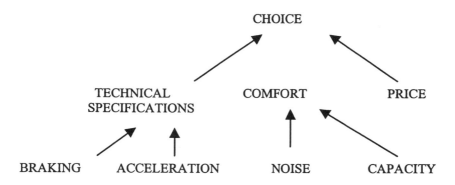

Figure 9.1 Tree diagram of criteria

This breaking down of criteria into sub-criteria is in fact similar to the system used in certain multicriterion methods such as the method of Saaty (1980, see sections 4.5 and 4.8 on hierarchization). In the above interpretation of vehicle characteristics, it is obvious that rules such as:

If NOISE(B) = HIGH and if CAPACITY(B) = MEDIUM
Then COMFORT (B) = MEDIOCRE

transforms a small part of the aggregation of the sub-criteria Noise and Capacity into a higher level criterion named Comfort. A rule is therefore a local aggregation procedure (Perny and Pomerol, 1999). Why local? Because the result of any aggregation will depend on the values obtained by the alternatives. In a Saaty-like procedure, aggregation can be characterized by a formula of the type:

$$0.4 \text{ Noise}(X) + 0.6 \text{ Capacity}(X) = \text{Comfort}(X)$$

All we then have to do is to assign values to Noise(X) and Capacity(X) to obtain the value of Comfort(X). To determine the basic values Saaty uses a special comparison procedure. Here we may assume they are given by the decision maker or have been evaluated by some expert system (Lévine *et al.*, 1990). In fact, ,as Lévine and Pomerol (1989) pointed out, each rule expresses a discrete functional dependence. Thus the fact that the choice depends on price, comfort and technical specifications can be expressed in a table (or decision table, see Benchimol *et al.*, 1990, p.134). Part of this table is shown in figure 9.2; for simplicity we have assumed that the values for comfort and technical specifications are limited to {BAD (B), MEDIUM (M), GOOD (G)} and the price to three ranges, Low (L) ($15.000 ≤ *Price* < $25.000), Acceptable (A) ($25.000 ≤ *Price* < $35.000) and High (H) ($35.000 ≤ *Price* ≤ $50.000). The choice is expressed as Refused (R), Acceptable (A) and Pending (P).

TECHNICAL SPECIFICATIONS	B	B	B	B	B	B	B	B	B	M	M	M	M	M	M	M	M	M
COMFORT	B	B	B	M	M	M	G	G	G	B	B	B	M	M	M	G	G	G
PRICE	L	A	H	L	A	H	L	A	H	L	A	H	L	A	H	L	A	H
CHOICE	P	R	R	P	R	R	P	P	R	R	R	R	A	P	P	A	A	P

Figure 9.2 Decision table

The complete table would required 27 columns to treat all combinations of values. Note that the first nine columns of the table can be summed up in four rules; this is what is called *induction* in some expert system generators. (see Benchimol *et al.*, 1990). Here are the rules:

If TECHNICAL CHARACTERISTICS(X) = BAD and if PRICE(X) = LOW then CHOICE(X) = PENDING

If TECHNICAL CHARACTERISTICS(X) = BAD and if PRICE(X) = HIGH then CHOICE(X) = REFUSED

If TECHNICAL CHARACTERISTICS(X) = BAD and if COMFORT(X) = GOOD and if PRICE(X) = ACCEPTABLE then CHOICE(X) = PENDING

If TECHNICAL CHARACTERISTICS(X) = BAD and if COMFORT(X) = BAD or MEDIUM and if PRICE(X) = ACCEPTABLE then CHOICE(X) = REFUSED

Aggregation between criteria can thus be done either with rules or tables. If graphs of functional dependence (or tree diagrams, Lévine-Pomerol, 1989) are systematically used, an entirely or partially symbolic multicriterion decision system may be devised. The expert system generator ARGUMENT (Saad, 1989), which allows rules to be written in tabular form as we explained just above, and to produce a tree diagram of associated criteria was devised in this spirit.

As we said, a table of values of the type shown in figure 9.2 represents a discrete function – in our example, Choice is a function of Technical Characteristics, Comfort and Price. Discretization means that there will be jumps in values, *e.g.* in Comfort, which is a staircase function dependent on noise and capacity.

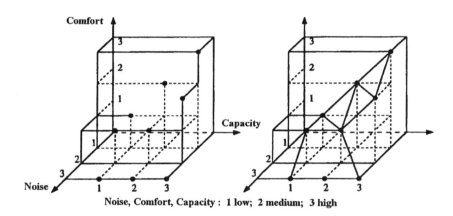

Noise, Comfort, Capacity : 1 low; 2 medium; 3 high

Figure 9.3 Discrete and linearized function

The jumps in the function cause problems in certain contexts, which is why Rajkovic *et al.* (1988) proposed a linearization of the function within the interval of values given by the table. In this way we obtain for the coordinates of the table (figure 9.3) a continuous aggregation coinciding with the values given by the decision maker. The reader will see from figure 9.3 that unfortunately, the linearization is non-unique. In the example of figure 9.2 we can see that it is the finite number of values that limits the discrimination possible; thus TECHNICAL CHARACTERISTICS = BAD and PRICE = LOW leads to CHOICE = PENDING regardless of comfort. If there existed intermediate values between REFUSED and PENDING it would probably be possible to take account of comfort. To achieve this Bohanec *et al.* (1991) designed a procedure allowing a complete ranking to be constructed in each class. For example, suppose three automobiles A, B and C satisfy TECHNICAL CHARACTERISTICS = BAD and PRICE = LOW; they will, *ipso facto*, be PENDING. If however we also know that COMFORT(A) = BAD, COMFORT(B) = MEDIUM and COMFORT(C) = GOOD, then it is reasonable to assume that within the class PENDING, the automobiles are ranked in the order: C preferred to B preferred to A. Bohanec *et al.* (1991) assign numerical values to their qualitative evaluations. Next they assign weights to the criteria by linear regression, so as to follow as closely as possible the decision maker's rules. A weighted sum is then applied, as in Saaty (1980), to evaluate all the alternatives. Most of the ties between alternatives are eliminated in this way. The result is formally identical to that given by Saaty's method, though in practice the weights and values of the leaves in the tree diagram are defined differently: the values are provided by the decision maker in symbolic form and the weights are calculated using rules also given by the decision maker. This idea resulted into DEX, an 'expert system like' aggregation system (Efstathiou *et al.*, 1986; Bohanec and Rajkovic, 1990).

Aggregating with rules requires a lot of information from the decision maker: rule by rule, he has to say how he aggregates the criteria. However there is the enormous merit that the aggregation thus defined is dependent on the observed values. Thus in our example, it is possible to express the fact that comfort and technical specifications are irrelevant if the price is too high and that, on the other hand, comfort is more important than technical specifications if the price is medium while the reverse is true if the price is low. In no fixed weighting method can these preferences be expressed – yet they are entirely natural. One way of getting over this is to vary the weights according to the level (Mandic and Mamdani, 1984, suggested multi-objective linear programming). The direct use of rules as described above seems to us to be the most natural solution.

We should however point out that the use of rules in decision theory to aggregate the preferences has been criticized as being too dependent on user and context (White, 1990). The rule approach is clearly not normative; it is actually very difficult to be normative in multicriterion decision, and constructivism (see section 9.3.1) is generally prevalent; here, it should appear perfectly normal for preferences to depend on the user, and the preferences expressed doubtless depend on context, presentation and even on the way knowledge is acquired. The way the rules themselves are written can introduce bias. Proponents of rules, to console themselves, can reply that utility functions are also dependent on context, method and presentation (see section 2.4.4).

Keeney (1988) points out, in connection with expert systems in general, that methods of choice by rules are an unfortunate mixture of facts and decision criteria, *i.e.* that rules express both the objectives, which describe the situation, and the decision maker's own choice elements. The resulting mixture can therefore be criticized as affecting clarity. One of the weaknesses of expert systems is that they ignore the factors involved in the decision. To avoid the mixing of facts and decision criteria, the respective functions must be kept quite separate, and this idea led to the design of MULTIDECISION-EXPERT (Lévine *et al.*, 1990). In fact, if things are to be done by 'rules', the factors involved in the decisions must be isolated and treated separately, with the decision maker writing a limited number of rules for real decision criteria without trying to include everything within a single expert system.

Note also that differentiating between 'factors entering into the decision' and facts that the decision maker considers as given is to a certain extent subjective, and in any case dependent on context. It is partly because of this that Keeney (1992) stresses 'objective-centered decision' rather than alternative-centered decision. According to Kenney (1992), the decision maker should first of all concentrate on his objectives, *i.e.* his criteria, his values and the objectives he wishes to achieve. The alternatives depend to a certain extent on the objectives in mind, and it is up to the decision maker to create alternatives that will take him to his objectives. To sum up, the criterion tree diagram and decision factors must be defined separately from, and preferably before, the alternatives. It is only after this that alternatives are evaluated (see MULTIDECISION-EXPERT in the next section).

Finally, let us note that the transformation of rules into numerical calculations along the lines of Bohanec and Rajkovic or even Saaty has evolved in a similar way to expert systems, with firstly the introduction of coefficients of similarity as in MYCIN, followed by fuzzy logic. Systems based on fuzzy logic or the theory of possibilities (Dubois and Prade, 1987) are systems in which fuzzy rules also perform aggregation. This idea is of course also present in multicriterion applications; see Perny and Pomerol (1999) for an overall survey of the question and many references.

9.2.2 Expert systems and multicriterion analysis

Section 9.2.1 was devoted to the use of rules as a preference aggregation procedure. There are of course other possible links between expert systems and multicriterion *analysis, which we shall deal with in this section. We shall not, however, include* here the integration of analysis modules with other models within an Interactive Decision Aid System (DSS); this will be dealt with in section 9.5.4.

a) Use of an expert module to evaluate alternatives

One reason why multicriterion decision has not enjoyed widespread popularity is that many decision makers find it difficult to construct a large enough set of alternatives, with each alternative properly evaluated and coherent with the others. In many cases, the analyst's job is to construct the decision matrix (*cf.* chapter 11). Why not use an expert system to do this? This idea was suggested by Lévine *et al.* (1986), and, as already mentioned above, resulted in a prototype, MULTIDECISION-EXPERT (Lévine *et al.*, 1990)

In this prototype, the decision maker or expert has to define the attributes. Each attribute, or criterion, appears as a collection of elementary facts relevant to each alternative. Thus the quality of service of a postal sorting machine will depend on the maximum height from which the packets are dropped, the risk of wrongly sorting a packet, and the risks of damage. The risk of wrongly directing a packet depends in turn on the time taken to stop the machine in case of error, the quality of security procedures and the likelihood of machine errors. This dependence can be shown on a tree diagram (figure 9.4).

Formally, the tree diagram of figure 9.4 is comparable to that of figure 9.1, but here the purpose is different; instead of thinking in terms of aggregation of low level criteria, we begin with the facts describing as objectively as possible a postal sorting machine. A fact base is associated with each machine (or alternative, since the choice is one of machine). This fact base is then treated by the expert system to produce an evaluation of machine i according to criterion j. A rule base R_j is then associated with each criterion j; this rule base treats all the fact bases B_i associated with machines i according to the architecture in figure 9.5.

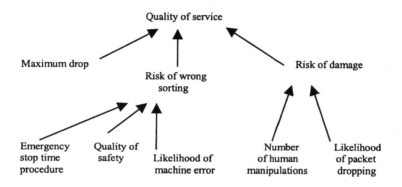

Figure 9.4 Evaluation of the criterion 'Quality of service' (from Lévine *et al.*, 1990)

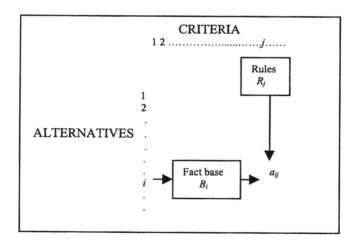

Figure 9.5 Architecture of MULTIDECISION-EXPERT (from Lévine *et al.*, 1990)

The result of applying all the rule bases to all the machine fact bases is a decision matrix (figure 9.6).

	Quality of Working Station	Maintenance	Using Facility	Cost of Operation	Quality of Service	Throughput of Sorted Packages	Price of Machine	Number of Delivery Directions
Machine1	Medium	Good	Very good	Rather expensive	Very good	Important	15.23	74
Machine2	Good	Very bad	Rather good	Very cheap	Very good	Important	15.70	60
Machine3	Medium	Bad	Rather good	Very cheap	Excellent	Poor	15.00	60
Machine4	Medium	Bad	Good	Very cheap	Good	Very important	15.55	60
Machine5	Medium	Very Bad	Very bad	Very expensive	Extremely bad	Very important	36.68	60
Machine6	Medium	Bad	Excellent	Rather expensive	Medium	Very important	22.90	61
Machine7	Medium	Excellent	Excellent	Rather expensive	Good	Important	19.58	61
Machine8	Medium	Very good	Bad	Extremely expensive	Very good	Poor	15.47	60
Machine9	Excellent	Bad	Very bad	Extremely expensive	Extremely bad	Important	13.99	60

Figure 9.6 Decision matrix , postal sorting machine choice (Lévine *et al.*, 1990)

Of course the difference between rule-based aggregation (section 9.2.1) and the construction of criteria from fact bases as just described is largely a matter of interpretation. In the former case, high level criteria are broken down into successively lower level criteria until a stage is reached where the decision maker is able to make an immediate evaluation. We thus have an approach from top to bottom followed by a calculation in the reverse direction. In the second approach, we consider that each alternative corresponds to a certain number of facts describing and qualifying it. These facts form part of the alternative. They express, not preferences, but an objective reality, *i.e.* a reality involving a maximum number of participants in the choice problem. From the facts, alternatives are evaluated according to a certain number of criteria characterized by fact bases, in a bottom to top direction. The criteria are important and significant enough for them to be considered as such and to deserve multicriterion treatment. *The basic question here is: when does a piece of information about an alternative change from being a descriptive fact to being an attribute relevant to the choice.* There is no standard answer to this question. The answer will depend on the context and on discussions with the people involved in the choice (see chapter 11). On the other hand, as we saw at the end of the previous section, the procedure described here has the advantage of clearly distinguishing the elements that the decision maker acknowledges to be decision criteria from those forming the description of one of the alternatives.

A third approach is to use an expert system to evaluate an alternative in relation to the whole of its attributes (O'Leary, 1986). This would only seem practicable for a small number of alternatives – two in O'Leary's paper. With an increasing number of attributes, the model is likely to become very confused since all the knowledge is contained in the same knowledge base and it is doubtful whether we could properly gather the knowledge on each criterion. This type of system is open to Keeney's (1988) criticism of being confusing (see section 9.2.1).

The problem of considering the various aspects of a decision or situation is not exclusive to multicriterion decision; designers of expert systems are also confronted with the problem. Take the example of credit for customers or businesses. When credit is requested it is granted or refused on the basis of several criteria. These will include the financial risk, the personal risk attached to the business, and the commercial risk attached to the market. in question. For cases where the criteria are essentially qualitative, several expert systems have been developed to evaluate separately each of the risks or, more generally, the attributes involved. The results from various knowledge bases are then combined into 'metaknowledge' to arrive at a conclusion. These systems, which are described in the literature on multi-expert systems (see Lévine-Pomerol, 1989 or Gleizes and Glize, 1990), are also systems which aggregate knowledge through a set of rules. Although the notion of criterion is not always very explicit in these systems, we can nevertheless say that each knowledge base evaluates the alternatives in relation to a domain (criterion?), while the system supervisor aggregates the knowledge from each module (the criteria) on the way to the final decision. Examples of such systems are to be found in the book on DSS (1989) by Lévine and Pomerol. We will go no further into them here since, given that most knowledge base systems are actually decision systems, and that a rational decision more or less always obeys a certain number of criteria, we may syllogistically conclude that the set of knowledge base systems is a subset of multicriterion decision! The role of decision in expert systems is nevertheless becoming increasingly invoked; see Keeney (1988), Langlotz and Shortliffe (1989) and White (1990).

b) The use of rules to construct or specify alternatives

In many real situations, the first step is not even to assess the alternatives but to understand what the alternatives are. This is generally the case for economical decisions where the alternatives actually are scenarios. For example, you do not decide between increasing or decreasing the price of the product but rather between two, or generally many more, scenarios: increase the price and begin an advertising campaign and differentiate your product (alternative A) vs. decrease the price and increase production capacity (alternative B) and so on. The alternatives contain different temporal actions and/or consequences. The first difficulty is to have a clear idea of what an alternative is and how to characterize it before assessing it! In our example, assume that the criteria are the net profit, the gross sales and the market share. It is not simple to define and evaluate alternative A vs. B; serious analysis is needed to characterize the results of A and B according to the chosen criteria.

Let us give another typical example. The problem is to assess the robustness of train schedules (Pomerol *et al.*, 1995). Given the railroad network (RN) and machines (M), we can define theoretical schedules (TS). Now, when an incident occurs, the train schedules are perturbed and it is obvious that the extent of the perturbation depends on RN, M and TS. With an unlimited number of railroads, an incident on one train would have no consequence on other trains, but this is not the case. An incident generally delays many trains, and the extent of the total delay depends on RN, M and TS. Assume that we want to assess an RN investment designed to improve the situation in case of incident. This means we will have to assess alternatives of the form (RN, M, TS) for robustness in case of incident. This is a typical case where human reasoning is unable to construct without help what we have called a *fully expanded alternative* (FEA). The fully expanded alternative associated with an alternative (*RN*, *M*, *TS*) is the real timetable incorporating all the delays caused by an incident. To 'propagate' the incidents the decision maker needs support. We have designed such a multicriterion DSS (Pomerol *et al.*, 1995). Here, an expert system plays the role of the dispatchers, *i.e.* it makes the decisions arising from an incident and its consequences. These decisions give rise to actions on the trains perturbed by the incident (typical actions include delay, cancellation and re-direction of trains). When the perturbed train schedule resulting from the preceding decisions is known, and only when this is so, we can assess the extent of the delays, the number of travelers affected and three more criteria. Within this framework, the expert system, or more generally the DSS, is used 'simply' to bridge the gap between the alternatives and the 'fully expanded alternatives'. We believe that this type of problem will become increasingly common as people address more complex problems. Any scenario is actually a sequence of intertwined events and actions, but by using robust actions designed to be good enough however events turn out, decision makers tend to reason in terms of a sequence of actions. This is why we have coined the term FEA for a sequence of robust (sub)actions (see Pomerol, 1999). The point is that computer support is necessary to deal with the complexity of scenarios and FEA.

In decisions about the price of a product, the reactions of competitors could be modeled by an expert system. This would then enable alternatives to be evaluated more easily in terms of profit, financial viability, etc. The use of an expert system to first of all model and then reproduce the sequence of actions and counter-actions of agents with differing interests is often appropriate. The example of evaluating of a piece of legislation according to the criteria of various pressure groups is to be found in Lévine-Pomerol (1989, p.249 *et seq.*) When responses and counter-responses are intertwined, use of an expert system is currently the most realistic way of evaluating the alternatives, each initial alternative giving rise to a scenario that it would be impossible to handle without support. Less common are applications in the domain of negotiation (see Lévine-Pomerol, 1991).

Another example appears in the paper by Du Bois *et al.* (1989). Here, a knowledge-based system of 'frames' is used to review the various pathologies which can affect a patient. Using the symptoms given and the existing information on the patient, the 'expert system' gives the possible pathologies. These pathologies are then subjected to multicriterion analysis by the PROMETHEE method (see section 7.4). For each pathology, the four criteria considered are the number of symptoms which correspond to the given pathology, the number of sicknesses associated with the observed symptoms, concordance of the pathology with the sex of the patient and concordance with the age. In this application, the expert system is thus used to construct alternatives, pathologies in this case, and the pathologies are then given multicriterion analysis. This example can be interpreted as a first step toward the introduction of case-based representation in MCDM.

In all the above cases, the knowledge base system models the expertise that enables the alternatives to be specified and constructed. The situations were all sufficiently complex for modeling in the form of knowledge bases to be considered the only practicable means. But in every case, this expertise, far from obviating the need for multicriterion choice, prepares the ground for it.

c) The use of logical rules and neural networks as a multicriterion choice method

In contrast to what happens in systems where rules are used to perform the initial aggregation, rules are used as elements of the method itself in the systems we now consider. Rules are used directly to eliminate, to classify and to order. This is somewhat reminiscent of the direct use of AI, as in PRIAM, but here we do not have heuristic searching.

The first of these systems, named PLEXIGLAS, is based on the resolution principle used by PROLOG. Suppose that you write a rule in PROLOG form, asserting that quality depends on comfort and technical specifications:

$$COMFORT(X) \land TECH.SPEC.(X) \rightarrow QUALITY(X)$$

PROLOG works in such a way that it will try to find X first of all with TECH.SPEC. as predicate. It will then proceed to COMFORT, introducing a lexicographical order relative to the criteria COMFORT and TECH.SPEC. Seeking some level of quality, we may introduce the rule:

$$COMFORT(X,5) \land TECH.SPEC.(X,3) \rightarrow QUALITY(X,4)$$

Using PROLOG to run this program in order to find all the X such that QUALITY(X,?) is greater than 4 amounts to starting a lexicographical search classifying the issues in the order in which the program finds them. This idea together with some consequences on classification was developed by Rommel (1989, 1991). The work of Rommel extends to the idea of comparing an initial ordering by the decision maker of certain alternatives with the ordering given by the system. In the UTA method (see section 6.6), the ordering within a subset is extended to a utility function that is applicable everywhere. Rommel's idea is equivalent to a learning process which tunes the system to the examples supplied by the decision maker. This method of learning is similar to that used in neural networks (see below).

Pasche (1991) has developed a system in which rules are applied to the preferences. Let J be the set of criteria and 2^J and the set of parts of J. We can define on 2^J a relation $>_j$ reflecting the importance of criterion coalitions using the following rules:

$$\forall \ I, K, L \in 2^J, \ I >_j K \ \text{and} \ K >_j L \Rightarrow I >_j L \qquad \text{(transitivity)}$$

$$\forall \ I, K, L \in 2^J, \ I >_j K \ \Rightarrow I \cup L >_j \ K \qquad \text{(amplification)}$$

$$\forall \ I, K \in 2^J, \ I >_j K \Rightarrow \ not(K >_j I) \qquad \text{(asymmetry)}$$

$$\forall \ I, K, L \in 2^J, \ I >_j K \ \text{and} \ I \cap L = \varnothing \Rightarrow I \cup L >_j K \cup L \quad \text{(additivity)}$$

PROLOG programming is particularly well suited to the weighty calculations introduced by the above rules, which produce $>_j$ from the partial preorder given by the decision maker. Once the relation has been completed it can be used to replace the weights in outranking methods such as ELECTRE III (see section 7.3 and Skalka *et al.*, 1992). It is actually sufficient to state:

a is strictly preferred to *b* if and only if $\{j \mid a \succ_j b\} >_j \{j \mid b \succ_j a\}$

Pasche (1991) developed these PROLOG-based ideas in his system EXTRA. Pasche's method, in which sets of criteria are compared, amounts to comparing different sums of criteria, and can thus be classed as an additive method. This is reminiscent of Churchman and Ackoff (see section 4.5.3).

In this section we have only applied the rules of classical logic. Numerous examples of non-classical logic use (modal logic, fuzzy logic, possibilistic logic etc.) are now appearing in the literature, and the reader is referred to Perny and Pomerol (1999) for an overview of the various links between logic and multicriterion decision. Here we include from that work a brief look at neural methods; neural networks open the way to non-linear aggregation, which is not often met in multicriterion analysis.

It is indeed quite possible to teach a neural network to perform aggregation. Wang and Malakooti (1992) introduce vectors to a neural network, together with the value of the aggregated utility function (the latter may if required be defined by the decision maker); the network 'learns' by giving weights to the links between neurons. Once the network has been configured on the vectors that have been used in the learning process, it can then evaluate any vector and therefore perform automatic aggregation. Numerous experiments have been done in this direction, particularly in the field of marketing using not only neural networks (see Wang and Malakooti, 1992; Malakooti and Zhou, 1992 and Boscarino *et al.*, 1993; Usher and Zackay, 1993; Kant and Levine, 1997), but also algorithms such as ID3 (Quinlan, 1979, 1983). Basically what the network does is to weight the criteria according to a subset of alternatives, replacing the linear programs in methods such as LINMAP and UTA. In addition, if the network has several layers, the aggregation is not linear; in other words, unlike statistical methods based on regression, the network can 'learn' non-linear functions. When preference independence is not satisfied, non-linear aggregation can be preferable (Chung and Silver, 1992).

Several dynamic network methodologies have been proposed for multiattribute categorization tasks (see Leven and Levine, 1996; Kant, 1995; Kant and Levine, 1997). These are based on the Adaptive Resonance Theory (ART) networks proposed by Carpenter and Crossberg (1987). This type of network consists of various interconnected modules encoding attributes of alternatives, categories and the goals of decision makers (prototypes of categories). Category nodes and attribute nodes are interconnected with the connections able to be modified by associative learning, leading to inter-level resonant feedback (for more details, see Leven and Levine, 1996). Assignment of an alternative to a category is only possible if the match between the alternative and the prototype of the category is above a certain threshold. When this is not the case, a new category is created and the unassigned alternative becomes its prototype. An interesting variation of the ART network named CATEG_ART has been developed and tested (Kant, 1995; Kant and Levine, 1997). This network is designed to implement a cognitive model based on Moving Basis Heuristics (see section 5.6.2) and automatically to produce categorization rules synthesized by polynomials> For example, the polynomial:

$$P(C) = X_1^2 X_4^3 + X_2^1 X_5^4 + X_3^5$$

which is made up of a disjunction of monomes; what this means is that, if an alternative has been placed by the decision maker in category C, this is because it satisfies ($X_1 = 2$ and $X_4 = 3$) or ($X_2 = 1$ and $X_5 = 4$) or $X_3 = 5$. Once the neural network has been trained in the reference set of alternatives, the assignment rules that have been created allow classification of new alternatives whose categories are unknown.

Some of the standard networks fit reference data by choosing optimal weights or coefficients within a predetermined mathematical model, and provide a direct assignment by the neural network as in Malakooti and Zhou (1992); Usher and Zackay (1993), however, propose a dynamic neural network methodology for multiattribute decision making. In their system, the network is regarded as a descriptive model aimed at approximating the behavior of a decision maker faced with a multicriterion problem. Consider a set of alternatives $A = \{a_1, ...,a_m\}$, where each attribute X_j has been evaluated by the decision maker and quantified by positive coefficients w_{ij} depending on the alternatives and criteria, $i = 1, ..., m$, $j = 1, ..., n$. We can then consider the following decision network architecture constructed from the prior knowledge of the decision situation (see figure 9.7, adapted from Usher and Zackay, 1993).

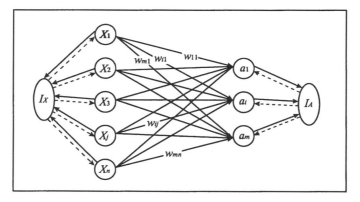

Figure 9.7 A network architecture for multiattribute decision making

This dynamic neural network consists of two connected sub-networks; the one on the left is composed of competing attribute assemblies representing the 'current state of the decision maker's mind' (the current focused attributes); the one on the right is of competing alternative assemblies representing the current potential decision. Full lines represent excitatory connections and dashed lines represent inhibitory connections. I_X and I_A are inhibitory assemblies that mediate competition in the left- and right-hand networks respectively. For example, when inhibitory neurons of I_x receive excitation from all the assemblies in the sub-network, they return inhibition which causes a decay the activity of some attributes. Arrows of the type (X_j, a_i) represent excitatory connections weighted by the coefficient w_{ij} (if w_{ij} exceeds a certain threshold, the a_i assemblies are activated each time attribute X_j is scanned).

After initial activation, the network is autonomous and evolves dynamically. For most of the time it converges towards a stable state. However, when dynamic variations of thresholds are allowed, the network may adopt periodic or chaotic behavior with some intervening stable periods. When stability is reached, the selection proposed by the network is made up of the activated alternatives.

Note that an inhibition parameter controls the average number of simultaneously activated attributes in the left-hand networks. It may be used to represent more or less demanding dominance rules for comparing alternatives (when only one attribute is activated at a time, the decision process can be compared to the 'elimination by aspect' process (see Tversky, 1972a,b). Another inhibition parameter attached to the I_A assemblies controls the level of competition among the alternatives and therefore the average number of alternatives that can be activated each time. Other parameters (noise factors) control competition and inertia among the alternatives. Depending on the values attached to these parameters, the network shown in figure 15.6 can simulate various decision behaviors. Among these are elimination by aspect, the dominance rule, conjunctive and disjunctive models and lexicographic decision rules (for more details see Usher and Zackay, 1993).

d) The use of rules to aid the decision maker in his choice, and miscellaneous applications

The first idea is to use rules to aid the decision maker to choose his alternatives. According to Reyes and Barba-Romero (1986) the rules for choosing investment alternatives could for example be in the form of a veto as follows:

If the social consequences would be undesirable and the turnaround time greater than two years then do not choose the investment.

In this example, we are using rules to restrict choice. Honda and Mimaki (1986), on the other hand, use rules in multi-objective linear programming to aid choice from among the Pareto-optimal solutions provided by their algorithm. Parisek *et al.* (1991) suggest using rules to express the decision maker's knowledge about the criteria. Non-local symbolic rules like those we have already seen could also be used to govern tradeoff between criteria.

If a decision maker is to be aided in his choice, he has to express his preferences in a realistic manner. These preferences will not necessarily be preorders, and pseudo-orders are often considered to be more realistic. We can in that case consider expressing complex preferences in the form of rules that do not necessarily follow monotonic logic. Thus, Tsoukias (1991) suggested expressing the decision maker's preference using logic rules in an inference network. As some preferences may become challenged, the network would incorporate the notion of TMS (Truth Maintenance System). Here, inferences would only apply if the 'IN' conditions were true and the 'OUT' conditions false; this would enable any deductions to be challenged. Preferences for alternatives would be denoted by expressions of the type $a \succ b$ if the decision maker were expressing a preference for a over b (justified by certain criteria) and if indifference were false; any expression of indifference would call into question this strict preference. The decision maker's preferences over all the alternatives would be expressed by the network in its entirety. The major difficulty lies in constructing a knowledge base requiring so much information. As the method has not been applied up to now, it is hard to say how useful it might be, but the system may well turn out to be extremely heavy and not provide the decision maker with much more information than he has fed in.

By now the reader may well have come to the conclusion that the most urgent necessity is an 'intelligent' decision support system to aid the decision maker in his choice of multicriterion method! The idea of using multicriterion IDSS for this was put forward by Jelassi (1986), and implemented by Ozernoy (1991) and Teghem *et al.* (1989). In the latter system, the choice of method is based on a decision tree that classifies decision parameters involved in the choice of method (discrete or continuous, uncertain or certain, etc.). The net result is an interesting classification of the various methods. The idea was also implemented by Hong and Vogel (1991) in a multicriterion IDSS that proposed various simple aggregation models (maximin, lexicographic, sum, etc.). The decision maker is steered toward one model or another through rules controlled by an expert system. Petrovic (1998) tackles the problem using another AI technique: case-based reasoning (CBR). Reasoning here takes the form of a search within a case-base for the case that is the closest (in the sense of a distance that has to be defined) to the case being studied. If the case is defined by several attributes, aggregation of these attributes is performed by the similarity function that measures the distance between the cases. The main difficulties are in defining a good representation of the cases that takes in the whole of the required context, and in defining an adequate similarity function. In the system developed by Petrovic (1998), a case is defined by the type of problem (choice, ranking, sorting etc.), the number of alternatives, the number of criteria, the weights etc. Then when a new case is introduced the most suitable multicriterion method is chosen according to a hierarchical similarity function.

Another attempt to merge CBR and MCDM goes back to 1992 with a system designed by Angehrn and Dutta (1992). In this system the CBR module is intended to act as an adviser, in the authors' words, 'to recognize whether a multicriteria approach was used successfully in other similar situations before, and provide information about the types of alternatives, criteria and preference structures used' (Angehrn and Dutta, 1992). The second role of the system is as a 'story teller'. Here the system allows the user to replay a previous case. The main component of the system is a case library which contains structured multicriterion problems with their solutions. This system remains one of the most complete examples of CBR structuring for multicriterion problems.

The examples we have given above of inter-penetration between multicriterion analysis and expert systems or, more generally, AI, do not, we are certain, exhaust the subject. Historically, multicriterion analysis has sought to involve AI for some time already, and AI workers are now starting to show an interest in multicriterion achievements. This is because available AI methods have popularized the notion of agents that are more or less autonomous (in the informatic sense, *cf.* Ferber and Ghallab, 1988). In these systems there are often decision to be made: which agent should perform which task? How should resources be assigned? Such decisions, which are rarely subject to a single criterion, could be made by the system according to certain multicriterion methods. In the same vein, it should be possible to use multicriterion evaluation functions in algorithms for heuristic searches. Thus Lirov (1991) and White *et al.* (1992) propose versions of A* using an evaluation function with several criteria.

9.3 Interactivity

9.3.1 The context of interactivity

To show just what is involved in interactivity within the multicriterion domain, we will begin with a story first told by Dupuy (1982); it was re-interpreted by Roy (1987a) and it is his version that we reproduce here. It involves a question-and-answer game played in three different ways.

Player B asks the questions and player A replies. There are three ways of playing. In the first way, player A thinks up a story. He then if he wishes tells B the beginning. B's task is to discover the story by asking yes/no questions. Player A must not cheat, *e.g.* by changing the story. After a time, B will probably have figured out the main thread of the story.

The second version of the game appears in the eyes of an outside observer to be exactly like the first; however, player A, though professing to have thought up a story, is actually making it up as he goes along, or else he did think of a story but deliberately gives wrong answers – which comes to the same thing. At this point, two sub-variants are possible: either player A replies according to a precise plan (*e.g.* 'yes' if the question ends with an 'e', and 'no' otherwise), or else player A answers whimsically with no precise plan. In either case the result of the game will be a more or less fantastic story.

The third way of playing is formally identical to the second, with the answers unplanned. The difference is that A now tries to take advantage of B's questions to invent a story that follows the pattern of B's logic and/or fantasies. From A's point of view the process is a creative one, and if B is aware that there is no predetermined story, the game leads to *a joint creative and constructive process.*

Interactive procedures can be compared to these games; in this book we have been referring to them by a term we prefer: progressive information procedures (see section 2.6), for consider this: on the one hand there is the system (sometimes the analyst) corresponding to player B asking the questions, and on the other hand there is the 'responder' *i.e.* the decision maker who supplies the information, like player A. What is the aim of the game? It is to find the decision maker's utility function (the ordering problem) or the alternatives the decision maker prefers (the selection problem).

In the standard description of progressive information methods the game is assumed to be of the first type, *i.e.* there exists an intangible pre-established story in the mind of the decision maker. The job of the interactive system is to discover this story, and the interactive process will be successful if it converges toward the right story. Progressive information methods can therefore be tested by setting a defined utility function in A and answering the system's questions according to the dictates of the utility function; the process should converge toward the utility function. Note however that *the convergence test is only of any use if it can be shown to be satisfied for an entire class of utility functions* (*e.g.* concave). For a single given function, satisfying the convergence requirement only shows that the procedure is correct for that particular case, and does not allow any conclusion to be drawn as to the value of the method (*cf.* below).

Unfortunately, although all this appears fairly satisfactory from a theoretical point of view, anybody who has actually practiced multicriterion decision with real flesh-and-blood decision makers will know that the reality is less clear cut and turns out to be much closer to the third version of the game than the first. Most often there is no predetermined story to be unraveled, but rather a story (*i.e.* a utility function or selection set) that is jointly constructed by two or more participants. Moreover in our experience it is precisely because he does not know a story but would like to construct one that is coherent and appealing to him that a decision maker will call in multicriterion specialists. There will then be a joint and voluntary process to construct the utility function or the set of selected alternatives. Bouyssou (1984), Roy and Bouyssou (1985) and Roy (1985) are absolutely right in distinguishing the search for the decision maker's utility function (game #1), a process they name descriptive, from the *constructive approach* that we have been describing.

If we accept the 'constructive' approach, the notion of convergence no longer has the well-defined meaning described above. There is of course nothing to stop us still applying the convergence test after initially setting a utility function, except that the very idea of the test would have no justification since the utility function would be arbitrary. For the test is only meaningful if it is applied to a class of utility functions, which presupposes that the decision maker's utility function does belong to the class in question; but the decision maker does not have a clearly defined utility function, and to assume, let alone verify, that it is of a certain type is meaningless. It is even essential that the approach be adapted to any modifications to the decision maker's preferences in the course of construction. For the only possible convergence is toward the satisfaction of a task accomplished when the interactive process has produced a logical, coherent story (read also: utility function or subset of alternatives which appear suitable to the decision maker).

One of the disagreeable features of the 'constructive' approach is that the result of the interaction depends on the process. There is no guarantee that the decision maker would obtain exactly the same result from one method or another. This fear has unfortunately confirmed in all experiments on constructing utility functions under uncertainty; se for example Hershey *et al.* (1982), Farquhar (1984); McCord and de Neufville (1983a and b, 1986); Vansnick (1986); Cohen *et al.* (1987); Essid (1990).

In the final analysis, all that we are left with for judging progressive information methods is value judgements that are more or less dependent on the user: ease of interaction, attractiveness and clarity of the interfaces, flexibility of use, a good ratio of amount of information required to results obtained, *confidence of the decision maker in the results, understanding of the logic of the system.* The last two points are highly correlated, as the work of Wallenius (1975) showed. For the theoreticians, there do remain a few firm criteria: the fact that non-dominated solutions are produced, and the number of questions to be asked before arriving at a result, to compare to the number $m \times (m + 1)/2$ of comparisons required to order $\mathcal{A}$.

Not an outstanding state of affairs; progressive information procedures should, nevertheless, be accepted for what they are: *procedures of explication-construction of the decision maker's preferences.* Finally what we have is the same situation as in AI: the so-called knowledge engineering, or know-how gathering, process is actually a cooperative explication-construction modeling process that takes place between the expert and the knowledge engineer (Lévine-Pomerol, 1989). Any pretence at forgetting the construction side in favor exclusively of eliciting of some underlying known object, is bound to end in nonsense.

9.3.2 Heuristic search and interactivity

Perhaps the most striking impact AI has had on decision aid in general and MCDM in particular is the increasing use of heuristic methods. The initial idea put forward by Newell and Simon (1972) is that human problem solving involves an exploration in a space of states, each state containing information about the solution. This exploration obeys the very simple rules of heuristic searching (see Newell and Simon, 1972). Heuristic searching is a systematic process of trial and error controlled by the evaluation of each state. This evaluation is usually performed by with evaluation function which formalizes the heuristic knowledge acquired and utilized in the search. However, in many cases it is actually difficult to formalize this knowledge, and the evaluation of each state becomes highly contextual and idiosyncratic. This is why decision making requires human intervention (Pomerol, 1992; Lévine and Pomerol, 1995).

The aim of a heuristic search is to find a sequence of well-defined operators that lead from the initial state to the target state (figure 9.8). On reaching a state (the current state) the explorer has to decide whether:

– to choose one of the daughter states as the new state (advance);
– to continue to develop new daughters (continue);
– to backtrack to an already developed and recorded state.

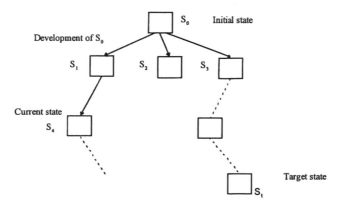

Figure 9.8 Heuristic search

This three-way decision pattern is the basis for any interactive decision process (see section 9.3.3). The general framework has now been appropriated by many researchers into optimization and decision aid, but we will restrict ourselves here to MCDM. Among the applications, the most direct is to apply the heuristic search process to that of choice among the set of all available alternatives (the choice set).

9.3.3 Heuristic search in the choice set

The set of all alternatives is considered as a state space. After defining the operators the decision maker can then proceed to a heuristic search in this space. The idea was used in PRIAM (Lévine-Pomerol, 1983 and 1986), and from this was developed the software application MULTIDECISION, incorporating the PRIAM algorithm (see Pomerol and Barba-Romero, 1993, chapter 9). In PRIAM, each state essentially contains the 'satisfaction levels' relating to the real or fictional alternative(s) present at that point, and the number of feasible alternatives that dominate this 'satisfaction level'. As such, the method is somewhat reminiscent of the elimination by aspects of Tversky (1972a and b). The decision maker evaluates, in view of the alternative proposed by the system, whether the alternative is satisfying and how promising continued exploration appears (see figure 9.9.below). On the basis of this evaluation, the decision maker decides whether or not to trigger an operator to have a new alternative proposed by the system. He can also decide to backtrack to a previously met alternative. The whole process is very similar to the A* algorithm (Nilsson, 1982).

How successful the decision maker's exploration is depends on how well he understands that he cannot increase all the criteria at once. In other words, if he wishes to gain on one criterion he must be prepared to make sacrifices on others; he will find himself on a learning process on the subject of tradeoff. However, the software must provide good operators that help the decision maker explore wisely and, more importantly, reach efficient points (the Pareto boundary). This type of exploration is inherently interactive and heuristic since, at each step, the decision maker is confronted with an alternative proposed by the software. This raises two questions: does the process lead to an efficient point? and has the decision maker sufficient information to control his exploration and evaluate each state attained by the system? To answer the first question, the idea in PRIAM is to provide the decision maker with an efficient point dominating the current point; this is done by extending the previous move up to the Pareto boundary.

To answer the second question on evaluating the current state, there are various possibilities. The system can display some alternatives in the 'neighborhood' of the current point to make the decision maker aware of other possibilities. It is also possible to give an aggregated real value of each alternative displayed by the system; this is a type of scoring system. In MULTIDECISION, some of the alternatives already met by the decision maker and considered as 'satisficing' are displayed, so that the decision maker can easily backtrack and explore various tradeoffs. With these facilities together with the possibility of knowing the distance to the Pareto boundary and hence knowing efficient points, exploration with MULTIDECISION is highly interactive and attractive. As a consequence, the idea has been utilized in other packages (Fiala, 1991, 1992). The idea of interactive searching is also used in AIM (Lotfi and Zionts, 1990 and Lotfi *et al.*, 1992) in one step of their method. PRIAM has also been adapted to continuous multiobjective programming (Pomerol and Trabelsi, 1987).

9.3.4 A general framework for the analysis of progressive information methods

Roy (see section 9.3.1) distinguishes between elicitation of the decision maker's preferences, and progressive articulation of these preferences. Our view is that in MCDM analysis it is the second situation that is prevalent. Thus we consider interactive methods as progressive articulation methods designed to obtain more and more knowledge from the decision maker in order to construct the decision maker's preferences jointly with him. Joint construction means that the decision maker and the system cooperate in the progressive elicitation of preferences.

In MCDM, progressive elicitation methods go back to the very beginning (*e.g.* Roy, 1968, Benayoun *et al.*, 1971 and Roy, 1974), and were developed quite independently from AI; however there are numerous similarities between heuristic searches and progressive articulation methods.

Current progressive articulation methods in MCDM always rely on one process: the system displays an alternative (the current alternative) to the decision maker, and then, according to the decision maker's reaction, tries to find a 'better' alternative. This is none other than a heuristic search process where exploration takes the form of a learning procedure. The principles of progressive articulation methods are shown in figure 9.9; see also Vanderpooten (1989, 1990a) for a comprehensive overview of interactive methods within a pure multicriterion framework. Among the heuristic search methods, Gardiner and Vanderpooten (1997) distinguish between the so-called search-oriented and learning-oriented processes, depending on the role of the decision maker. For them, both processes involve heuristic searching, but in the 'search-oriented' process the decision maker is assumed to have a pre-determined preference structure to unveil, whereas in the 'learning-oriented' process the decision maker has his preferences elicited through the interaction. Unlike Gardiner and Vanderpooten we do not adopt this distinction because we believe that:

1) the first case does not really exist in multicriterion decision (*i.e.* the decision maker never does have a pre-determined preference structure in mind);
2) the observer has no means of distinguishing between search-oriented and learning-oriented behavior.

In figure 9.9, the current state is generally an alternative or satisfaction level that the system displays to the decision maker. Once the decision maker's reaction is known, the system determines another alternative which becomes the new current state. The words 'continue', 'advance' and 'backtrack' refer to heuristic searching (see section 9.3.2). By 'satisfaction level' we mean a vector with a value assigned to each attribute (which need not necessarily be an alternative), this vector indicating the level of satisfaction the decision maker would like to get for each criterion. When the information provided by the decision maker does not constitute a feasible satisfaction level such as an ideal point, the system generally calculates the alternative(s) which is or are closest to this satisfaction level by using a *temporary aggregation function* (Vanderpooten, 1990b).

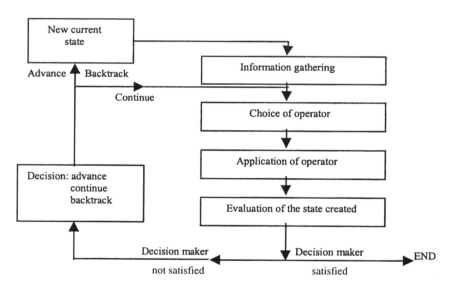

Figure 9.9 The principle of progressive articulation methods

This interpretation of progressive articulation methods provides a link between the search for a compromise in MCDM, and problem solving. Interactivity in MCDM can be studied without any reference to AI (see for example Vanderpooten, 1990a), but we think that heuristic methods provide a useful framework for thinking about interactivity and man-system cooperation. Firstly, the role of evaluation is highlighted in heuristic searching, and evaluation is exactly what the decision maker has to perform when a new current alternative is displayed by the system. Secondly, an important point introduced by Simon's work is that this heuristic searching leads to a 'satisficing' issue. While much of the work relating to interactive MCDM has been directed toward seeking some kind of efficient alternative by using a real and valued aggregation function, it is clear that this is not a problem for decision makers, who only need a satisficing compromise between two more or less contradictory criteria.

The various stages shown in figure 9.9 are not of course peculiar to multicriterion decision, and they can be found in identical form in multiobjective linear programming. In this connection, Larichev and Nikiforov (1987) proposed segmenting the operations into seven stages of which the main one in our case is the generation by a defined method of a current state and its subsequent evaluation by the decision maker.

For a method to be given the label 'progressive information', certain of the five stages in figure 9.9 must not be automated. In general it will be evaluation and decision that will be left to the decision maker. These two stages correspond essentially to the steering aspect of heuristic searching. Most of the methods we deal with next fall into the pattern of Figure 9.9 .

9.4 Interactive multicriterion methods

In this section we give a general overview of interactivity in discrete multicriterion methods. In fact the first interactive multicriterion methods in general use were designed for continuous problems. The STEM method (Benayoun *et al.*, 1971) was designed for linear programming with several criteria, while that of Geoffrion, Dyer and Feinberg (1972) was aimed at concave differentiable utility functions. For a description of the interactive aspect of these methods, see the articles by Larichev and Nikiforov (1987) and Vanderpooten and Vincke (1989). Many ideas in the domain of interactivity are transferable from the continuous to the discrete, and the reader will therefore find many allusions to continuous methods in this section. Let us now examine the various points where interactivity can occur.

9.4.1 Information provided by the decision maker

The basic idea is from cognitive psychology (see section 3.4 and 8.3, and Montgomery (1983)). Every individual is assumed to reason in terms of aspiration levels, which in its most banal form is equivalent to envy. These aspiration levels normally represent an individual's wishes that are difficult to achieve, similar to the ideal point in multicriterion decision or Roy's (1974) focal point, or the objective in 'goal programming'. Thus the decision maker has to indicate an aspiration level $y = (y_1, y_2, ... y_n)$ where y_j is the utility the decision maker would ideally like to obtain relative to criterion j. In multicriterion methods, aspiration levels can change from one iteration to another (the distance method of Dinkelbach, 1971; the evolving focal point of Roy, 1974; the AIM method of Lotfi *et al.*, 1991, etc.). With this frequent variation we are obviously a long way from a psychological interpretation.

In the PRIAM method we introduced satisfaction level, taking our inspiration from Simon (1955). What this satisfaction level expresses is a realistic level that the decision maker would be perfectly happy with at a certain point in time, given the available information. The 'reference level' of Stewart (1991) is a similar notion. In conception, *the satisfaction level is a mobile level that depends on the current state*. Simon (1955) points out the high mobility of 'satisficing' according to what has already been obtained and the speed with which it was obtained. It seems to us more reasonable to rely on these satisfaction levels, which can be increased at iteration $n+1$, if the level at iteration n was obtained easily, than to envisage modifying an aspiration level which in general stands no more chance of being achieved at the end of the search than at the beginning (an ideal point for example). In other words, aspiration level is to desire as satisfaction level is to reality.

Sometimes a third type of information is considered; this is the *rejection level* (Vanderpooten, 1990b), or *veto level*. This is a level y such that any alternative that does not dominate y is rejected. In methods other than progressive information methods, this level is the anti-ideal or nadir $m = (m_1, m_2, ..., m_n)$. Note however that in this case all the alternatives in $\mathcal{A}$ will dominate m: there will be no actual veto effect. In interactive methods the rejection level can increase progressively (as in the STEM method and the method of Vanderpooten, 1989). In all these methods where the information is given in the form of levels, it is obvious that the decision maker can restrict himself to merely modifying the levels relating to a subset of criteria at each iteration.

Another level to consider is the satiety level, where the decision maker becomes indifferent between two otherwise equal alternatives but which reach or exceed the satiety level relative to one criterion; this amounts to saying that when he chooses, he considers only criteria for which the level of satiety has not been reached. This satiety level corresponds more or less to the 'satisficing threshold' of Lotfi *et al.* (1991); however, we should be careful not to confuse the idea of satiety with the previously defined satisfaction level which, in our opinion, expresses Simon's idea of 'satisficing' much better.

Figure 9.10 shows the various levels from rejection to satiety for a criterion to be maximized. The levels 'rejection' and 'satiety' may be considered as the decision maker's psychological constants, while the levels 'satisfaction' and 'aspiration' will change – at least, in certain software applications – according to the alternatives that are proposed and the utility level already reached (the current state).

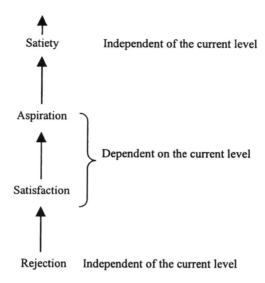

Satiety   Independent of the current level

Aspiration

Dependent on the current level

Satisfaction

Rejection   Independent of the current level

Figure 9.10 The various levels of decision

Another kind of information that the decision maker may be asked to provide is what Vanderpooten (1990b) called indirect information. This means that they do not have a direct bearing on utility levels. First, there is information about the criteria. This is generally the *ordinal or cardinal weights*, which will vary during the interactive process. Weighting information will be obtained from the decision maker by one of the many processes referred to in chapter 4. In some methods, the decision maker is made to vary the weights through cleverly designed displays and through sensitivity analysis (TRIPLE C, Angehrn, 1989; GAIA, Brans and Mareschal, 1991). Information about weights can thus be obtained by indirectly by asking questions on substitution rates between criteria (Geoffrion *et al.*, 1972; Sadagopan and Ravindran, 1986). The reader will remember (see sections 4.1 and 6.3) that substitution (compensation) can take place between criterion i and criterion j if the loss of one utility unit on criterion i is compensated by the gain of k utility units on criterion j. We can ask the question: "How much do we have to gain on criterion i to keep the same satisfaction level despite the loss of on unit on criterion j?" For the decision maker, these questions are difficult and quickly become tedious.

Finally, an interactive process can be aimed toward *the direct gathering of information on the ranking of a subset of alternatives*. In chapter 5 we saw how to aggregate information on the ranking of pairs of alternatives. If we can obtain the ranking for a subset of alternatives, then we can construct a decision maker's utility function that is consistent with the ranking (using for example the UTA method; see 6.7). In many cases, the utility function will be constructed after determining weights that are coherent with a ranking for a subset of alternatives (the alternatives adjacent to the solution obtained from the nth iteration in alternatives comparison methods; see 8.2). The solution from the $(n+1)$th iteration will maximize the sum weighted by the weights found in the nth iteration. This procedure is often performed using linear programs. Information obtained on certain alternatives is also used to eliminate the lowest ranked alternatives or those which are dominated by a given alternative.

The MCRID method (**M**ultiple **C**riteria **R**obust **I**nteractive **D**ecision) by Moskowitz *et al.* (1991) combines a request for ordinal weights and the pairwise comparison of 'good alternatives' determined by the weighted sum where the weighting intervals have been obtained from the first session of questions.

9.4.2 Choice, construction and application of the transition operator

Several types of operator are possible, and the one chosen will depend on the information that has been gathered. It may be an operator that modifies the current state or the current solution; this is the most common type (as in the methods of Wierzbicki, 1980, 1983, 1999; PRIAM (Lévine-Pomerol, 1986); and the method of Korhonen and Laakso, 1986). In some cases the operator does not act on the aggregation but allows the user to enrich his information; highly interactive methods all have this feature (MULTIDECISION, TRIPLE C). These are the *interactivity operators* and they modify the information displayed on the screen.

We referred above to operators that modify the current solution; in general it is *not the user who chooses these but the system that determines them*. They are usually the result of an optimization. Broadly speaking, the system chooses the feasible alternative or level that is closest to the aspiration or satisfaction level indicated by the decision maker; it is only the type of distance that varies. We name $x = (x_1, x_2, ..., x_n)$ the achievable current level and $y = (y_1, y_2, ..., y_n)$ the level of satisfaction or aspiration expressed by the decision maker at iteration n. In PRIAM and VIMDA the new current state is of the type $x + t(y - x)(t > 0)$. The direction $(y - x)$ is sometimes called the *reference direction*. In pure distance methods (Dinkelbach and Isermann, 1973, 'goal programming' by Charnes and Cooper, 1961 and the other methods described in 8.3), the new state is a solution of the program:

Min $\{d(y,z)$ / for the z feasibles$\}$
where d is a distance or norm of the type we have already met.

Many other methods take their inspiration from Wierzbicki (1980, 1984) in that they use what Wierzbicki names an 'achievement' or *'scalarizing function'*; Vanderpooten (1990b) uses the name *'temporary aggregation function'*. The simplest of these functions is where we consider the sum weighted with provisional weights and display either the best alternative(s) resulting from the maximization of this weighted sum, or the evaluation intervals, if each weight can vary within the interval; this is the case in the first step of MCRID (Moskowitz *et al.*, 1991). Following the decision maker's reactions, the weights are modified and the process continues.

The temporary aggregation function is more general than Wierzbickis' scalarizing function, in that, to each triplet (w,z,y), it associates a numerical value $s(w,z,f(y))$, where $f(y)$ is a point dependent on the aspiration level, z is any point in the choice set and w is the weight vector. The operator associated with this function takes us from the current solution x and information y to the new solution x' which is the solution to the program:

Min $s(w, z, f(y))$ under the constraint $z \in A_n$

where A_n is a subset of A that depends on iteration n. Wierzbicki (1983) has suggested several desirable properties for temporary aggregation functions; in particular he suggests that solutions to the above program, *i.e.* Arg (Min $\{ s(w, z, f(y)) \; / \; z \in A_n \}$) should be more or less Pareto optimal. This property is not satisfied for all $f(y)$ and any function s, but it is satisfied for $f(y) = y$ and for the Tchebycheff norm (Changkong and Haimes, 1983). On the other hand it is well known that in 'goal programming', if the objective is not high enough, the nearest point, in Euclidean distance, is not efficient (see figure 9.11). The search for points 'at the closest distance' in continuous problems is in fact based on linear or quadratic programs. The situation is not so simple in discrete mode where the programs are inherently enumerative. Where there are many alternatives, their ranking by computer is a decisive aspect of the efficacy of the method. In this connection, the use of quad trees in multicriterion decision by Habernicht (1982, 1992) is very interesting. The quad tree enables those alternatives that dominate a given alternative y (*e.g.* an aspiration level) to be found very easily. Using these ideas, Sun and Steuer (1996) have developed an interactive method named INTERQUAD. In the method the quad tree is used very successfully to find the closest point for the Tchebycheff norm of a given aspiration level y. INTERQUAD can of course be made interactive by varying y.

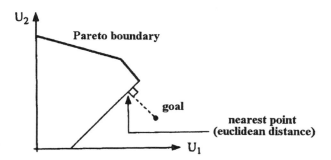

Figure 9.11 The nearest point to the goal is not on the Pareto boundary

The conditions on $f(y)$ and the semi-order relation in $|R^n$ for Arg (Min $\{ s(w, z, f(y)) \; / \; z \in A_n \}$) to be Pareto strong or weak are to be found in Wierzbicki, 1983) and Skulimowski (1987). This property appears to us to be fundamental in a non-interactive method such as standard goal programming (and also the interactive goal programming of Spronk (1981)); on the other hand it is much less desirable in an interactive method where the decision maker is actually informed that he is not on the Pareto boundary and that he moreover has at his disposal operators enabling him to move toward this boundary.

In any case we feel that *in an interactive method, the idea of systematically steering the decision maker toward the Pareto boundary is a bad one*. For it immediately results in the decision maker reasoning in terms of compensation: what do I have to lose on one criterion to have a chance of gaining on another one? We feel it preferable for the decision maker to increase his satisfaction gradually through a sequence of points that may be dominated, and of which only the last one or last few are Pareto optimal. There will thus be progressive improvement accompanied by *learning*. At each iteration, the decision maker will refine his preferences without any painful back-tracking or compromising. This is the type of chain we have introduced in PRIAM in continuous mode (Pomerol and Trabelsi, 1987).

Following Wierzbicki's scheme, weights are either set or deduced from the information y given by the decision maker. In the most commonly used relations, the higher the weight the closer the aspiration level is to the ideal. We thus have a choice between:

$$w_j = 1/(M_j - y_j) \tag{1}$$

or the normalized formula:

$$w_j = [1/(M_j - y_j)][1/\sum_k (1/(M_k - y_k))] \tag{2}$$

We naturally assume that the aspiration level y is such that $M_j - y_j > 0$.

The most widely used temporary aggregation functions are those of the *'weighted Tchebycheff distance'* type:

$$s_1(w, z, y) = \text{Max}_j \{w_j \mid z_j - y_j \mid\} \tag{3}$$

or else of the *'augmented Tchebycheff distance'* type:

$$s_2(w, z, y) = s_1(w, z, y) + \alpha \sum_j w_j \mid z_j - y_j \mid \tag{4}$$

in the latter formula, α is a control parameter.

Wierzbicki's ideas were first put into practice at the IASA at Laxemburg (Austria) by Kallio, Lewandowski and Orchard-Hays (1980, SESAME package) and Grauer *et al.* (DIDASS application, 1984). In DIDASS (**D**ynamic **I**nteractive **D**ecision **A**nalysis and **S**upport **S**ystem), the following function is used:

$$s(w, z, x) = \text{Max}\{\beta \, \text{Max}_j w_j \mid z_j - y_j \mid, \sum_j w_j \mid z_j - y_j \mid\} + \sum_j \alpha_j \mid z_j - y_j \mid \text{ in which the}$$

parameters α and β are to be adjusted.

In AIM, the weights are defined by $w_j = (y_j - m_j)/(M_j - m_j)$ and the function s by the weighted Euclidean distance between z and y or by the function s_2. In Roy's *evolving focal point method* (1974), which pre-dates Wierzbicki's work, weights are defined by formula (1) while the minimized function is

$s_3(w, z, y) = s_1(w, z, y) + \sum_j \alpha_j z_j$. The set A_n on which minimization is performed changes at each iteration, as in the STEM method.

In Steuer and Choo's method (1983), weights are according to formula (2), while the function s is identical to the one in the evolving focal point method. In the method of Korhonen and Laakso (1986), weights are assigned at the start. We let

$$f(y) = x + t(y - x) \quad \text{and} \quad s(w, x, f(y)) = \text{Max}_j[(f(y_j) - z_j)/w_j]$$

As we stated above, the alternatives m_h of Arg (Min $\{ s(w, z, f(y)) \ / \ z \in A\}$), corresponding to angular points, are displayed to the decision maker who makes his choice for the next iteration. In Korhonen and Karaivanova (1999) an algorithm is given for projecting a reference direction onto the Pareto set.

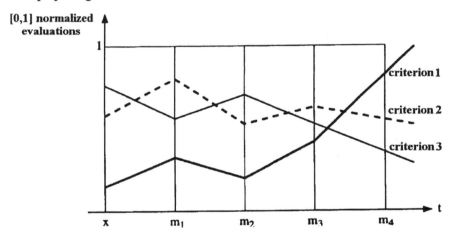

Figure 9.12 Pareto points displayed in the method of Korhonen and Laakso

Vanderpooten's method (1989) combines minimization of a function s of the type s_2 with reduction of the choice set. Weights are given by formula (1). In the TRIPLE C method (Angehrn, 1989), the temporary aggregation function $\sum_k w_k (| z_k - M_k |)/(| m_k - M_k |)$ serves to produce a temporary ranking that the decision maker is to react to.

When the information volunteered by the decision maker concerns weights, the operator function consists of seeking and displaying the alternatives which are suitable for a temporary aggregation function compatible with the given weights. The decision maker's reactions to the alternatives presented trigger a modification of the weights and the process continues. The methods of Zionts (1981) and Moskowitz *et al.* (1991) are of this type.

To sum up, from the information provided by the decision maker, there are two main schools for the definition of the operator that will transform this information into a new state which will in general be an alternative or a satisfaction level submitted to the decision maker. The first and most widely followed school optimizes a temporary aggregation function, to quote the apt expression of Vanderpooten (1990b). This aggregation will, in most cases, be a more or less improved distance. The result of optimization is displayed to the decision maker as being the following state. The second less popular school reasons geometrically on directions in 'criterion space' $U(\mathcal{A}) \subset |R^n$. This idea, which avoids intermediate aggregations, is used in PRIAM; for the cognitive reasons we have already mentioned, the number of criteria is limited to around seven. For more than seven criteria, the geometrical methods with hierarchized criteria would have to be adapted; see Mili (1984).

9.4.3 Evaluation

Once the operator has been applied, a new state emerges. New information is now available to the decision maker. In Wierzbicki type methods this will usually be the compromise solution from s. It would actually be more useful for the decision maker to carry out a complete evaluation of the state of the search, including the current state. This presupposes that he has at his disposal a complete overview together with really informative displays. Efforts are being made in this direction at the current time in certain methods which display not only the compromise the system has just come up with but also a view of the surrounding terrain. In TRIPLE C (Angehrn, 1989), the whole of the ranking obtained is displayed; in MULTIDECISION, TRIPLE C and AIM, the anti-ideal and the ideal are permanently displayed for comparison. In Stewart (1991), the compromise position x is displayed with the ideal point and aspiration level y, and is furthermore expressed as a percentage. In other words, for each criterion, the operator sees the ratios $|x_j - m_j| / |M_j - m_j|$ and $|x_j - m_j| / |y_j - m_j|$; AIM also uses this idea.

BASIC DISPLAY						
Objective	STU	EXP	AGE	INT	TES	Proportion
Ideal point	7.00	8.00	25.00	10.00	10.00	
Next better	5.00	6.00	34.00	6.00	7.00	0.22
Current goal	4.00	5.00	35.00	5.00	6.00	0.33
Proportion	0.89	0.78	0.89	0.89	0.89	
Next worst	3.00	2.00	38.00	4.00	5.00	0.89
Nearest Sol.	6.00	8.00	30.00	7.00	9.00	Georgia
Nadir point	3.00	1.00	38.00	4.00	5.00	

F1: Help; F2: Next Scrn; F3: Explore; F4: Reset CG; F5: ReCalc;
Up goal: Down goal: Next obj. v. obj. ESC: Exit

Figure 9.13 An AIM screen display

The AIM method (Lotfi, Stewart and Zionts, 1992) described in section 8.3.4, which is essentially based on the ideas of Wierzbicki, displays the alternatives close to the compromise solution. Zionts and Wallenius (1976, 1983) had already used the idea of adjacent alternatives in continuous mode. In the method of Korhonen and Laakso (1986), a set of non-dominated alternatives leading in the direction of the decision maker's aspirations is displayed on a single screen (*cf.* figure 9.13 above). The same way of displaying alternatives on a single screen had already been used by Belton (1985) to achieve an effect dialogue with the decision maker.

It is in any case preferable to facilitate the decision maker's learning process by presenting him with a set of alternatives rather than a single one, even a compromise one. In MULTIDECISION, those alternatives can be displayed that dominate the proposed current level, and the previous steps can also be displayed. At this stage it is vital to be able to display as much information as possible, and interesting ideas have been proposed. One of the most synthetic displays is that in GAIA (Mareschal and Brans, 1988 and Brans and Mareschal, 1991), which is an extension of the PROMETHEE method using a clever display on a single screen of alternatives, criteria and weights (*cf.* 9.5.1).

In progressive information methods, this stage of evaluation is never automated, and it is always the decision maker who decides on the following step: to end the search or to continue. When the decision maker decides to continue, he can choose between backtracking to a previous state, advancing from the compromise state that has just been proposed, or trying another operator. These three possibilities actually exist in PRIAM. Stewart (1991) also uses backtracking, while storing all the states of the system within a database. In the other methods the decision maker only really has the choice between ending the search and advancing from the latest compromise.

Finally, note that there does of course exist the possibility of automatic evaluation. Suppose we have to maximize n criteria represented by the utility functions U_j. The function $U = \sum_j w_j U_j$ (w_j being positive or null weights) is an evaluating function since we can follow a heuristic search consisting of going from one state to another that produces a better value of U. *Any aggregating function can thus play the role of an evaluating function*. The trouble is, for it to actually play this role it has already to be known, which is not the case in progressive information methods! This explains the usefulness of temporary aggregation functions.

9.5 Incorporation of multicriterion methods in DSS

In this section, we deal with multicriterion decision aid systems. The main software applications will be described in the next chapter; here we shall simply outline the main features of these systems and the emergence of multicriterion DSS, which contain expertise that can go well beyond multicriterion decision. In a recent forward-looking article, Dyer *et al.* (1992) considered that the coming years would see the development of multicriterion DSS as meeting the real needs of decision makers. A critical presentation of multicriterion DSS can be found in Pomerol (1993).

9.5.1 The quality of interfaces for knowledge gathering and dialogue with the decision maker

In the section on progressive information methods, we distinguished between the interactivity side, *i.e.* everything involving the quality of the interfaces and/or the interaction, and the search process controlled by the decision maker. It is this aspect of the quality of the interfaces that we deal with now; of course this is not specific to multicriterion decision, and concerns any interactive system. Given the vastly superior visual faculties of human beings over machines, one of the important aspects of interaction is the exchange of data in visual form. Designers have come up with some very sophisticated interfaces. In some problems of network management, DSS can be designed almost entirely around graphic exchanges; see Angehrn and Lüthi (1990) and Angehrn (1991). In previous sections (figures 9.12 and 9.13) we showed several screens displaying alternatives and various other information. One of the specific things we are interested in is displaying the vectors $U(a) = (U_1(a), U_2(a), ..., U_n(a))$ to the decision maker in a visually attractive form.

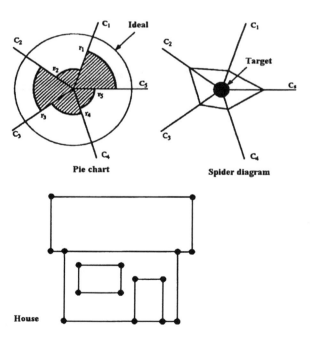

Figure 9.14 Visual representation of vectors

Angehrn (1989) in TRIPLE C uses a pie chart, while Korhonen (1991) tries a representation in the form of more or less harmonious houses (see figure 9.14). Each characterizing point of the house, of coordinates (α,β), is attached to a parameter v. The normal value of α, α_1, corresponds to the value v_1. The parameter α is assumed to vary within the interval $[\alpha_1,\alpha_2]$, whilst v varies within the interval $[v_1,v_2]$. For the observed value of v, the parameter α will have the value: $\alpha_1 + ((\alpha_1 - \alpha_1)/(v_1 - v_1))(v - v_1)$. The same thing is done for β. In figure 9.14 the control points will correspond in our case to the coordinates of the vector that we wish to represent. In multicriterion decision, for a given alternative, each characterizing point of the house will correspond to the utility relative to a criterion. The positions of the characterizing points, and the shape of the ideal house, are chosen arbitrarily. The ideal house will correspond to the ideal point. An alternative different from the ideal point will, on applying the above rules, lead to a distorted house. According to the degree of distortion, the decision maker can 'visually' accept or refuse the choice proposed.

A spider diagram is used by Kasanen, Östermark and Zeleny (1991). This has the advantage of displaying the decision maker's goals in the form of a target with the ideal point at the center. If the length of a ray expresses a percentage of $r_j = |x - m_j| / |M_j - m_j|$, figure 9.14 indicates, for five criteria, what r_j represents in each case.

The displays used in GAIA (Geometrical Analysis of Interactive Assistance) deserve separate treatment. This system, developed by Mareschal and Brans (1988), is base on a matrix F of flows associated with PROMETHEE, of dimension $m \times m$, defined by $F = (f_{ij})$ where $f_{ij} = \sum_k [1/(m-1)][S_j(a_i,a_k) - S_j(a_k,a_i)]$ (see section 7.4 for the notation). Each alternative a_i is then characterized by the vector $f_i = (f_{i1}, f_{i2}, ..., f_{in})$. According to the authors, these vectors give a good synthesis of information on the alternatives. The idea of GAIA is essentially to display the information contained in the matrix F by means of a projection on the plane of the main components; this is a standard method of data analysis (see section 8.5). In the method named BIPLOT, Lewandowski and Granat (1991) suggest the same idea, which they credit to Gabriel (1971), using it to improve interactivity in DIDASS, a multicriterion DSS.

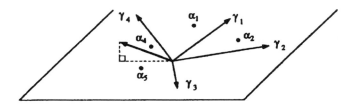

Figure 9.15 GAIA Plane representation

In GAIA, projecting values f_i on the plane of the main components provides information on the position of the alternatives in relation to the criteria which are represented by the projection γ_j of the n vectors $e_j = (0,0, ..., 1, 0, ..., 0)$ of $|R^n$, the 1 being the jth. coordinate (see figure 9.15). Through the properties of projections of main components, close vectors γ_j display criteria which are in concordance (γ_1 and γ_2), and opposing vectors indicate criteria in discordance (γ_1 and γ_3). The greater the length of γ_j, the more discriminating is the criterion (high dispersion of f_{ij} for a given j). In the figure, the projections α_i of f_i represent the positions of alternatives. For a given α_i, angular proximity of a γ_j is proof of the strength of the alternative a_i according to criterion j. Thus a_1 and a_2 are good relative to criteria 1 and 2 (figure 9.15). Finally, if we project the weight vector w on this plane, its direction will indicate the most favored criteria. The alternatives which are good for these criteria will naturally stand a high chance of being ranked first by PROMETHEE II (in our case, a_4 followed by a_5). The reader will appreciate the interest of this method of display, even though the ranking cannot be deduced directly. In GAIA, the weights can also be varied interactively, and the alternatives can be seen coming more or less close to the PROMETHEE II optimum.

A point that is specific to multicriterion decision is the difficulty of the questions addressed to the decision maker which we have already referred to in connection with various methods. Empirical studies (Wallenius, 1975; Buchanan and Daellenbach, 1987; Mote *et al.*, 1988) and comparative studies (Larichev *et al.*, 1987; Vanderpooten, 1989) show that decision makers often run into difficulty in replying correctly to complex questions, especially about substitution ratio. Here the quality of the interfaces depends on the questions being presented within the framework of a real explanatory support; the display can also play a part in indicating the immediate effect a given response will have.

9.5.2 DSS and multicriterion systems

DSS have popularized the notion of 'what if?' or sensitivity analysis (see Keen and Scott Morton (1978), Keen (1987) and Lévine-Pomerol (1989)). A method does not have to be of a progressive information type for sensitivity analysis to be felt useful for the decision maker. Among sensitivity studies we can distinguish (Lévine-Pomerol, 1989) those concerning the data (the first level) and those concerning the models (second level).

In the vast majority of cases, the designers include sensitivity analysis on data. This enables the decision maker to vary:

– the weights or coefficients of importance (TRIPLE C, PROMCALC and GAIA);
– the coefficients of concordance and/or veto in outranking methods (ELECTRE IS, Roy and Skalka, 1985) and IDEAS (Vetschera, 1989)).

In the various applications, the results are presented with more or less clarity. The program answers a set or subset of questions of the type (from the easiest to the most complicated):

– 1 Does the order change if the weights (or parameters) are changed by such and such an amount?

– 2 By how much can the given parameters vary without affecting the final order?

– 3 Can zones be given for parameters that leave the final result unchanged?

Most software applications reply to question 1 by supplying the new result. The validity limit of the weights (question 2) is given by PROMCALC and DEFINITE (van Herwijnen and Janssen, 1989). Question 3 is dealt with by TRIMAP (Climaco and Antunes, 1989). Similarly, some software applications allow the weights to vary within an interval and provide the aggregated result in the form of an interval of possible values; this is in particular the case with MCRID and DECISION PAD.

As far as we know, only one software application (TRIPLE C) deals with the following relevant question:

– 4 What value should be given to the parameters to obtain, if it is possible to do so, a given order?

This question, to which it is often possible to give an answer in cardinal methods such as the weighted sum (see section 4.1.3), without changing the semi-order defining the criteria, is, of course, such as to undermine the confidence of any user, which is why it is generally avoided!

Another popular function of DSS is 'reporting', producing the clearest and most complete reports possible. This function has been more or less developed in the various multicriterion software applications. The most complete appear to be DECISION PAD and EXPERT CHOICE.

The change from one model to another (second level exploration) is an idea that is less commonly implemented. Some systems such as AIM (Lotfi *et al.*, 1991) have, however, used the idea to propose to the decision maker a method that combines various models. Here, however, there is not really any exploration. Making several aggregation methods available is an idea taken up by various authors (Jarke *et al.*, 1984; Barba-Romero, 1987 and 1988; van Herwijnen and Janssen, 1989; Teghem *et al.*, 1989; Balestra and Tsoukias, 1990; Hong and Vogel, 1991) although no thought is really given to the metamodel which would enable the change from one method to another to be modeled. Teghem *et al.* (1989) does construct a decision tree enabling one method to be chosen rather than another (*cf.* section 9.3.2 above).

9.5.3 Multicriterion DSS

According to Simon (1977), there are four stages in decision:

– intelligence (information gathering)
– design

– choice
– review (as a learning process).

A complete DSS for multicriterion decision should incorporate all four stages. Let us examine them in more detail.

The first stage of data acquisition will involve database management systems (*cf.* Lévine-Pomerol, 1989). A multicriterion DSS should therefore offer the possibility of storing data on alternatives in a more or less standardized form so that the decision maker can retrieve information and update it as necessary, while the system itself can use it as a facts base, for example (see section 9.2.2a). This fundamental link with databases is treated in depth by Jelassi *et al.* (1984), Jarke *et al.* (1984) and Jelassi *et al.* (1985). In these studies, the system database is a fundamental component in decision support. This enables various types of information selection and retrieval: choice of criteria, choice of alternatives, choice of methods. Several authors, and in particular Jelassi *et al.* (1985) and Balestra and Tsoukias (1990) highlight the possibility of being able to manipulate various types of criterion (quantitative, qualitative, to be maximized, to be minimized, modal, deterministic, threshold etc.). Most DSS packages are heavily dependent on databases (see for example Stewart, 1991).

The second stage in decision is design. It is undoubtedly here that DSS in general and multicriterion DSS in particular are weakest (Pomerol and Brezillon, 1997). This is despite the fact that it is extremely useful to broaden the decision maker's thinking. The DSS can help in constructing and defining the alternatives, and a few systems have been produced in this area (see section 9.2.2). What should be incorporated is the creation of new alternatives or criteria, either from scratch or from existing ones. In DEFINITE, for example, mixed alternatives, formed from several existing alternatives, can be introduced. It should also be feasible to have regrouping or elimination of criteria when they are not sufficiently discriminating or are too highly correlated; see Roy (1985), Korhonen (1986), Roy and Bouyssou (1987), Bouyssou (1989) and Roy and Bouyssou (1993) for some ideas on correlation and discrimination in criteria.

A widening of the decision maker's thinking is actually envisaged in several software packages, even though it does not result in any enlargement of the choice set. The way this is done is by presenting several alternatives in the neighborhood of the current alternative so as to prevent the latter from being the focus of the decision maker's thinking. In AIM (Lotfi *et al.*, 1991) neighboring alternatives that outrank the current one are displayed, while in Korhonen and Laakso (1986) and Korhonen (1988) Pareto optimal alternatives 'close' to the decision maker's direction of improvement are displayed.

It is important to avoid focusing on a very small number of alternatives, but it is also important to avoid using too simplistic a model for an alternative (Pomerol and Brezillon, 1997), for the alternatives that are envisaged by decision makers are often very complex because they are time-dependent scenarios. The decision maker is actually choosing between various scenarios in which uncertainty has to be considered. To visualize the way the results of initially choosing an alternative evolve as events unfold it is often necessary to use some support from knowledge-based systems, as in the train schedule example (9.3.2). It was to be able to take into account the role of more realistic alternatives involving scenarios that we introduced the notion of fully expanded alternative (*cf.* section 9.2.2 and Pomerol, 1998, 1999).

Another idea concerns choice among criteria and constraints. In section 1.3 we saw that the distinction between goals, criteria and constraints depends closely on context, though this dependence does not stop them playing a key role in the result. In a program by Korhonen *et al.* (1989), it is possible to go from one to another, and the decision maker can be supported here in his modeling process.

There is no need for us to say a lot about choice since until recently this has been the only thing that has really been dealt with by systems designers. To a certain extent, aggregation, which is the central concept of multicriterion decision and of this book, is only concerned with choice. At the beginning of the book we stated that under the classical paradigm, the criteria and the alternatives are fixed, in other words that the design phase has been completed at the point where most multicriterion methods begin.

There remains Simon's final stage, 'review'. What this ought to involve is a look back to the decision and a learning process, but this is not exactly the strongest point in currently available DSS! Yet support for the decision maker could play a fundamental role here. Imagine a DSS which, unlike the human brain, can reconstitute without any bias all the circumstances and reasons for the choice made (stored in the database). This playback could be used to identify the right decisions that were made, and also the parameters or criteria that were not considered which subsequently turn out to have been behind a failure. The system could then correct the cognitive bias of the user. Some of these functions had been suggested in DECIDEX (Lévine *et al.*,1987). In an unstructured environment, important and relevant information can go unnoticed, and Smith (1989) shows how easy it is to miss such information even if it is perfectly accessible.

A database memory is a cumulative record of past information, and a multicriterion DSS could thus become a memory in the field for which it has been designed; it is with this in mind that the total integration of multicriterion decision within a decision support system has been undertaken.

9.5.4 Integrated multicriterion DSS

Purpose-designed multicriterion DSS have now appeared. Eom (1989) examined 203 DSS of which 23, *i.e.* 13%, involved multicriterion; a modest proportion, but the movement only really started after 1980, and the idea was clearly taking hold. There are two approaches to designing these DSS: either the multicriterion method is built into the DSS and the decision maker cannot avoid using it, or there is free access to multicriterion decision support, and the decision maker decides whether or not to avail himself of it.

In the first of these two approaches are to be found numerous applications of Wierzbicki's methods, including DIDASS (Grauer *et al.*, 1983) and those of Steuer, Korhonen and Laakso. In these DSS which feature practically none of the functions we have just seen, the 'multicriterion choice' part is the main ingredient; around it there are merely a few spreadsheet functions to record databases.

The most ambitious DSS focus on the 'database' and 'report' aspects that we developed above, and these include Stewart (1991), with a DSS designed for the choice of research and development projects, and DEFINITE for the choice of public amenities. One special version of DEFINITE, designed for the choice of agricultural land destined for other uses, uses aggregation by distance from the ideal and has a highly visual map display of the results; this is a form of report (Janssen and van Herwijnen, 1991). Another example is BANKADVISER (Brans and Mareschal, 1989), designed around PROMETHEE, where the user has multicriterion functions for evaluating companies, together with the possibility of managing a portfolio of cases (client database).

These types of DSS can try to combine various types of modeling and expert systems. An example of this, the choice of a legislative law, is to be found in Lévine and Pomerol (1989). In some cases the expert system part takes preference over the multicriterion part; an example is the programming support system for measures to help the unemployed (Labat and Futtersack, 1992).

Multicriterion DSS of the second type, which prepare multicriterion decision as much as executing it, leave the user with the possibility of using one method or another. This is true of MULTIDECISION-EXPERT which aids the management of knowledge relevant to the alternatives while preparing multicriterion choice. Another example is the DSS on the robustness of train schedules (see section 9.3.2). In Bayad and Pomerol (1992), the DSS prepares a decision to reinforce a medium voltage electricity distribution network. These DSS produce a decision matrix containing a small number of satisfactory alternatives, but it is the decision maker who carries out the multicriterion choice. A possible option is: design a multicriterion DSS which expressly does not perform aggregation. This is the option that was chosen for the electricity network reinforcement problem, with the aim of retaining motivation and fostering reflection in the decision makers involved. DSS here considerably widens research and analysis possibilities for the decision makers while leaving the final choice up to them.

In many cases, the system should simply have the job of eliminating the largest possible number of alternatives and leaving the decision maker with the final choice between satisfactory alternatives. This seems to us a forward-looking idea in which the motivation and responsibility of the decision maker are conserved. The open-ended metamodel will then be kept for the user's decision (Pomerol, 1990).

The ever-increasing complexity of decisions, motivation and interests will undoubtedly lead designers more and more in the direction of multicriterion DSS; the trend detected by Eom (1989) can only intensify.

9.6 Conclusion

We share the opinion of many workers that the future of multicriterion analysis lies in interactive methods. We hope we have amply demonstrated that from a purely rational point of view there is no such thing as an optimum method of aggregation.

Many methods are moreover heavily dependent on the reliability of the information they handle (*e.g.* ordinality and cardinality). Now, the quality of the information provided by the decision maker depends on the quality of the interaction between him and the system, and there is a genuine mutual enrichment taking place between the method and the decision maker. This brings us back to the constructive point of view of section 9.4.1. *We feel that this philosophy is more realistic and within it multicriterion analysis is a matter of exploration and learning.*

Exploration means that the decision maker really can explore the choice set and get to understand its multidimensionality. Exploration was introduced with PRIAM and in more recent methods such as AIM and TRIPLE C. Angehrn (1990) states that exploration should be stimulated. This stimulation comes from 'what if' type sensitivity analyses, from which the decision maker can appreciate how simple it is to modify the parameters of the analysis. Using computer methods, exploration can be presented in the form of a game.

At the present time the weakness of exploration is that it does not extend to modeling. To enable and encourage the decision maker to build several models for a single problem, simple methods of entering data and visually displaying alternatives and criteria are being introduced. However, the truth is that in the building of models the help of the analyst is very necessary, for two main reasons. The first and most important is that the complexity of reality cannot be simplified and in real problems results in complex alternatives (see for example the problem of robustness of train schedules in section 9.3.2).

The second reason is the tendency of the human mind to focus on a very small number of alternatives (*cf.* section 11.3.1). Many decision makers restrict themselves to the possible alternatives, which they view as data, instead of considering the choice set as a set to be built up, for example in terms of goals (Keeney, 1992). Very much remains to be done in this area of decision support. A first requisite is probably to construct a database of models of problems. With this in mind, 'case-based reasoning' is a promising avenue already being explored by Angehrn and Dutta (1992).

Case-based reasoning requires cases to be stored in structured form, which will also enable decisions to be reviewed. Here we enter the sphere of learning. In our domain there are two types of learning which can take place. First there is the improvement of the decision maker's performance in exploring the choice set. In most multicriterion methods learning is seen as screen-based through various types of visual stimulation, as the reader will have appreciated. Methods based on distance from the ideal are rather dangerous here, since they lead very rapidly to Pareto optimal alternatives, and this does little for the learning process. In our opinion the Pareto boundary should be approached more slowly, varying the operators as in MULTIDECISION. Learning by the system itself also takes place during progressive adjustment of the aggregation function in terms of the decision maker's expressed preferences. UTA and similar methods involve this type of learning by the system. In this field, Rommel (1991) and neural networks (see sub-paragraph 9.2.2 c) open the way to the use of artificial intelligence.

Learning has taken place when the decision maker has well understood the effect on the system of the information he provides, but this is far from being the case in every method! However, this goal is being borne in mind by an increasing number of designers of recent methods. Learning also presupposes that the questions will not be too difficult, and questions of the 'aspiration or satisfaction level' type appear the simplest and are generally more appreciated than questions on tradeoff rates which remain esoteric for most decision makers. On the other hand, questions on weights are generally well accepted, even though experience has shown that the responses are not very stable.

The second type of learning is at a higher level and concerns the quality of the decision: why did such and such a decision turn out to be the right one; did we have the right alternatives? The right criteria? and so on. Functions designed to improve the decision presuppose that multicriterion decision will be built into a complete DSS, since the data must necessarily be stored (in a database) to enable subsequent review. We come back to the problem, still in its infancy, of learning in DSS.

Informatics and AI open the door to multiple prospects for exploration and learning. And the field of multicriterion decision, far away as it is from that of optimization, and absolutely dependent upon interactivity, represents an ideal application.

10 SOFTWARE FOR DISCRETE MULTICRITERION DECISION

10.1 Introduction

This chapter is a continuation of the preceding one since it is devoted to software tools for discrete multicriterion decision (DMD). We shall be mainly concerned with software packages designed to perform the methods described in chapters 4 through 8. (We use the term 'software package' for commercially produced software, which can be purchased as a finished product).

Though DMD software packages are less numerous than in other domains such as expert systems and statistical econometrics, there are already several dozen available, some of which have attained a highly professional degree of development. This chapter is divided into two distinct parts; in the first part, there is a section on each of the four software packages LOGICAL DECISIONS, PROMCALC, EXPERT CHOICE and QUALIFLEX, all of which in our opinion deserve detailed treatment for various reasons including the facts that they are commercially available, applicable to real problems, representative of their methodology, widely used and well known.

In the second part of the chapter, section 10.6, we conclude with a summary of the characteristics of all the DMD software packages which we have encountered. Although this software is often full of interesting ideas, we devote less space to it, either because it does not satisfy the criteria mentioned above or because we have not had the opportunity to thoroughly test it ourselves. Inevitably the information in this chapter will have started to go out of date by the time the book is published; we hope the reader will be indulgent and indeed search the specialized literature for updates. This chapter will at least be a starting point to get acquainted with software available. Finally we should mention that this chapter contains no advertisements and that our choices and evaluations have been made completely independently. Note too that the vast majority of software packages are protected by trademark.

10.2 LOGICAL DECISIONS

10.2.1 General information.

LOGICAL DECISIONS for Windows (which we shall refer to as LDW, this being the name used in the software itself, and the data file extension) is a software package for multicriterion decision based on the methodology of multi-attribute additive utility as seen in chapter 6. It is available from Logical Decisions (1014 Wood Lily Drive, Golden, CO 8041 USA) at $494 ($300 for universities). There is an Internet site (www.logicaldecisions.com) from which a 1.8Mb file can be downloaded; this contains a very complete demonstration of version 5.0 available from February 1999, except that models and results in the demonstration cannot be printed or saved. We used the demonstration to evaluate the software.

10.2.2 Theoretical background

Modeling and analysis of a problem using LDW is in three distinct stages: structuring a decision, assessing preferences and reviewing results. The LDW manual very laudably stresses the importance of separating the objective part of the analysis (structuring) from the subjective part (assessing). In other software this distinction is not made so clear.

In the structuring stage alternatives are first created, followed by criteria, which are called measures in LDW terminology. This done, the alternatives and criteria can be given suitable names for the benefit of the user, as we have done ourselves. The worst and best levels are defined for each quantitative criterion, and for qualitative criteria are given text labels which LDW then ranks by default; the user can of course modify this ranking if desired. Evaluation of the alternatives is then carried out in a natural way without any transformations to hinder the decision makers' understanding.

The criteria can be organized in two hierarchical levels, the upper level being called goals. Among other functions, a goal performs the aggregation of its offspring using preferences which will be expressed in the next stage. LDW is a good example of how powerful a good representation of criteria and super-criteria can be in analyzing and modeling a problem.

In the stage 'assessing preferences', the MAUT method (chapter 6) is used. All the information on preferences which is to be progressively constructed is contained in the preference set. Several preference sets can co-exist within the same model, as the decision maker may wish to consider different points of view, or several decisions makers may be involved. The decision set in use at any given moment is clearly displayed on the screen and in the ensuing results. To each decision set is attributed a default value (equal weights, linear utility function for each criterion, etc.) so that if the decision maker supplies nothing more precise, the system nevertheless produces results which can then be refined.

To evaluate and normalize the utility of each attribute (or 'common unit' in LDW terminology) we need information from the decision maker. This information can be supplied in five ways: SUF, AHP, AHP-SUF, adjusted AHP and direct AHP.

The SUF (single utility function) method is probably the best known; the user supplies the utility function of the criterion in question graphically or numerically (this is illustrated in figure 10.1 below). There are several ways, even leading to non-monotone preferences, for the user to quantify the qualitative evaluations (using text labels).

For the weights, there are seven different methods of evaluation: trade-off, SMART, graphic SMART, SMARTER, direct entry, pairwise weight ratio and AHP. All of these methods assume that there are no interactions between the criteria and the utilities are additive. If this is not the case the user can choose various non-additive procedures; these are by no means simple, and require a sound grasp of the underlying theory.

The final stage in the analysis is of course to examine as thoroughly as possible the abundant information given by LWD. The most immediately obvious result is the complete ranking of all the alternatives in graphical-numerical form, but there is much more: stacked bar ranking, results matrix, compare alternatives graph, scatter diagram, etc. The sensitivity of the results to weights can be appreciated through an interactive display which gives the global evaluation and the ranking of weight variations, controlled on the screen by mouse.

Among the other useful features of LWD is the possibility of setting satisfaction cut-offs, thereby performing a satisfaction pre-analysis (see section 3.4). Another is that the user can introduce uncertainty into the evaluations. Here, he can choose the uncertain evaluations from among six statistical distributions, the system will ask for the parameters of the distribution. Thus the decision maker can carry out any number of trials and vary the random seed numbers in the Monte Carlo simulation used by LDW in the processing of uncertain evaluations. User uncertainty is thus taken into account; for example, in the graph displaying the final evaluation of the alternatives, a value interval is superimposed on the definitive results to show the dispersion.

10.2.3 Example of use

We used LDW to create a model of our candidate selection example. Using the three-stage process described above, we began by describing alternatives and criteria. The process is quite user-friendly, though a little tedious owing to the multitude of windows to be filled out. We rather arbitrarily fix the worst and best levels of criteria to be maximized at 0 and 10 (except for criterion #3) and the levels for the single criterion to be minimized (#3) at 20 and 40. We can now easily enter our evaluations into the decision matrix.

Once we have structured the data, we can evaluate the preferences. To illustrate the SUF (single utility function) feature, suppose the criterion EXPerience does not have a linear utility function. Let us say that four years' experience are preferable to two, but that the utility of experience subsequently drops off. Figure 10.1 shows the LDW screen where, using the mouse, we have interactively set the concave form of the utility.

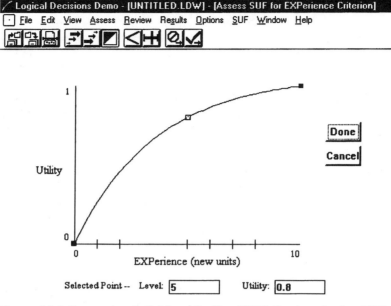

Figure 10.1 Interactive definition of utility (SUF) for the criterion EXPerience.

Weights are given by direct evaluation, as we have done throughout with this example, and are set as usual at 5, 5, 2, 4 and 4. The basic results are shown in figure 10.2.

Alternative	Utility	
Georgia	0.758	
Donald	0.678	
Blanche	0.669	
Helen	0.651	
Albert	0.610	
Emily	0.540	

Figure 10.2 The final ranking of the alternatives.

We referred above to weight sensitivity of the results, and this is illustrated in figure 10.3 In the upper window are the results for each alternative while the lower window shows the weights used. Using the mouse, the weights can be changed and the result on the alternatives is instantly seen in the upper window.

Dynamic Sensitivity of OVERALL Ranking

Alternative	Utility	
Albert	0.610	
Blanche	0.669	
Donald	0.678	
Emily	0.540	
Georgia	0.758	
Helen	0.651	

◀

Drag bar end or click on weight to adjust

Member	Weight	
AGE	10.0	
EXPerience	25.0	
INTerview	20.0	
STUdies	25.0	
TESt	20.0	

Figure 10.3 Interactive analysis of sensitivity to weights.

10.2.4 Computer aspects

The demonstration version of LDW is easy to install. It takes up 5.6Mb in its directory on the hard disk. A single file (uninstaller) will be located in the directory C:\WINDOWS. The software is Windows-style user-friendly. There are a number of personalizing options: user-redefinable names, font selection, black and white display and printing for finer detail, various printing options for the alternatives in the final result.

There is a wide range of functions, as we have tried to show in our description; these are complemented by standard computer functions such as data import/export in various formats, and screen zoom.

10.2.5 Overall assessment

LDW is a package which is heavily biased towards the weighted sum, with a strong theoretical foundation and a whole arsenal of techniques to use. There could even be too many, and it is easy for the user to get lost among all the possibilities on offer; however, default parameters and options are always there to fall back on (Zapatero *et al.*, 1997, back up this criticism).

One other criticism that could be leveled is that the process of creating alternatives and criteria is rather laborious, lacking a spreadsheet-type matrix that could be filled out directly.

Apart from this, LDW is user-friendly and very complete at an affordable price. This is all that is required for working by MAUT on discrete problems.

10.3 PROMCALC

10.3.1 General information

PROMCALC (PROMethee CALCulations) was devised by B. Mareschal and J.-P. Brans in the Free University of Brussels and is marketed by the university itself (U.L.B., Center for Statistics and Operations Research, Pleinlan 2, B-1050 Bruxelles, Belgique). The latest version that we have tried is 3.4; it is priced at $500 and is supplied as an unprotected disk accompanied by a publication (Brans and Mareschal, 1994) describing the methodology; there is no actual user manual.

10.3.2 Theoretical background

PROMCALC is the computer version of the PROMETHEE method (see 7.4). The software can handle up to 150 alternatives, 150 criteria and 3600 evaluations. The preference indices of the alternatives are calculated as the weighted sum of the pairwise preferences for each criterion or pseudo-criterion. From these indices the input and output flows are obtained (see definition 7.16). A (partial) outranking relation is thence defined, and this can be shown graphically (in the PROMETHEE I method). If we consider the total flow, we arrive at a complete pre-order (definition 7.18) (in the PROMETHEE II method).

Other interesting features of the software include:

a) Statistical analysis (descriptive and correlation) of the decision matrix data.

b) A λ-μ filter (see Karkazis, 1989, for a description of this concept) allowing dominance analysis to be performed; this eliminates non-useful alternatives. The parameters λ and μ loosely correspond to concordance and discordance thresholds.

c) Weight sensitivity analysis.

The software GAIA (Geometrical Analysis for Interactive Aid) by the same authors is a useful complement to PROMCALC; it provides visual information on criteria concordance and discordance, and on the relative proximity of alternatives. In 9.5.1. we saw the principles involved in GAIA for analysis into principal components, together with the main properties of the 'GAIA plane' projection. The weight vector, or 'PROMETHEE decision joystick' is interactively modifiable by the decision maker, and the GAIA plane projection shows the position of those alternatives which are well placed in the aggregation based on the weight vector in question. The relative proximity of alternatives can also be seen and evaluated.

10.3.3 Example of use

Once again we shall use the candidate selection problem for ease of comparison by the reader. Figure 10.4 shows our model as it appears on the screen. We have avoided using pseudo-criteria – one of the strong points of PROMETHEE – in order that comparisons can be made with other software. Thus our criteria are all of the simple type (type I in figure 10.4). The model having been defined, we can perform the various analyses mentioned above (correlation, dominance), or else go on directly to calculating preference indices and input and output flows ϕ^+, ϕ^- (figure 10.5).

```
 ┌═ PROMCALC & GAIA   U.3.4  STUDENT ══════════ C.S.O.O. - U.U.B. - 03/1994 ═┐
 │                      C..1       C..2       C..3       C..4        C..5     │
 │      Criterion      STUdies   EXPerience    AGE     INTerview     TESt     │
 │       Min/Max         max        max        min        max        max     │
 │        Type            1          1          1          1          1      │
 │       Weight         5.00       5.00       2.00       4.00       4.00     │
 ├═ Actions ─────────────────────────────────────────────────────────────────│
 │ A..1  Albert          6.0        5.0       28.0        5.0        5.0      │
 │ A..2  Blanche         4.0        2.0       25.0       10.0        9.0      │
 │ A..3  Donald          5.0        7.0       35.0        9.0        6.0      │
 │ A..4  Emily           6.0        1.0       27.0        6.0        7.0      │
 │ A..5  Georgia         6.0        8.0       30.0        7.0        9.0      │
 │ A..6  Helen           5.0        6.0       26.0        4.0        8.0      │
 │                                                                           │
 │                                                                           │
 │                                                                           │
 │                                                                           │
 │                                                                           │
 │                                                                           │
 │                                                                           │
 └───────────────────────────────────────────────────────────────────────────┘
 F1 : Help  -  F7,F9 : Actions  -  F8,F10 : Criteria  -  Ins/Del  -  ESC : Stop
```

Figure 10.4 The PROMCALC decision matrix for candidate selection

Preference Flows

Actions	Leaving	Rank	Entering	Rank	Net Flow	Rank
A..1: Albert	0.33000	6.0	0.57000	6.0	-0.24000	6.0
A..2: Blanche	0.51000	2.0	0.45000	2.0	0.06000	2.0
A..3: Donald	0.45000	3.0	0.50000	3.0	-0.05000	3.0
A..4: Emily	0.37000	5.0	0.53000	4.0	-0.16000	5.0
A..5: Georgia	0.70000	1.0	0.16000	1.0	0.54000	1.0
A..6: Helen	0.40000	4.0	0.55000	5.0	-0.15000	4.0

Figure 10.5 Calculation of flows in PROMCALC

From the flows in figure 10.5, PROMCALC deduces the partial outranking relation of the PROMETHEE I method (figure 10.6, screen copy). Note that A6 (Helen) and A4 (Emily) are incomparable, whilst A5 (Georgia) dominates.

If we want a complete preorder a single keystroke takes us to PROMETHEE II with the results shown in figure 10.7.

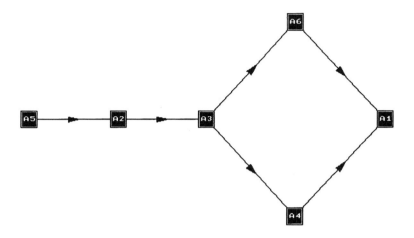

Figure 10.6 PROMETHEE I outranking relation

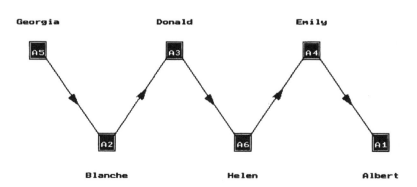

Figure 10.7 PROMETHEE II complete preorder.

    If we compare this preorder Georgia ≻ Blanche ≻ Donald ≻ Helen
≻ Emily ≻ Albert with the one obtained by simple weighting (see chapter 4), *i.e.*

Georgia ≻ Donald ≻ Helen ≻ Blanche ≻ Albert ≻ Emily, we note that Georgia is ranked first in both cases whereas the rankings of the other candidates are modified (with Blanche going from fourth right up to second place). The reader who has followed us up to here will not be surprised to learn that it is hard to explain away this difference, the philosophies behind the two methods being so different; probably, Blanche is 'lucky' in pairwise comparisons (PROMETHEE), though she is just average in all the criteria.

Weight Stability Intervals

Criteria	Weight	Interval		%	% Interval	
C..1: STUdies	5.00	< 0.00,	5.20>	25.00	< 0.00,	25.74>
C..2: EXPerience	5.00	< 4.83,	6.83>	25.00	< 24.37,	31.30>
C..3: AGE	2.00	< 1.50,	3.25>	10.00	< 7.69,	15.29>
C..4: INTerview	4.00	< 2.75,	4.25>	20.00	< 14.67,	20.99>
C..5: TESt	4.00	< 3.50,	6.50>	20.00	< 17.95,	28.89>

Figure 10.8 PROMCALC weight sensitivity analysis

Let us now look at the highly useful weight sensitivity analysis. In figure 10.8 we see, for example, that for the criterion STU (studies), and for a weight value of 5.00, the stability interval is [0, 5.20]. This means that if the weight of STU varies between the bounds of this interval, with the other weights unchanged, the PROMETHEE II ranking does not change. This is clearly very high value information given the uncertainty that usually surrounds the choice of weights.

Finally we show our candidate selection example projected on the GAIA plane (figure 10.9). The horizontal and vertical axes, respectively u and v, are hypercriteria resulting from separation into main components. The percentage of information from the flow matrix retrieved by the projection is $\delta = 73\%$. The alternatives are located on the plane according to affinities. The 'decision joystick' is shown as a small line which here is close to the v-axis. This joystick indicates the criteria which are most favored by the weighting (according to the angular proximity of the joystick). Those alternatives which are not opposed to the direction of the joystick will thus be relatively highly ranked, as are Georgia (A5), who is the closest in angular distance, followed by Donald (A3) and Blanche (A2).

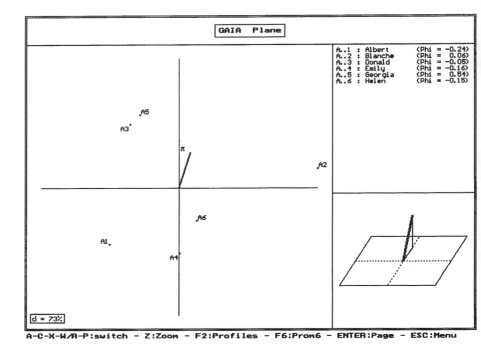

Figure 10.9 GAIA plane projection of the candidate selection example.

10.3.4 Computer aspects

PROMCALC does not require any special hardware and though a printer is not indispensable, it is obviously useful. The program takes up 256 Kb of memory and can be run immediately without prior installation or configuration. Relatively user-friendly, it functions through pleasingly designed menus of useful options. It is easy to introduce models and modify data through an optional 'spreadsheet' type format. Calculations are fast and graphic information is quickly obtained.

The drawbacks are the sparseness and inflexibility of written reports and the absence of on-line help, aggravated by the lack of a manual.

10.3.5. Overall assessment

PROMCALC is an effective and reasonably well-designed software package, reliable and convenient in use. Though poor in supporting documentation and written reports, it has good graphics. The GAIA software is a very useful complement available as a built-in option in version 3.4. Its 'pairwise comparison' methodology often gives different results from simple weighting.

10.4 EXPERT CHOICE

10.4.1 General information

EXPERT CHOICE is a package based on the AHP (Analytic Hierarchy Process) method of T.L. Saaty. It is distributed by Expert Choice Inc. (5001 Baum Boulevard, Suite 650, Pittsburgh, PA 15213, USA). Here we shall describe the version ECPro 9.0 for Windows (a trial version); version 9.5 can be downloaded from the home page on Internet (www.expertchoice.com). Also available on Internet is information about AHP and an on-line sales service for other products from the company. A downloadable program named Slide Show gives an amusing explanation of the mathematics of AHP.

10.4.2 Theoretical background

EXPERT CHOICE authentically reflects Saaty's AHP method for computer. The standard and most comprehensive work that is both popular and scientific is Saaty (1995), but there are also excellent introductions to AHP in Zahedi (1986), Harker and Vargas (1987) and Saaty (1995). Without going into details, the two main features of AHP (see section 4.6) are as follows.

First there is the structuring of the problem as a hierarchical tree diagram. From the root of the tree, or 'Goal', spring several layers of criteria (up to five in EXPERT CHOICE), divided into sub-criteria. In EXPERT CHOICE, each node can only have nine daughters. The leaves represent alternatives. In the basic setup up to m alternatives and n criteria can be handled, where $m \leq 9$ and $n \leq 9$; if the criteria are arranged in hierarchical form, up to 9^5 criteria and nine alternatives can be processed.

The second feature is the use of pairwise comparisons, both between criteria to estimate the decision maker's desired weights and between alternatives to evaluate each one relative to each criterion. Comparisons are made on a numerical scale from one to nine or on a qualitative scale as suggested by Saaty (see section 4.6). These can be displayed graphically. From the evaluations obtained AHP, and of course EXPERT CHOICE, calculates an evaluation for each alternative by simple weighting, and hence a total preorder. The package contains several useful extra features including:

a) Real time graphic analysis of sensitivity to changes in the weights of the decision maker's criteria (the other criteria being modified in proportion to their original values).

b) The 'What-if' function allows the sensitivity of the final result to be assessed in relation to modification of the preference between two given alternatives.

c) Through the 'Ratings' function the software limit of nine alternatives can be extended: a set of alternatives is defined and evaluated as if it were a single alternative. Once the best set of alternatives has been chosen, the best alternative within that set can be found. The process can be repeated any number of times as required.

d) The package has a diagnostic function for the most likely causes of inconsistency resulting from pairwise comparisons when the inconsistency coefficient is greater than 0.1. AHP, as we saw in section 4.6, will accept a certain degree of inconsistency.

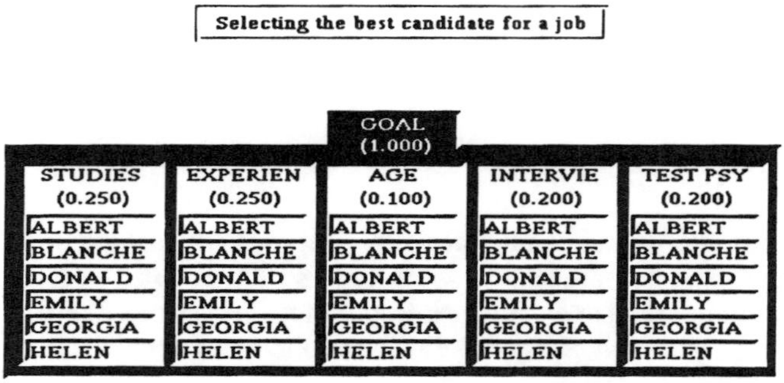

| Selecting the best candidate for a job |

| GOAL (1.000) | | | | |
STUDIES (0.250)	EXPERIEN (0.250)	AGE (0.100)	INTERVIE (0.200)	TEST PSY (0.200)
ALBERT	ALBERT	ALBERT	ALBERT	ALBERT
BLANCHE	BLANCHE	BLANCHE	BLANCHE	BLANCHE
DONALD	DONALD	DONALD	DONALD	DONALD
EMILY	EMILY	EMILY	EMILY	EMILY
GEORGIA	GEORGIA	GEORGIA	GEORGIA	GEORGIA
HELEN	HELEN	HELEN	HELEN	HELEN

Figure 10.10 Candidate selection model with EXPERT CHOICE

10.4.3 Example of use

To run our candidate selection model it has to be adapted in the following way.

a) Create a hierarchy. Here we shall simply consider the root 'Goal' and its five daughters, each one corresponding to one of our criteria. The leaves are our six alternatives as usual. The resulting hierarchy is shown in figure 10.10, where there are also shown values of the weights and evaluations of each alternative for each criterion.

b) The weights are ascertained from pairwise comparisons between criteria. From the responses, given in the wished proportion for our example as summarized in figure 10.11, the system deduces the weights and displays them graphically.

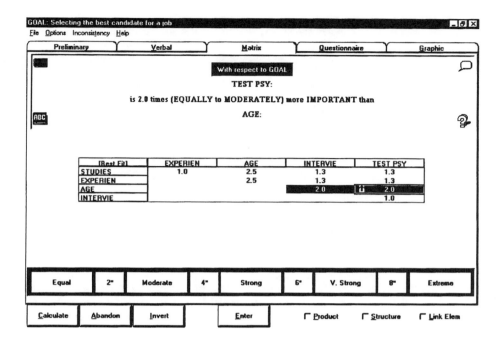

Figure 10.11 Comparisons between criteria in order to assign the weights in EXPERT CHOICE

c) Each alternative is assessed for each criterion in the same way, by pairwise comparisons. The process is exactly the same as the one for the weights that we have just seen. In our case, since values are available we enter them directly using the sub-command 'Data' in the command 'Assessment'.

When the assessments have been made, the final result is clearly displayed and can be printed (figure 10.12). In this example, the final results are almost identical to those obtained in chapter 4 by simple weighting after normalizing by dividing by the sum. This is not surprising since EXPERT CHOICE is based on the weighted sum, and the slight discrepancies observed may be attributed to rounding errors.

Figure 10.12 The final results in the candidate selection example

Finally, to give some idea of the kind of sensitivity analysis that can be interactively carried out on-screen, figure 10.13 shows the five sensitivity graphs available: Performance, Dynamic, 2D-plot, Gradient and Differences. In each of these the decision maker can modify the relative size of a weight and immediately observe all the various consequences.

10.4.4 Computer aspects

Minimum configuration is quite modest: a 386 processor at 33 MHz and 8 Mb of RAM and 10 Mb on the hard disk, and a printer (highly recommended but not indispensable). No other software is required.

The trial version has no date limit and has nearly all the functions of the software proper, except that only three levels of hierarchy can be handled. The words 'Trial version' appear on all printed sheets. Installation is quick and easy and the software can be personalized through various options (color, printer type, fonts etc., using 'Setup'). These parameters are set as default.

The software is controlled through menus with informative help screens available at all times. The design is quite user-friendly, but it is recommended to use the online tutorial to get familiar with certain features.

The interfaces are well thought out with effective displays and easy navigation through the tree hierarchy of the model in use. It is easy to enter or modify a model and the pairwise comparison process is intuitive and simple to carry out, which is vital given the large number of comparisons that have to be made.

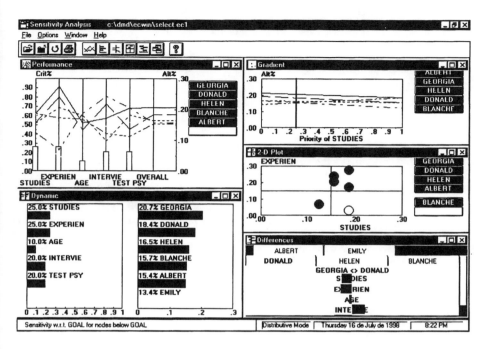

Figure 10.13 Weight sensitivity analysis with EXPERT CHOICE

Hard copy can be printed out at any stage of the process (hierarchy model, comparisons, final result, sensitivity analysis). With EXPERT CHOICE the tiresome but necessary task of producing reports becomes effortless. The results can be saved on a file for subsequent use.

Commands are flexible and relevant; errors are easy to avoid or correct. Help is reasonably comprehensive though sometimes rather wordy. There are various ways of importing and exporting data. The five types of sensitivity analysis can be consulted simultaneously on a single screen in their five windows whenever any one of them is modified (see figure 10.13). The resulting analytical power would have been hard to envisage only a short time ago.

10.4.5 Overall assessment

As a DMD tool, EXPERT CHOICE is powerful, flexible and user-friendly, well documented and theoretically sound. Its only limitation stems from its very reason for existence: AHP. It forces the user to adapt to a hierarchy structuring, to make many pairwise comparisons and not to exceed nine alternatives (unless entered into 'Ratings') and 9^5 criteria, and of course to accept the theoretical assumptions of simple weighting. All these considerations result from AHP methodology and not from EXPERT CHOICE itself.

In spite of the above limitations, AHP and hence EXPERT CHOICE has considerable advantages: reliable, intuitive and robust assessments through pairwise comparisons with consistency check and appropriate representation when there is a natural hierarchy of criteria.

10.5 QUALIFLEX

10.5.1 General information

The QUALIFLEX version 2.3 package is available on 3½" disk accompanied by a book entitled 'QUALIFLEX. A Software Package for Multicriteria Analysis' by authors J. van der Linden and H. Stijnen, published by Kluwer Academic Publishers in 1995, referred to below as Linden and Stijnen (1995). The package price is $1173.50. The book, which replaces an earlier version by Ancot (1988), is 32 pages long (hardly longer than the software manual) and has a short appendix describing the theoretical basis of the algorithm. The user is advised that only a single backup copy may be made; the software may, however, be installed on several machines provided only one is in use at any given time.

10.5.2 Theoretical background

QUALIFLEX is based on the algorithm of the same name described in section 8.4 on permutation methods. The basic idea, it will be remembered, consists of comparing the various possible rankings of alternatives with the information in the decision matrix. For each ranking we calculate a value index characterizing the matching between the ranking and the values in the decision matrix, and the ranking with the best index is selected. We can also calculate a distance between each alternative and the best one. To do this the positions of two consecutive alternatives in the final ranking are exchanged and then observe the differences between the value indices of these hypothesis and that of the best one. It is then assumed that the total distance is the sum of the partial distances and finally normalized between 0 for the best alternative and 100 for the worst.

The qualitative (QUALI) and flexible (FLEX) assessments (as indicated by the name of the software) are highly ordinal; this makes the method very convenient and robust for the decision maker (see example 10.5.3). The evaluations are internally transformed into cardinal form by calculation of the barycenter of the multidimensional subsets formed from the set of evaluations that are compatible with the values given by the decision maker.

Paelinck (1983) succeeded in formalizing the basic algorithm of the method as a special case of a quadratic assignment (see section 8.4). Ancot (1988) and Linden and Stijnen (1995) further refined it and it is this algorithm which ensures the very effective functioning of QUALIFLEX.

The maximum problem size that QUALIFLEX can handle is $m = 500$ alternatives and $n = 20$ criteria, largely exceeding the capacity of earlier versions.

One important feature of QUALIFLEX is its ability to treat the assignment of weights to the criteria in various different ways: fixed weights, ranking of criteria or interval weights. As a result the decision maker has to provide respectively: a cardinal value, a preorder on the weights and a cardinal value interval for each weight. In the latter cases, the analysis is repeated with 1000 sets of arbitrarily constructed weights compatible with the ordinal preference or the intervals given by the decision maker. The final results are the mean and the standard deviation of the rankings and distances obtained by each alternative on the set of 1000 simulations. The following example will make this less abstract.

10.5.3 Example of use

Setting up our habitual example of candidate selection is a laborious exercise because QUALIFLEX requires that the 'structure' of the model be previously created, *i.e.* the field characteristics of each alternative and the number and type of criteria. For each alternative there are then entered the name, abbreviation and the assessments for each criterion. If the criterion is cardinal, a positive assessment must be given; if it is ordinal a string of + and – signs will indicate the preferences. All assessments are assumed to be monotone increasing, in other words they must if necessary be transformed so as to enable maximization. Figure 10.14 summarizes our model (the alternatives are designated by their initials) and age is qualitative.

```
+--------------------------------------------------------------------+
|     Print data                      | QUALIFLEX version 2.3        |
|              All alternatives       | (C) NEI Software 1994        |
+--------------------------------------------------------------------|
| NR   ALTIDEN STUDIES  EXPERIENCE   AGE    INTERVIEW  PSYCHOTEST     |
| ------------------------------------------------------------------ |
|  1      A       6         5        +++        5          5         |
|  2      B       4         2        ++++       10         9         |
|  3      D       5         7        --         9          6         |
|  4      E       6         1        +++        6          7         |
|  5      G       6         8        ++         7          9         |
|  6      H       5         6        ++++       4          8         |
|                                                                    |
+--------------------------------------------------------------------+
```

Figure 10.14 Candidate selection model with QUALIFLEX

The model can optionally be saved and printed, then under the option 'Analysis' entered in the main menu for evaluation of the weights. There are various options available as described above. Here we shall detail the most flexible possibility of defining the weights in ordinal form. In the corresponding window we enter the ranks of the criteria STUdies and EXPerience, which is 1, of INTerview and PSYchometric test, which is 2, and of AGE, which is 3. With this information together with the random method described in 10.5.2, QUALIFLEX displays the final results in various forms (graphic table, ranked alternatives) as shown in figure 10.15.

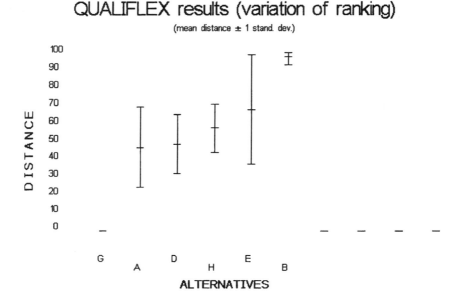

Figure 10.15 QUALIFLEX results for the candidate selection example

With the second type of weight (interval weights), the decision maker assigns cardinal values to the weight intervals and the results appear in a similar form to the above. The third method is to assign a precise cardinal value to the weights (assuming this is possible). In our case these will be the usual values of 5, 5, 2, 4 and 4 for criteria C_1 to C_5. Results are obtained as shown in figures 10.16 and 10.17.

```
+----------------------------------------------------------------------+
| QUALIFLEX results (standard weights) | QUALIFLEX version 2.3         |
|          sorted by distance          | (C) NEI Software 1994         |
+----------------------------------------------------------------------+
|  NR      ALTIDEN      Prio1         Dist1                            |
|----------------------------------------------------------------------|
|   1         G           1            0.0                             |
|   2         H           2           83.2                             |
|   3         D           3           84.6                             |
|   4         B           4           89.0                             |
|   5         A           5           96.7                             |
|   6         E           6          100.0                             |
|                                                                      |
|                                                                      |
+----------------------------------------------------------------------+
```

Figure 10.16 QUALIFLEX results with cardinal weights

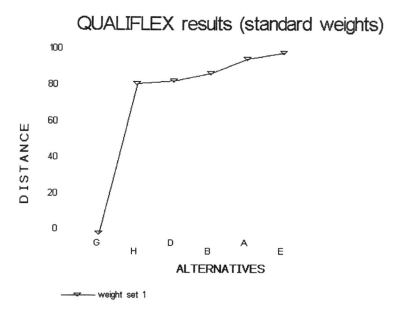

Figure 10.17 Graphic display of results with cardinal weights.

In the above two figures we note that the final ranking is G ≻ H ≻ D ≻ B ≻ A ≻ E, which is almost exactly the same as that obtained by weighted sum in chapter 4, the only difference being the inversion in the order of Donald and Helen. But in spite of this we should stress that the methods are very different in nature. The fact is that the results in figure 10.15 for ordinal weights are rather different from those in the following figures, especially for Albert, even though first is still the robust alternative G. The precision that we required in the cardinal weights may make us think that the results in figures 10.16 and 10.17 make better use of the information produced, provided the stated precision is real.

10.5.4 Computer aspects

QUALIFLEX runs in DOS. A color screen is necessary together with a printer. No software other than DOS is required. Installation is simple, and the various files take up 1 Mb of memory. A single BAT file executes the program. The package is user-friendly with easy navigation between the various options on-screen (the main problems and serious errors in the previous version have been corrected). Once the working method has become familiar it is relatively simple – fortunately, because there are no help screens and the manual is not designed as a reference but as a tutorial, and is not very illustrative.

Creating models is a tedious job, as we have already mentioned, but various input and output options are available for the model in text, spreadsheet or database format. Overall performance is effective and the graphic interface is both powerful and attractive.

10.5.5 Overall assessment

QUALIFLEX has two great advantages making it attractive as a DMD method. The first is its low requirement of information from the decision maker, particularly in its acceptance of ordinal utilities. The second is its great flexibility in defining weights and its powerful weight sensitivity analysis. On the other hand, an inherent drawback in permutation methods is the difficulty in interpreting the results produced. Then there are the assumptions that we have to make about weighted sums.

QUALIFLEX is the computerized version of the method of the same name using an elegant and effective calculation algorithm by Paelinck and Ancot, based on a quadratic assignment model; it features attractive graphics and is, in sum, the professional exploitation of an interesting methodology and algorithm; it is in the way models are created that there is still room for improvement.

10.6 Brief review of DMD software

In this section we give all the reliable information we have been able to obtain on the software in question. In very many cases we had, at least, the possibility to test a demonstration version, and our comments refer to this test. We would like to thank all those designers and companies who made their software available to us. Whenever possible we have given a postal, email or URL address where the reader may obtain additional or updated information – for this is obviously a field in which things tend to move fast!

The information is divided into two sub-sections. The first (10.6.1) is on packages that can be thought of as 'live', *i.e.* they can be purchased from the supplier or obtained from the authors. This is the most detailed of the two sections of course. However, we should not ignore many other pieces of software that at one time or another have played a role in DMD, even though they may now have been abandoned, or become obsolete or hard to find. Thus it was with a sense of history together with some nostalgia that we decided to include sub-section 10.6.2, where the reader will find succinct information on these early programs.

10.6.1 Alive DMD packages

AIM
Methodology

This is a progressive information method where the decision maker supplies aspiration levels, and the search is based on an operator that has been defined by a temporary aggregation in which the distance to the aspiration level is minimized (see sections 8.3.4 and 9.4.3). Furthermore, ELECTRE I is used to search for the alternatives in the neighborhood of the temporary optimization.
Source

Designed at the State University of New York at Buffalo by Zionts' team. Production and sales: V. Lotfi, School of Management, The University of Michigan, Flint, MI 48502, USA. (lotfi_v@umich.edu).

Comments

Aspiration-level Interactive Model. Version 3.0 (DOS) handles 50 alternatives and 10 criteria. The criteria are maximized or minimized, or else the user can define a goal to be attained. Dominance analysis and use of interactive veto levels.

References

Lotfi and Zionts (1990), Lotfi and Teich (1991), Lotfi *et al.* (1992).

CRITERIUM DECISION PLUS

Methodology

Uses the AHP and SMART methodologies .

Sources

Infoharvest Inc. P.O. Box 25155, Seattle, WA, 98125-2155, USA. fax: (206) 729-1854

URL: www.infoharvest.com

Price: $495 (version 2.0). Version 3.0 has recently been released (October 1998).

Comments

Accepts uncertainty in weight assessment, enables weight sensitivity analysis, pre-analysis of satisfaction and a good many inputs and outputs (data, graphic displays, DDE etc.), in a Windows environment. Maximum of 50 alternatives and 7 levels of hierarchy for criteria.

References

Saaty (1995), Edwards (1977).

DECIDE

Methodology

DECIDE includes Borda, Condorcet, ELECTRE, PROMETHEE and GAIA methods, as well as other classical algorithms used in operational research. DECIDE is a Windows Dynamic Link Library callable from virtually any current programming language.

Source

DECIDE is currently in its development phase by its author Christophe David (Universite Libre de Bruxelles, Belgium). The latest beta version can be downloaded from http://ourworld.compuserve.com/homepages/christophe_david. The installation program asks for a (free) password that must be asked by email to Christophe David.

Comments

The above mentioned page also gives on-line access to the latest version of the DECIDE documentation. The DECIDE package comes with a user friendly installation program that will guide the user through the installation process, and includes samples of C sources and a Microsoft Excel workbook with Visual Basic for Applications code that demonstrates how you can use the DECIDE functions.

References

None.

DECISION PAD
Methodology
   Weighted sum.
Source
   Apian Software, Inc. 400 N. 34th St., Suite 310, Seattle, WA 98103, (USA); email: sales@apian.com; URL: apian.com. Price: $395 in October 1995.
Comments
   Version 2.03 is the latest, in DOS environment. It handles up to 250 criteria and 250 alternatives. Moreover 60 decision makers (referred to as evaluators) in a group can each have their own decision matrix. Evaluations can be given on various scales, predefined or user-built.
   Weights can be determined in several ways (ordinal, cardinal by direct evaluation, pairwise comparisons) and in various modes (numeric, symbolic, graphic). Criteria can be hierarchized into two levels and a satisfaction level can be defined for each criterion.
References
   Kepner and Tregoe (1981), Pomerol and Barba-Romero (1993).

DECISION GRID
Methodology
   Some non detailed Concordance method.
Source
   Visual Decision Inc. (Complete information at www.decisiongrid.com, where a limited demo version can be downloaded). Version 3.5 Professional costs 500$ (February 1999).
Comments
   Easy to use and with quite common features. An originality, given its concordance methodology, is the possibility of using interval or range evaluations. Sensitivity analysis for the evaluations but not for the weights.
References
   None.

DEFINITE
Methodology
   This is a multicriterion DSS combining several DMD methods.: ELECTRE II (section 7.3), REGIME (section 7.5.1), EVAMIX (subsection 7.5.1), weighted sum and others.
Source
   R. Janssen, Free University, 1007 MC Amsterdam, Netherlands. Janssen's book (1992) comes with a restricted demonstration version.
Comments
   DSS developed by van Herwijnen and Janssen for the Dutch Treasury for public investment planning. It is the only known package to use the REGIME method.
References
   van Herwijnen and Janssen (1989), Janssen *et al.* (1990), Janssen (1992).

ELECCALC

Methodology

Uses pairwise comparisons of alternatives with ELECTRE II. The comparisons are made on a subset of the choice set. In ELECTRE II weights and indices of concordance and discordance have to be defined. From arbitrary starting values, the program allows the user to follow interactively the parameters that are compatible with the decision maker's preferences on the reference set. These parameters are then used for the final ranking.

Source

L.N. Kiss and J-M Martel, Faculté des Sciences de l'Administration, Université Laval, Québec, Canada G1K 7P4. There is no commercially available software.

Comments

Interactive visualization of the parameters to be estimated. High quality implementation.

References

Kiss *et al.* (1992).

ELECTRE

Methodology

ELECTRE I interactive (ELECTRE IS), ELECTRE III-IV and ELECTRE TRI (see section 7.3).

Source

B. Roy and various collaborators from LAMSADE, Université de Paris-Dauphine, Place du Maréchal De Lattre de Tassigny, 75775 Paris 16, France. URL: www.lamsade.dauphine.fr. Price 7500FF to 9000FF in 1998.

Comments

All the possibilities for interaction and data handling in the various versions of the ELECTRE method. Version 3.1 of ELECTRE III-IV is for Windows.

References

Roy (1968a), Roy (1978), Roy and Skalka (1985), Skalka, Boyssou and Vallée (1992), Mousseau *et al.* (1999).

ERGO

Methodology

The so called Weighted Average Composite Index. For each of the alternatives, ERGO computes a Composite Index measuring how close its evaluations are to the user's vector of weights. The Composite Index is then combined with the weighted average.

Source

Commercialized by Arlington Software Corp., Montreal, Canada. Information and demos at www.arlingsoft.com. Price: 2000$

Comments

ERGO states that "Weighted Average is recognized as being highly inadequate.....", and can lead to less-than-optimal decisions well over 50% of the time". Instead, the Composite Index correction gives "accurate results and reliable decisions". Powerful graphical interface.

References

None.

EXPERT CHOICE
See section 10.4

HIPRE3+
Methodology
Basically AHP (see section 4.6), with considerable extension.
Source
Raimo Hamalainen and his collaborators in the HUT (Helsinki University of Technology). A demonstration version can be downloaded from Internet (http://www.hut.fi/units/SAL), and it is possible to work directly with the version WEB-HIPRE of the software.
Comments
HIPRE3+ (**HI**erarchical **PRE**ferences) runs in DOS, but is highly interactive. Allows modeling in 20 hierarchical levels with 50 elements per level. Weights are assigned either directly or by pairwise comparisons, and the user can easily define various utility functions for the criteria. Features complete sensitivity analysis. The parallel package HIPRE3+ Group Link is a Windows version for group decision work. See also the INPRE package below.
References
Wesseling and Gabor (1994).

HIVIEW and EQUITY
Methodology
Weighted sum.
Source
Enterprise LSE, London School of Economics, Houghton Street, London WC2A 2AE, UK. email: enterprise@lse.ac.uk URLs: www.lse.ac.uk/enterprise and www.tidco.co.tt/seduweb/research/axelk/ parks/parks.html
Comments
HIVIEW and EQUITY assisted decision processes have the feature of giving a balanced view over a wide range of criteria where hard data (e.g. financial) is integrated in a rigorous manner with subjective data. HIVIEW is the most suitable of the two when the problem is one of choosing between different management options. EQUITY, on the other hand, is designed to help in resource allocation, budgeting and prioritization modeling.
References
Phillips (1990).

INPRE
Methodology
AHP (section 4.6) with pairwise comparisons of intervals.
Sources
As for the HIPRE3+ package .
Comments
INPRE (**IN**teractive **PRE**ferences). This software handles pairwise comparisons of intervals, allowing greater flexibility in the expression of preferences that are generally imprecise. Uses an internal linear programming algorithm.

References
Salo and Hamalainen (1995).

LOGICAL DECISIONS
See section 10.2

MACBETH
Methodology
Decision maker's preferences (criteria utility funtions and weights) are extracted by means of qualitative binary comparisons and finally obtained by a linear programming algorithm (see section 4.7). Basically MACBETH is intended to help the decision maker to build consistent cardinal utilities.
Source
Prof. C. Bana e Costa (U. Técnica de Lisboa) (cbana@alfa.ist.utl.pt) and Prof. J-C. Vansnick (U. Mons-Hainaut) (Jean-Claude.Vansnick@umh.ac.be). A commercial version may be obtained from the last author.
Comments
Measuring Attractiveness by a Categorical Based Evaluation TecHnique. Inconsistencies the decision maker's estimations may be detected, and suggestions given for solving them. An external to MACBETH weighted average ranking method is used by the authors for the final multicriterion aggregation. Methodology and package in continued development since 1992.
References
Bana e Costa and Vansnick (1997, 1998).

McVIEW
Methodology
Uses a procedure of progressive elicitation of the decision maker's preferences based on a mixture of two methodologies: comparison of pairs of alternatives that are easy for the user to compare, and cardinal evaluation of preferences in the form of weights or aspiration levels; an optimization algorithm is used to make the two sets of results consistent with previous comparisons.
Source
Professor Rudolph Vetschera, Management Center der Universitaet Wien, Bruenner Strasse 72, University of Vienna, 1210 Vienna, Austria. (rudolph.vetschera@inivie.ac.at). There is no commercially available package.
Comments
The data, decision maker's preferences and possible inconsistencies are displayed on the screen in the form of main components. The preferences can then be modified and the consequences immediately seen on the screen, the process being repeated until the decision maker is satisfied.
References
Vetschera (1994).

P/G%
Methodology
  Weighted sum
Source
  Professor Stuart Nagel. Decision Aids Inc., 1720 Parkhaven, Champaign, IL 61820, USA. email: s-nagel@staff.uiuc.edu URL: uxl.cso.uiuc.edu/~s-nagel.
Comments
  Uses the method of part/whole percentaging, which is simply a weighted sum method with a great emphasis on evaluations' normalization. Written in the language PLATO but there is also a LOTUS version in macros. Handles up to 15 alternatives and 15 criteria.
References
  Nagel and Long (1986), Lotfi and Teich (1991).

PROMCALC and GAIA
  See section 10.3.

QUALIFLEX
  See section 10.5.

SMC
Methodology
  Weighted sum.
Source
  Professor S. Barba-Romero, Dpto. Fundamentos de Economia, Universidad de Alcalá, 28802 Alcalá de Henares, Spain. email: fesergio@alcala.es
Comments
  User-friendly package in the form of a spreadsheet under Windows. Direct weight evaluation or by pairwise comparisons. Alternatives evaluated on quantitative or qualitative scales. Several normalization methods are available. The package allows pre-analysis of satisfaction and dominance. Several weight sensitivity analysis.
References
  Barba-Romero (1998, 1999), Barba-Romero and Mokotoff (1998).

UTA PLUS
Methodology
  UTA (see section 6.7).
Source
  LAMSADE, Université Paris-Dauphine, Place du Maréchal de Lattre de Tassigny, F-75775 Paris CEDEX 16. URL: www.lamsade.dauphine.fr/logiciel. html
Comments
  UTA PLUS was developed for Windows by M. Kostkowski and R. Slowinski and is in the classic PREFCALC tradition (see section 6.7)
References
  Jacquet-Lagrèze (1990).

VIMDA
Methodology
Interactive comparison of alternatives in graphic mode.
Source
NUMPLAN, Helsinki, Finland. email: NumPlan@NumPlan.fi. URL: www.numplan.fi.
Comments
Visual Interactive Method for Discrete Alternatives is based on the multiobjective optimization method of Korhonen and Laakso, adapted for the discrete case (see subsection 8.2.2, point 3). Handles up to 500 alternatives and 10 criteria.
References
Korhonen and Laakso (1986), Korhonen (1988).

V.I.S.A.
Methodology
Weighted sum.
Source
Professor Valerie Belton, University of Strathclyde, UK. URL: www.cqm.co.uk. Price: $495.
Comments
Visual Interactive Sensitivity Analysis performs graphic interactive analysis of weight sensitivity. Criteria can be hierarchized.
References
Belton and Vickers (1990).

WEB-HIPRE see HIPRE3+

10.6.2 Historical DMD packages

The DMD software packages in this section all played their part at one time or another in the development of discrete multicriterion decision, but they are now either unavailable or have been made obsolete by progress in computer technology. We simply give a bibliographic reference for each one.

ARIADNE Sage and White (1984), Goicoechea and Li (1994), Olson (1996).
BEST CHOICE PCWeek Reviews, June 25, 1990 p.119-21.
CONJOINT LINMAP Carmone and Schaffer (1995).
DMDI Barba-Romero (1988).
DISCRET Majchrzack (1985, 1989).
DIVAPIME Mousseau (1995).
ELECTRE RUSCONI/MBA Schärlig (1985).
IDEAS Vetschera (1986, 1988, 1989).
IMAP II Malmborg *et al.* (1986).
MAPPAC and PRAGMA Matarazzo (1986, 1988, 1990, 1991).
MARS Colson (1989).
MATS-PC Brown *et al.* (1986), Lotfi and Teich (1991).

MCRID   Moskowitz *et al.* (1991).
MicroMULCRE   Danev *et al.* (1987), Lockett and Isley (1989).
MINORA   Hadzinakos *et al.* (1991).
MULTIDECISION   Lévine and Pomerol (1986), Pomerol (1985).
PCPDA   Kirkwood and van der Feltz (1986), Lotfi and Teich (1991).
PREFCALC   Jacquet-Lagrèze (1990) (see UTA PLUS).
PRODALT   Khairullah and Zionts (1987).
RADIAL   Buede (1996).
REMBRANDT   Lootsma (1992a), Olson (1996).
TRIPLE C   Angehrn (1989).
ZAPROS   Larichev *et al.* (1993), Olson (1996).

11 MULTICRITERION DECISION IN PRACTICE

Introduction

The previous chapter was about software for multicriterion decision; in fact we could almost have called it multicriterion choice, for the vast majority of this software carries out aggregation, and hence choice, at the expense of other types of support which, as we have stated several times, ought to embrace the whole process of decision making. This overall support is probably more typical of multicriterion DSS (Decision Support System) than of multicriterion analysis programs. Leaving aside now questions of theory and software, in this chapter we deal with problems and advantages of multicriterion decision in actual practice. We begin by recalling the salient features and advantages of multicriterion modeling; we shall also be examining the various phases and individuals involved in practical multicriterion decision. We go on to discuss modeling, aggregation and finally fields of application.

11.1 The role of multicriterion decision in descriptive models of human decision

It has long been known and recognized that more or less conflicting criteria can co-exist in the mind of any individual. However, it was not until around 1955 that Simon was the first to give a scientific basis to this fact, introducing the idea into his *model of limited rationality* (Simon, 1955), and giving status to a 'satisficing' solution which is not necessarily optimal. This idea took some time to gain acceptance; though in 1968 in an editorial, Roy stated that 'operational research should be de-optimized', Keen was still justified in 1977 in referring to the sixties as the 'Elizabethan age' of optimization, in other words single-criterion optimization. The reader will by now have understood that the notion of 'optimization' does not really have a sense in multicriterion analysis. Along with the golden age of optimization came the Grand Illusion that, following the very real successes of linear programming, operational research methods would give substance and meaning to the notion of optimal decision, and enable this to be reached. It is this last assertion which is the most arguable: if we are in a situation in which optimal decision does have any meaning, then it is probably not too difficult to identify that decision; this task is in fact the real domain of operational research. Keen (1977) points out that this optimism prevailed equally in economics (maximizing utility) and politics (rationalizing budgetary choices). In France at the same time Roy (1976, 1977) was developing his arguments on 'de-optimization', an idea which was beginning to find some echoes. Roy's criticism focuses on three main points: *the need for global knowledge of the choice set, its stability (with the choice set being considered as one intangible piece of data), and the existence of a complete pre-order in the choice set.*

Towards the end of the seventies, the 'Olympian' model of maximization of expected utility (MEU), as Simon named it (1983), started to weaken and Roy (1987) started talking about the emergence of a new paradigm. Several strands can be identified in this new way of thinking; the first is that of limited rationality through the use of different criteria at different times as the action-decision interaction develops. Here we shall go no further into limited rationality, as we have touched on it several times already; the reader should refer to Simon (1983).

The second of these strands is multicriterion modeling, or paradigm (Roy, 1987b). This, according to Roy, involves accepting that:

1) several more or less conflicting criteria (conflicting at least on a local scale) are relevant and legitimate in driving the organizational system in question;

2) the term 'optimal running of the system' is meaningless, the aim being rather to find *successive equilibria*, each equilibrium position resulting from a compromise or arbitration between criteria.

The successive compromises have to be discovered one by one through a system that relies on the ability of its members to devise compromises. The notion of absolute optimum loses all meaning; however, this does not imply that certain decisions are not better than others, but rather that *'the best decision from all points of view'* is a myth. Several successive compromises are possible which, determined as they are by different criteria, are not necessarily irrational: they represent the normal evolution of equilibria to be found in any living organism, this evolution being brought about by changes in the environment, in the balance of forces, power, information etc. within the organization. These modifications are moreover often the result of imbalances arising in the previous equilibria!

As we saw in the section on DSS (sub-section 9.5.2) within the optimization paradigm we focus almost exclusively on the choice phase of the decision process. This focusing only makes sense if we have faith in the idea of the 'best decision'. In fact, if we want optimization, it ought rather to be optimization of the process of arriving at the successive compromises. As Simon (1983) puts it, *it is much more a question of procedural rationality than of substantive rationality.* Keen (1977) also points out that in going from one paradigm to another several changes in point of view occur, including going from one to several criteria., recognition of the role of *decision-making context*, introduction of the notion of decision process (already considered as fundamental by Simon) and abandoning the irrelevant notion of convergence and the incremental process.

Keen's analysis leads on to a vaster paradigm than that of multicriterion decision, that of *'apprehensive man'*. This individual is simultaneously worried about the future (apprehension in the sense of fear), able to see globally rather than to analyze in detail (apprehension in the sense of perception) and able to seize opportunities (apprehension in the sense of prehension).

Apprehensive man has confidence from his prior experience of situations he has apprehended, and on the basis of this confidence he will make his decision, which will not be optimal but rather, conservative and based on satisfaction levels. He is much more of a coping person than an optimizer in the face of the uncertain; Lindblom (1959) calls this 'muddling through'; it is much more intuitive than calculating. The consequences of the paradigm of the apprehensive man on decision support closely match what is observed in practice. Like Keen (1977), we shall list six points, emphasizing the consequences on practical multicriterion analysis.

1) *The analysis is dominated by apprehension.* In complex decision situations the decision maker does not feel he has to believe, much less to follow the results of unidimensional analysis. This is an argument for multiple criteria but is also a warning that even with multicriteria, in the mind of the decision maker there is a high risk of an analysis not apprehending the entire situation. *The analyst must therefore remain modest, be ready to present divergent findings and not be offended if his work is only one among several elements contributing towards the final decision.* No surprise should be felt by the analyst if the recommendation from the system is not considered as Gospel.

2) The decision process is interactive and recursive; evaluations and modifications are often local and *it is rare to see a decision process in which overall substantive rationality is conserved from start to finish.* Progressive information multicriterion methods generally accept these assumptions. The decision maker can evolve, change his point of view and go back again. But this is not enough; a good multicriterion DSS should be usable over time to help the decision maker during quite a long period. It should also enable him to get over apprehension through unbiased formalizing and memorizing which improve perceptive learning.

3) *Conflicts of value and interests are inevitable; they are a fact of life in any organization and are theoretical impossible to resolve. Yet both organizations and individuals get along perfectly well despite this.* This is because at any given moment one or other of the interests prevails. The compromise can be set in time or can be by separating domains. With multicriterion DSS this pre-supposes that the criteria are not fixed, and can change over time. Creating new alternatives or eliminating old ones can make it easier to reach a compromise. DSS should therefore be used iteratively, with negotiation about or between each run.

4) Decision makers are much more apprehensive than calculators. One consequence of this is that decision makers are reluctant to use complicated multicriterion methods. *Any method or model too remote from the decision maker's calculating abilities or imagination is very difficult to impose in practice.* On the other hand, and more optimistically, this means that when a model is accepted, it is likely to be really worthwhile in making available to the decision maker information he would not have been able to compute himself.

5) *Optimization is a theoretical concept of no help in situations full of uncertainty, in conflicts, in ill-defined multi-dimensional situations which do not lend themselves to numerical modeling.* This argument favors multicriterion decision up to a point, for the latter does assume the presence of some structuring. It does not, however, offer any help in evaluating uncertain alternatives. On the other hand, it can be of help in defusing conflicts, as we shall see in this chapter.

6) Keen (1977) adds a final 'axiom' which, he goes as far as to say, may constitute the key statement: there can be no universal definition of optimality. Everything depends on the organizational context, the thoughts, secrete motives and aims of the decision maker. Again, multicriterion decision is obviously not favored by this point of view. Multicriterion decision does at least allow a decision to be arrived at when a large number of dimensions are involved, without forcing optimization.

We have devoted a fair amount of space to this concept of the apprehensive man in order to show clearly how the multicriterion paradigm fits in with the general pattern of paradigms involved in the practice of human decision making. The points presented in orderly fashion above will be found dispersed throughout this chapter. To maintain a clear overall understanding, the 'apprehensive' aspect of man and the concepts of 'limited rationality' should be borne in mind at all times. These ideas are the natural accompaniment to the practice of multicriterion decision.

11.2 People and timing in multicriterion decision

11.2.1 People

To keep matters simple, let us state the roles of the people involved in actual multicriterion decision situations. We have already had to distinguish between decision maker and analysts (section 2.1). Within a company we consider, in order of appearance, the champion(s), the decision maker(s), the analyst or working party and finally, if multicriterion DSS or software package is being used, the users. In the basic paradigm, the user and the decision maker are one and the same person, but in practice things may be different.

The *sponsors* are by definition those who initiate the multicriterion analysis and provide the necessary resources for it. At the present time it is rare for company managers spontaneously to consider multicriterion analysis; private companies are still steeped in the monocriterion paradigm, which matches up well with unity of decision and objective – the dominant vision within much of senior management. Multicriterion decision generally slips in, not at the decision and strategic planning level, but at the stage where surveys are being made for the dossier to be drawn up, where it is hard to avoid multiple criteria. These surveys will include choice of site (new depots, new production or distribution units), choice of material investments (machines, premises etc.), choice of financial investments with the balance between at least two factors: risk and return. Yet even in these studies, and in particular in the choice of financial investment, competition is sharp between multicriterion methods and methods involving maximization of utility expectancy, with the older, expectancy methods, still being favored; these methods have been well tried and tested and are part and parcel of the dominant paradigm of optimization. The situation is a little better in the case of material investments as qualitative arbitration between price and quality (robustness, ease of use, ease of maintenance etc.) must be performed. Basically, sponsors are most likely to turn to multicriterion analysis when the *situation inevitably involves multiplicity of qualitative criteria.*

To sum up, multicriterion is seldom used in private companies, and is mostly used in technical dossiers with qualitative criteria. Fortunately there is a second context within which well-informed sponsors are now beginning to turn to multicriterion decision: the situation of conflict. In section 2.6 we referred to the problem of negotiation; when parties such as management, unions, polluters, public authorities etc. strongly diverge in their objectives or tastes – in a word, in their preferences, it is difficult to deny the multicriterion aspect of the decision to be reached. In these situations, compromise is the only possible outcome of the conflict, and this compromise can only be reached by taking into account the criteria of all parties. We have noted (section 2.6) that discussion centered on the evaluation of each party's criteria, on methods of aggregation, on methods of building the various possible, or 'constructible' compromises is much preferable to discussion centered on each party's favorite alternative. In this situation, multicriterion analysis is needed *to provide methods and a structured support for discussions rather than to provide aggregation and/or ranking.* The compromise reached will obviously reflect the relative strengths, threats and skills of the various protagonists; it will not come out of any method of aggregation. This is even more the case in negotiations of any complexity, where compromise never features among the basic alternatives on the table at the beginning. Here we hit a weakness (which can be overcome by various iterations) of the classical multicriterion paradigm which assumes that the alternatives are given; this is not always the case in practice. and almost never the case in situations of conflict.

Recent work (Fang *et al.*, 1988) do suggest that the multicriterion approach can be effective in certain negotiations. This is a very promising field of application, where the analyst's skill as a catalyzer of discussion within the multicriterion paradigm is as important as the supporting multicriterion method itself. Sponsors interested in this type of approach are mainly from the public sector; large investments in public or semi-public projects (such as subways (Roy and Hugonnard, 1982), airports (Keeney and Raiffa, 1976, Martel and Aouni, 1991). electricity supply, freeways, dams) have thus been subject to multicriterion studies. In these cases, the public authorities find themselves confronted with various single-interest pressure groups and multicriterion analysis serves among other things to show that decisions reached in this way are the 'best possible compromise' between these interests. This justification or legitimization aspect is underlined by Roy (1999). Multicriterion analysis is both a negotiating tool, as explained above, and an indicator of the rationality and impartiality of the decision maker who is here the willing seeker of the best compromise from the aggregation. For the decision maker, it is a way of shedding himself of responsibility by re-introducing, through the function of aggregation, the die-hard myth of maximization of the collective utility function. So much the better for him if it works!

We turn now from the sponsors to the decision maker(s). In many investment situations, the decision maker is also the sponsor, initiating the analysis to enlighten his choice. In situations of conflict the analyst is advisor to one of the protagonists, namely, the decision maker. But the analyst may also be the arbitrator.

One interesting situation is where the decision maker is not the direct sponsor. This is the case, for example, in organizations where planning departments have the job of preparing decisions. Here, the multicriterion problem is entirely one of selection and the department will be well advised to propose a small set of possible solutions from various angles (for example, how important each criterion is or should be). It is strongly recommended that the decision maker be left with a fairly open choice of solutions and related assumptions. This situation is frequently met in politics (*cf.* Lévine-Pomerol, 1989, p.249 *et seq.*, in a practical application of assessing a proposed privatization law). But in other contexts too, and increasingly, we feel it is a good idea to provide the decision maker with a small set of selected alternatives (a dozen or so), presented in terms of their multiple criteria, and to have him consider these alternatives. Multicriterion DSS should all allow this function, which has the advantage of 'not saturating the decision space of the decision maker' (de Terssac, 1992). For various reasons (*cf.* 11.3) we think it is undesirable – particularly where the decision maker is not the sponsor – to push analysis right through to a final ranking.

This is also the situation when an organization wishes to make a multicriterion DSS available to some of its staff who, though potential decision makers, are obviously not the sponsors. Here too we feel strongly that aggregation and final ranking of the alternatives should not be the first objective of the DSS. It would appear more advantageous to use it to facilitate multicriterion modeling while partially leaving choice to the decision maker; in fact if this is not done, the decision aid vanishes since the DSS itself becomes the 'decision maker' (*cf.* Lévine-Pomerol, 1989).

In some cases it is not the decision maker who uses the DSS, and here again we are in a situation where preparation for the decision is distinct from the decision itself. The qualities to be expected of multicriterion DSS are no different from what we might expect of any modeling aid; the only specific requirement is for DSS to remain within the multicriterion paradigm and to be able to manipulate alternatives according to several different attributes.

Lastly there is the analyst, or the analyzing team. The analyst is often, but by no means always, brought in as a multicriterion specialist. When the sponsors are convinced of the necessity for multicriterion analysis, and, fully aware of the various methods, they make their choice – fine. The analyst's job will then be to begin a modeling process based on their intuition, and more or less encouraged by them.

Two scenarios are possible: either the sponsors (sometimes the decision makers) want the analysis to support them in their responsibility for making the choice, or the sponsors are more interested in the modeling than in the final choice, which will be governed by a different logic. In the former case, the decision maker often hopes that the analyst will introduce rationality into the choice or selection process and that this 'rationality' will lessen his responsibility. The reasoning is as follows: 'As you can see, the case for choosing alternative X is clear: this is the solution the computer (or the analysis) has come up with.' This is not necessarily a comfortable position for the analyst who knows there is no rationality outside the decision maker capable of arbitrating between Pareto optimal alternatives. It is sometimes an extra reason for leaving the choice open, which can be explained as follows: 'If you consider risk is less important than return, then choose alternative X, while if you think the opposite, choose alternative Y.' For the analyst a response like this has the advantage of throwing the decision maker back on to his responsibilities while at the same time allowing him to benefit from the possibilities of multicriterion modeling which can, among other things, quantify or give a content to the common-sense response.

Given that it is sometimes possible to choose how important criteria are, so that a pre-defined alternative becomes optimal at the end of the aggregation process, the professional conscience of the analyst is sometimes severely tested, especially in situations of conflict. The ideal thing in this case is for the status of the analyst to be clearly understood: either he is working for one of the protagonists, or he has been chosen to arbitrate by all parties involved in the confrontation. Both are acceptable provided they are clear and balanced. Let us hope that in the future, those involved in multicriterion negotiations will have the wisdom to seek advice and to really pose their problems in terms of more or less opposed preferences.

Another case in point is when the sponsors are mainly interested in the modeling itself, or the multicriterion paradigm; they will already have fully realized how complex and multidimensional are the problems with which they are faced. Here we may hope for constructive participation in the modeling process, allowing a real discussion on criteria and alternatives. It is here that multicriterion analysis is at its richest, enhancing the decision maker's thinking and interactively broadening decision perspectives.

11.2.2 The various stages of multicriterion decision aid

Multicriterion decision usually starts of with a discussion with the sponsor(s). In other words, the analysis starts with an informal exposé of the reasons why multicriterion analysis is thought desirable. Note that this presupposes that the sponsor is already familiar with the technique; actually it is very often a member of the case study team who, confronted with a decision aid situation, thinks that this deserves multicriterion treatment for various reasons.

Whatever the case, the first thing the analyst does is to acquaint himself with the decision problem to be analyzed. This first contact with the problem is important in evaluating the feasibility of the work and identifying the people who are going to be concerned, the problem itself and the kind of decision problem involved. Thus, as with the design of knowledge-based systems, the analysis begins by an identification phase (*cf.* Benchimol *et al.*, 1990). The main objective of the thinking which takes place at this time is to understand and above all to accept the terms of reference of the analysis. All the people involved at this stage (sponsor(s), decision maker(s) and analyst) must agree on a clear definition of the problem and of the decision context. This is vital, above all if multicriterion analysis has been recommended by the analyst. At the end of this exploratory phase the people who will make up the modeling team will be named. The decision maker will generally be a member of the team, and will aid the analyst in his modeling task.,

The second stage of the analysis begins with a survey of the alternatives and construction of criteria. Here, basically, the multicriterion decision model which will be constructed and finalized by the production of a decision matrix. The construction of the model will be managed by the analyst, backed up by the working team with its various skills relative to the decision within the setting of the organization. As soon as the model has become sufficiently meaningful, even if it is still under construction, it must be shown to the sponsors and the decision makers for discussion and validation. Both the criteria and the alternatives can change. A glance at the performances of various alternatives can be enough to produce a substantial improvement in the model by creating and introducing new alternatives, by changing criteria and/or improvement of evaluations. This phase of the analysis, if it is carried out well , greatly enhances the quality of the decision. It is one of the most important stages in multicriterion decision aid.

The next stage requires the active presence of the decision maker as it will have a bearing, either on the choice of aggregation method (a method without progressive information), or on the progressive gathering of information relevant to the decision maker's preferences (a method with progressive information). Even in the case of a method without progressive information, the analyst will have to mobilize the decision maker to determine weights or obtain other prior information on the latter's preferences (*e.g.* pairwise comparisons, reference point).

There is little to say on the functioning of the model once it has been constructed. It will feature either an interactive process as described in chapter 9, or a computer program which will duly produce the result from the method used. It is important however to stress that a decision aid process should not end without a careful analysis of the results obtained. A discussion of the results might involve the decision maker, the analyst and possibly people affected by the decision. For the analyst, this is the time to state the limits of validity of the model and above all to have the decision maker perform various sensitivity tests (weight variation, concordance or veto thresholds, change of evaluation etc.). When the first results come out of the computer, these can be thought about and discussed, all of which will prove highly useful, especially in the domain of strategic decision. Where a DSS is used, there should be functions enabling the decision maker to carry out this post-aggregation study alone. In the absence of such functions, it is up to the analyst to foster this type of reflection.

11.2.3 The general set up

In figure 11.1 we show the main stages in multicriterion decision. The analyst, or the person with that responsibility, is present at all stages. The decision maker (or the conflicting parties) are also involved at nearly every stage, either for information gathering or to validate the work of the analyst. The sponsor, if that is not the decision maker, is mainly involved in the first two stages and in the final result (recommendations and feasibility study).

Although the stages described may appear to follow a strictly sequential linear flow pattern, most of them can be recalled recursively and there will normally be many opportunities for questionings which will trigger a return back to previous stages. We show on the diagram several examples of this. *This type of interactive process, which contributes to the richness of the decision aid, is in any case the rule in negotiation situations.*

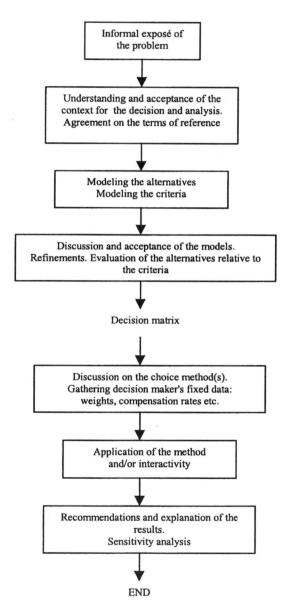

Figure 11.1 The various stages in multicriterion decision

Several multicriterion decision specialists have proposed more or less similar schemes to that shown in figure 11.1. The review by MacCrimmon (1973) contains the earliest thoughts on this theme. Decision aid methodology progresses with Roy (1975) and Jacquet-Lagrèze (1978, 1982). The reader will find an extremely thorough analysis of multicriterion methodology in the Roy (1985). The most recent articles often tackle the problem through the use of DSS, attempting to define (if not actually to produce!) DSS capable of aiding the decision maker at all stages of multicriterion analysis as described above. This approach is to be found in Jelassi *et al.* (1985), Jacquet-Lagrèze (1982) and Balestra and Tsoukias (1990). The reader can also consult the literature on multicriterion DSS referred to in chapter 9.

Notice that the various stages are not all equally visible, and do not even share the same validation possibilities. MacCrimmon (1973), in his scheme for interaction, drew attention to the fact that in our process there is cohabitation between constructed models resulting from the analysis process on the one hand, and models that are invisible but are assumed to be in the decision maker's mind. The former are the models on which discussions between the decision maker and the analyst will be based. The latter must serve to validate the method. This means that the multicriterion method serves to reveal the decision maker's preferences, not only to the outside world but also to himself. This interpretation of multicriterion decision aid as a process of revelation and construction of the decision maker's preferences is a very lively one; it forms the basis of the design of some methods such as AUT (see section 6.7). This philosophy is summarized in figure 11.2.

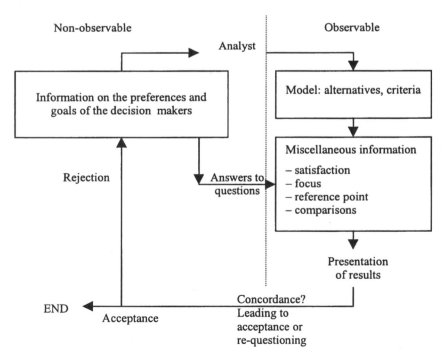

Figure 11.2 The process of revealing preferences

We do not set much store by this scheme of interaction in an industrial context, as it depends too much on the existence of a single decision maker, the revelation of whose hidden intuition will lead to a good decision. This idea is a bit too much like a story book to be true. Our experience has shown that either the decision is the fruit of intuition and the private domain of an individual, and then no recourse is had to any decision aid, whether multicriterion or not, the most we can hope for being a written report; or alternatively the decision process involves several people (even if only one of them finally makes the decision), and then modeling leads to the creation of an object which existed in nobody's mind beforehand. In this case the very idea of revelation is of no interest to anybody; what gives meaning to the process is the construction of a consistent model which the participants consider to be rational, and which they accept. We are getting away from the idea of validating a decision by confronting it with the intimate convictions of a decision maker, and getting closer to the idea of justifying a decision by shared and rationally supported arguments. In all cases, we keep coming back to the idea that the multicriterion paradigm leads to a much richer and realistic structuring than the monocriterion paradigm, and that the resulting model is much closer to a consensus.

11.3 On modeling

We will now assume that the decision problem has been properly identified and agreed upon by the various agents involved: sponsors, decision makers and, if they are not the same people, users. During this preliminary discussion everybody will have understood what is at stake and what are the conditions of use of the aid system. Just as in the case of knowledge-based systems, discussion about the users and conditions of use are generally very useful in assessing the feasibility of the system. It is here too that divergent views among the agents quickly emerge and the analyst will see straight away if these divergences can be overcome to reach agreement among the parties on the role of the system and whether its ultimate utilization may be problematic in view of these initial contradictory views. At the end of this first round of discussion, the various agents will in principle be in agreement on the problem formulation (see section 2.6), *i.e.* they will know whether they want a complete ranking of alternatives, classification, a selection or merely a negotiated compromise. According to the choice, the subsequent operations will be subject to some variation, and in particular, the multicriterion method chosen will be different. Whatever the case, the next stage is to model the alternatives.

11.3.1 The alternatives

According to the 'simple' multicriterion paradigm, the alternatives will have been set before the start of the method. This strong assumption which, if it were true would be such as to weaken the multicriterion paradigm substantially, is one of the three main criticisms leveled by Roy (1976, 1977, 1999) at operational research style optimization (*cf.* section 11.1 above). Here we must distinguish between interactive methods where modification of the choice set is supported and even encouraged, and methods where the choice set is assumed to be untouchable. Whatever the case, it is in practice seldom that the alternatives are known and presented to the analyst set in stone. The alternatives represent the decision maker's possible choices. In the most favorable cases, *e.g.* the choice of machines, the supplier will have checked a series of more or less precise specifications, and the analysts and decision maker therefore have their choice well circumscribed. In many other contexts, and especially in situations where notions of call for tender and specification list are meaningless, such as the choice of budget or a new line, the alternatives have to be constructed with the help of the decision maker or by an ad hoc working group. This is the dominant situation whenever strategic planning is involved.

When the alternatives are to be constructed, the starting point is usually a small number of ideas from the decision maker. These ideas must then be enriched to arrive at a respectable choice set. This enrichment is twofold: firstly, the choice must be enlarged, and secondly each alternative must be clothed with information.

The process of increasing the number of alternatives recalls a well known function of the DSS, that of enlarging the choice of the aided decision maker (see 9.5.2). This also falls within the remit of the analyst. By bringing an outsider's vision into the organization, he can urge the decision maker and the working group to explore other solutions. Mintzberg (1975) observed that actual decision makers very quickly eliminate a large number of possible alternatives to limit their choice to a very small number of alternatives that are examined in detail. This may be related to the cost and difficulty involved in analyzing the alternatives. We shall come back to this in the next section. A more psychological interpretation (Mitchell and Beach, 1990) is that intuitive choice is the result of an initial filtering process (a compatibility test) to rapidly eliminate a large number of alternatives on the basis of their 'value image', before seriously thinking about the small number of alternatives that have passed the compatibility test. Clearly this pre-selection is a fundamental part of the choice if the resulting number of remaining alternatives is very small. The analyst should consequently review this process of intuitive elimination by a kind of intellectual midwifery where the decision maker has to justify his rejection of various deliberately naive propositions put to him by the analyst. This can be done in a systematic way by making a mix of previously envisaged alternatives. The multicriterion DSS DEFINITE (van Herwijnen and Janssen, 1989, sub-section 9.5.2) contains such a function. Alternatives are actually rarely structured in a way which enables this type of procedure to be carried out, but an approximation to it can be accomplished informally by 'creative' methods such as brainstorming. Although this lies outside the sphere of multicriterion decision, we feel that it is a key task of any consultant responsible for modeling.

One restricting factor that can arise in the task of enlarging the field of alternatives is a reluctance on the part of the decision maker to have too large a choice set, which would involve looking at other alternatives. Alternatives are indeed rarely simple, and are often actually multiple alternatives or even whole scenarios involving sub-alternatives. Consider for example the job of deciding the route of a freeway or an electric power line. Clearly, an astronomical number of routes can be envisaged, every one of which must be studied and quantified – and in terms of several attributes. Here we touch on our second point; alternatives must not merely be suggested, but also clothed with information. In other words, *each alternative must be studied*. Now, the resources available for such studies are not limitless, a fact which will limit the imaginations of the decision maker and the analyst. The example of the freeway routing is obvious, but the resource problem is encountered in more or less all multicriterion decision contexts. The situation is the same for financial or productivity investment. Here, each basic alternative will generate several scenarios, depending on what assumptions are made about reactions from the environment, and these scenarios will all have to be analyzed willy-nilly. Each scenario will then be considered as an alternative, with the probability of being chosen as one of the attributes. Another way of going about things is, rather than splitting up the alternative according to various scenarios, to consider the probabilities of each scenario and to calculate the utilities as expectancies. Whatever approach is taken, the work involved is clearly neither negligible nor routine. The analyst's job is thus to construct well-informed alternatives, with the aid of an DSS. We gave an example in section 9.2.2 for assessing train timetables. Constructing and handling scenarios is a well-nigh impossible task without aid, so here too DSS can be used. An example of a suitable DSS devised for handling scenarios is DECIDEX (Lévine *et al.*, 1987). In making complex decisions, it is becoming ever more obvious that the decision maker must be aided in constructing alternatives rendered highly complex through the involvement of chains of external reactions. Multicriterion DSS designers have much food for thought here. This aid can take the form of file storage, as we have seen, the files subsequently going through a filtering process to produce the right information for evaluation according to the agreed attributes to be carried out. In MULTIDECISION-EXPERT this filtering is achieved by and expert system (see 9.2.2).

Many types of aid are possible, depending on the context; in the field of strategic choice or negotiation, constructing an alternative is always a matter of constructing scenarios that depend on responses from adversaries or competitors, and this task is sometimes feasible using an expert system; see Lévine-Pomerol (1991).

As a last point, it is obvious that the construction and, especially, analysis of alternatives is not completely divorced from discussion on the criteria since each alternative has finally to be assessed in relation to these same criteria, and the information we are trying to collect on each alternative is therefore highly oriented towards the type of assessment proposed.

11.3.2 The criteria

Roy (1975, 1985), Bouyssou (1989) and both Roy and Bouyssou (1993) have for several years been closely concerned with analyzing the qualities of a good system of criteria.

According to Roy (1985), *a family of criteria is coherent* if it satisfies the following the three following properties:

Exhaustiveness

None of the attributes used to discriminate between alternatives has been forgotten. Put more formally, this means that the family is exhaustive if there exist no pairs of alternatives (a,a') that are tied according to all the criteria in the family and such that the decision maker can without hesitation say that $a \succ a'$ or $a' \succ a$; the latter would mean that the decision maker is in possession of a choice criterion which has not been taken into account.

Consistency

The decision maker's global preferences should be coherent with the preferences according to each criterion. This means that if a and b are two alternatives between which the decision maker is indifferent (and especially if they achieve the same score for each criterion), then the improvement in a according to one criterion and/or the degradation in b according to another criterion does indeed imply $a \succ b$ for the decision maker.

Non-redundancy

A family that satisfies the properties of exhaustiveness and consistency is non-redundant if removing one single criterion leads to the rest of the family no longer satisfying the requirements of exhaustiveness and consistency.

Out of the three above properties, the one that the analyst must bear in mind at the time of modeling is exhaustiveness. This is because it is vital to see, from discussions with the decision maker and by tests on his preferences, whether the selected criteria do indeed express all of the attributes actually considered in the decision. The quality of any decision aid depends on this.

The property of consistency is generally satisfied with a rational decision maker, and in any case it is not the easiest of properties to test during modeling since it presupposes good knowledge of the decision maker's global preferences and because it is precisely because the decision maker has difficulty in expressing his global preferences that the analysis has been undertaken.

The property of non-redundancy is desirable but the drawbacks which result from it depend on the method of aggregation used; the greater the cardinality of the method, the greater the drawbacks. There are no drawbacks in methods such as PRIAM. In fact the risk in redundancy is to attach too much importance to a criterion which happens to feature two times in more or less closely similar forms.

The reader will also have realized that the idea of independence between criteria is irrelevant; criteria are seldom uncorrelated, and in real problems there are naturally links between criteria such as those between price and quality. These links must be taken into account in aggregation methods, especially when determining the weights. *The choice of criteria will essentially be governed either by descriptive considerations if there is a single decision maker, or, in a negotiation context, by considerations dictated by the nature of the conflict.* In the former case, the criteria must take into consideration the dimensions that the decision maker thinks are relevant and which he is capable of actually evaluating. The analyst should take as criteria the questions of how good the model is and how familiar the decision maker is with the given attributes. Once the family is exhaustive it is up to the analyst and the decision maker to make any rearrangements to the criteria to be able to appreciate the weighting and to avoid unduly overweighting a sub-group of more or less strongly related criteria. In a negotiation situation, the criteria must be representative of the interest groups involved.

In choosing criteria, the decision maker should be warned against having too many. Certain methods involve comparisons between alternatives, and it should be borne in mind that the limited cognitive ability of the human brain makes it extremely difficult to make any meaningful comparisons between vectors with more than six or seven components, or perhaps a few more with good visual representation (see section 9.5.1). This effectively reduces the number of criteria to around seven. In any case, in methods with global comparisons of alternatives, these comparisons must be limited to seven items. If it is considered absolutely necessary to go beyond this number of criteria, the difficulty must be divided up by giving them a hierarchical structure, a possibility which is afforded by several packages such that EXPERT CHOICE (see section 10.4), or else by getting aid from an expert system (see sub-section 9.2.2). In purely aggregative methods without progressive information, one is obviously less limited. The difficulty will however be found to recur during weighting or whatever process is used in its place. It is our opinion that no serious analysis can be performed with more than around twenty real decision criteria (not to be confused with the facts, see sub-section 9.2.2), otherwise the cognitive load is so high that no stable and meaningful information can be expected from the decision maker; it becomes all too easy for the unscrupulous to have him say anything they want and thereby to obtain any desired result.

11.3.3 The weights

As we have just seen, it is difficult to separate the two operations of thinking about the criteria themselves and, immediately afterwards, thinking about the relative importance of the criteria, *i.e.* the weights. In chapter 4 we saw how difficult it can be to evaluate weights in a reliable fashion, especially if they are to be cardinal. In fact, as the multi-attribute clearly shows (chapter 6), *the weights should strictly speaking be determined together, at the same time as the utilities related to each criterion.* There are thus two relationships that must be respected: a global view where we have the weights depend on the whole set of criteria and the relations that may exist between these criteria, and then the relationship between the weights w_j and the scale U_j that is used to measure the utility of each alternative. We shall examine these two aspects of the same question separately.

First of all, it is clear that *the weights must be a function of the family of criteria* and that the analyst must be careful not to end up overweighting any attribute. The problem actually arises for practically every family of criteria, because since the latter are not independent, there is often a more or less strong correlation between them. For example, in planning a freeway route, the nuisance level is strongly and inversely correlated with the cost; In most automobiles, comfort is correlated with safety; many other examples can be found. But does this mean we should no longer consider both the criterion of comfort and that of safety? Certainly not, if they enable us to discriminate between automobile of fairly comparable quality. What we must do when we give weights to comfort and safety is to make sure their sum is not too high compared to the weight given to price (for example).

Using a hierarchical modeling of criteria is often an effective way of handling complexity (see section 4.8). Above about seven criteria, we think it very advisable. However, here too we must be careful to avoid overweighting subdivided criteria a tendency that is generally observed in experiments that have been carried out (*cf.* Weber *et al.,* 1988). Empirical studies strongly confirm the usefulness of a hierarchy of criteria in getting a good idea of their importance (Stillwell *et al.,* 1987 and Borcherding *et al.,* 1991). Remember, though, that a hierarchical representation is not unique and that the final result can be affected by the hierarchical structure adopted (Borcherding *et al.,* 1991). Thus in choosing an automobile, one may be unsure whether to include the quality of the braking system, which could in itself be a criterion, in the higher categories of 'safety' or 'comfort'. What must definitely be done is to count braking quality twice without appropriately dividing up its weight.

The influence of the whole set of criteria on weighting is not the only problematic factor for the decision maker. We have already mentioned the effects of context (see sections 2.4.4 and 4.5.2), and of the indisputable influence of the method used in determining the weights (Schoemaker and Waid, 1982). The nature of the alternatives available also has an influence. If an individual is given a set of rather bad alternatives, the weighting proposed is generally different form that adopted by the same individual for a set of good alternatives (Goldstein, 1990 and Mousseau, 1992b). This will be no surprise to the reader who knows that preference additivity assumes preference independence, *i.e.* that weights are independent of the level of utility of the alternatives. In many practical situations, of course, it is difficult to accept such rigidity; fortunately, there are some interactive methods in which this problem can be overcome by not fixing the weights; this is no small merit. In our opinion, this rigidity goes a long way to account for the discrepancy observed (Schoemaker and Waid, 1982; Goldstein, 1990; Borcherding *et al.,* 1991) between methods which, for a given decision maker, determine the weights from global comparisons of alternatives by regression (as with LINMAP and UTA), and direct methods such that the ratio method (see section 4.5.2), direct comparisons, compensation, the swing method and evaluation by price (see section 4.7); and the direct methods are not necessarily the worst, especially for evaluation by price if the situation lends itself to this (Borcherding *et al.,* 1991).

Leaving aside the numerous factors likely to perturb the decision maker in his assessment of weights, we come to the second type of difficulty, stemming from the links between weights and utilities. We have already pointed out the difficulties in determining weights in terms of the utilities adopted for each criterion (see section 4.3 and chapter 6). What is especially difficult is to arrive at a reasonable evaluation of cardinal weights. Normally the weights w_j depend on the scales chosen to evaluate U_j values. Now, empirical studies (Nitzsch and Weber, 1993) have shown that decision makers often fail to grasp this link; they tend to consider that weights have an absolute meaning and, when changing scale, they fail to make the necessary corrections. Even if analyst and decision maker are heading towards the use of a simple weighted sum, we feel that discussion about scales should not be dissociated from discussion about weights. To avoid such a situation, we are strongly inclined to recommend that weights be determined through one of the alternatives comparison methods (section 4.7), or, if the analyst is sufficiently available, the compensation method (example 6.44). In all cases there must be consistency checks.

We end this section by stating that, though the practical and theoretical difficulties surrounding weight determination are often considered as the Achilles' heel of multicriterion decision, we do not share this view; we have not tried to evade the problems, for potential users have a very fertile methodology at their disposal, and naivete and innocence must not be allowed to dissuade them. On the other hand, we must stress that *knowledge must not paralyze practice*. It is vital at one moment or another to discuss the importance of criteria and, if necessary – and this is not mandatory in all methods – to cardinalize the weights; this must be done without complexes, and with all the means necessary to do the job properly, with hindrance from as few precautions as possible. If there are more than about seven criteria, precautions can include the use of an AHP type hierarchical structure (section 4.6), in conjunction with a descending hierarchy type of weight attribution method (section 4.8); several different structures can be tried, with the one that the decision maker feels to be the most natural being selected. The overall coherence of these weightings can then be tested by the compensation method and by questioning, as in the MAUT problem; here, example 6.44 can usefully be studied.

If these checks are carefully carried out, one can be reasonably confident in the weights obtained and use them without restriction in the chosen method of aggregation, *provided the result of aggregation is accompanied by a sensitivity analysis proving that the ranking obtained is stable to small weight variations*.

11.3.4. Constructing the criteria

In many applications, it is not enough to agree on the criteria which will govern the decision; the alternatives have to be assessed according to these criteria. We assume that there are a sufficiently large number of alternatives in the set and that the decision maker or analyst can find the necessary information. What we shall now look at is how we go from verbally defining a criterion such as 'the quality of an automobile' or 'the noise nuisance of an airport' to actually evaluating the various available automobiles or possible airport sites.

Bouyssou (1989) draws attention to the difficulty of this transition. He points out that it involves modeling the criterion and cites the example of noise pollution. How shall we measure it? In decibels, or in number of inhabitants affected? Suppose we adopt the second solution. What noise level determines whether an inhabitant be counted or not? And just who are the inhabitants – those who are living there now, or those who will be living there in the ten years ahead? If there are several runways, do we simply take their sum or do we take the actual number of days each runway is used? Do we take into consideration the way airplane engine noise changes over time? If we choose decibels weighted by the number of inhabitants, this weighted mean is not very fair since those who suffer most are drowned in the mass of those who have a very low noise tolerance; we should therefore choose some particular way of calculating a mean: but which way? Should we take into account the time at which the noise occurs? And if so, how? These are only some of the questions modelers have to deal with in constructing the 'noise pollution' criterion. All things being equal, the difficulty will be identical when constructing an automobile quality criterion; compared to the case of noise pollution, the gain in certainty in the case of the automobile is offset by a loss in precision because the elements assessed are highly qualitative. Included in the criterion will be luxuriousness, seat comfort, driving position, noise (at different speeds?), visibility and esthetics. We can well imagine the kind of discussion there may be toward a precise definition of the criterion.

To come back to the airport, if h_i is the number of inhabitants in a noise zone and w_i a weight dependent on the level of noise i, Bouyssou suggests three types of criterion (and this is far from exhaustive):

$$g_1 = \sum_i w_i h_i$$

or

$$g_2 = \sum_i w_i h_i (1 + \alpha_i)^d$$

(where α_i is the annual population growth rate in zone i for the d years to come)

or $g_3 = \sum_k [\sum_i w_i h_i (1 + \alpha_{ik})^d] P_k$

where P_k is the probability of growth scenario k occurring. It is easily to see that the complexity resulting from such a criterion is almost without limit. The analyst has to fight against this complexity in order to preserve the clarity of the criterion. *The decision maker must at all times be able – at least in principle – to understand what the criterion involves and what assumptions have been made in constructing it. It is these assumptions which set the limits of the model.*

To model the criteria the analyst will try to reach a combination of points of view. In his construction he must of course take into account the available data; it would not be reasonable to include future inhabitants among those suffering from noise pollution if nobody had a very good idea about how urbanization was going to progress.

We have already spoken about the difficulty of defining complete alternatives, *i.e.* accompanied by their consequences. This is particularly difficult because of the uncertainty stemming from the behavior of people or events external to the model. If we are working with probabilized uncertainty we can evaluate the alternatives through expectancy. One of the criteria might be, for example, gain expectancy. But since expectancy is an average quantity, it is not a very good indicator of risk and one might therefore consider a second criterion of the variance type. This in selecting a portfolio, portfolio a, with a lower gain expectancy than portfolio b, will be preferred by a prudent decision maker because the variance of a is much lower than that of b. According to more recent models (Jaffray, 1988, Cohen, 1992), it should be possible to introduce the attribute of certainty indicating, for each alternative, what is certain to be obtained. Several suggestions can also be found in Colson and Zeleny (1980). Thus *multicriterion decision is capable of taking into account various components of risk and not merely an average as in monocriterion decision.*

To summarize, we have seen that modeling criteria is a fundamental operation in multicriterion decision. In carrying out the modeling we must remain as close as possible to the decision maker's preoccupations, while avoiding the naive and the incoherent. Modeling is subject to two limits: on the one hand, clarity and acceptability for the decision maker and on the other hand, the difficulty of collecting information on the alternatives. Because of the latter point, the processes of constructing the alternatives and the criteria cannot be considered as separate.

11.4 From aggregation to choice

There is no obligation in multicriterion decision to end up with an aggregation. This point has been repeatedly stressed. Our hope is that after reading the previous sections the reader will appreciate the richness of the multicriterion paradigm and the richness of thought which is necessary to achieve good modeling of the alternatives and criteria. The combination of alternatives and criteria, plus the evaluation and thence the decision matrix, constitute a mine of information for arriving at a decision. If there is a small number of alternatives and fewer than six or seven criteria, the decision matrix becomes an aid which a trained decision maker can use without the necessity of introducing any method. As we have already stated, contamination by the paradigm of the optimal solution has caused many multicriterion methods to be attached more to what happens once the decision matrix has been obtained. *What happens beforehand is at least as important.* This is why, in multicriterion DSS, we advise a modular construction separating what happens before obtaining the decision matrix from what happens afterwards, so that the DSS can function with any multicriterion method and not only a purpose method.

Once the aggregation stage is reached, we consider that, except in the case of ranking, it is a mistake to head straight for aggregation. Instead, interactive methods should be favored, where the decision maker can explore, think and learn. *Whether it is interactive or not, aggregation is meaningless without at least a modicum of sensitivity analysis. the reader will no doubt have noticed the amount of arbitrariness in the determination of cardinal utilities and weightings, but this should not be cause to reject multicriterion decision; on the contrary, it should be seen as a source of richness provided it is well understood and accompanied by sensitivity analyses necessary to judge the robustness of the most interesting solutions.* Within the multicriterion paradigm, choosing alternative *a* because it happens to come out marginally better than *b* through some abstruse aggregation function, makes no sense. Then, either the choice must be left to the decision maker, or other criteria must be considered or else the evaluation must be reviewed. What usually happens is a new bout of reflection, after which it may well be decided that both *a* and *b* have their advantages and disadvantages, that there is nothing to choose between them and that the choice is irrevocably a political one, in the general meaning of the term, *i.e.* a set of moral and cultural values which cannot easily be reduced to attributes. In fact it would be true to say that *the choice between two (nearly) Pareto-optimal alternatives is always a political one, with the role of the multicriterion paradigm being to reveal the choices which are generally hidden beneath the technicalities of maximizing some function representing the choice of a few individuals* rather than the public interest or the higher interests of the organization concerned as we are sometimes led to believe.

Provided we do not invest more than reasonable faith in the aggregation, multicriterion decision appears to correspond quite well to Keen's apprehensive man (see section 11.1). It clearly exposes the conflicts between criteria without preventing the decision maker from acting, often in a strongly intuitive way. This is why, in highly complex situations of uncertainty, it is always preferable to give the decision maker a selection of alternatives, each one good in its own way, without going as far as final aggregation; this would deprive the decision maker of his 'apprehensive man' reflex. In any case, if given a final aggregation, the decision maker would not believe in it. DSS like GAIA, which enable the user to visualize the regrouping of alternatives which are satisfying for various criteria, are a real help. It is true that statistical classification methods based on data analysis can also be useful. In a general way, once we have a numerical decision matrix, we can use any method of data analysis to show up sub-groups of alternatives. Aggregation could then consist of looking for the 'best alternative' in each sub-group.

Even if the decision is in the end apprehensive, in multicriterion analysis everything which has happened up to the moment of choice will have been, for the decision maker, of great educative value. This is especially so in the case of negotiation. Here, multicriterion analysis is *first and foremost a framework for dialogue*. Discussion on the alternatives: does everybody agree on what alternatives are possible, and on their consequences? In general, no. Here too computer aid can be used to design the alternatives and develop the scenarios. Discussions can also lead to the construction of new alternatives: blends of, or compromises between, the first-proposed alternatives.

There may also be argument on the criteria. Here the situation is relatively simple since in the first stage, the multicriterion methodology accepts all possible criteria. When agreeing a set of criteria (see section 11.3.4), the objectives of each participant are discussed, a very enriching process. On reaching the stage of evaluating the criteria, the strengths and weaknesses of the various arguments come to light. The decision matrix will reflect all these discussions. It then remains to agree on the alternative or alternatives to choose. The discussion can then be on the method of aggregation, which will usually translate into discussion on the weights or on the compensations. Finally there is the 'give and take', inevitable in any compromise. The most important thing is that the final discussion be soundly based on foundations understood and accepted by all parties. If necessary, discussion can turn back to the choice of alternatives, because after initial discussions *it is often possible to define new alternatives* which better match the attributes that the various parties consider primordial. Negotiation and diplomacy need patience and imagination, and it is the latter that is involved in both the construction of new alternatives and the timing of that process. Imagination can also be used in the domain of procedures and here too multicriterion decision offers an impressively wide range of possibilities.

Multicriterion decision aid gives organizations a framework for structuring complex decisions carrying strong implications for the future. We must stress that interactive methods involve both simulation and a learning process; the decision maker who uses these methods is forced to think about the various factors in the decision, and this can lead to learning. While explicit learning is not yet an option that is widely available in multicriterion DSS, the very way the applications are meant to be used means that a learning process is more or less implicitly present. What is really lacking in DSS is the ability to go back on a decision and to carry out *a posteriori* analysis. This type of learning with feedback is now a feature of some available DSS, but to gain ground it will need the encouragement of decision makers. We have to admit that as far as decisions (especially strategic decisions) go, learning is not (yet?) a major priority for real live decision makers or their organizations. This leads us to ask the question: how do users react to the paradigm of multicriterion decision?

11.5 Reactions of decision makers to multicriterion decision

Multicriterion decision is a technique that is still not widely used by the decision makers themselves. Those who do use it are mainly concerned with preparing cases, and their role is limited to making proposals. They have a consultancy role and it is often they who are involved in constructing the model (alternatives and criteria). Their understanding of the multicriterion decision paradigm is generally good, and they very often advocate its use.

This then is our first observation: that the choice of multicriterion modeling almost always comes from the consultant rather than the decision maker or sponsor. Because of the culture and domination of the 'optimizer' paradigm which was mentioned in the first section, the multicriterion paradigm and methods are not the first things to spring to the mind of executives. To them, multicriterion methodology has its defects: it not only requires analysis of more alternatives than the 'assumed optimal decision', but this analysis must be detailed, looking at the alternatives from several points of view. This is likely to increase costs and has a destabilizing effect. We shall be returning to the latter point, which is perhaps the main defect in multicriterion decision. Finally, there are the two weaknesses of simplicity and of complication.

Simplicity because if we stop at the stage of the decision matrix, multicriterion methodology becomes more a matter of modeling than of mathematics. Modeling is actually not simple, but depends only marginally on formalized methods, at least in our present point of progress. Thus it is still dependent on the skill and experience of the modeler, and since it does not come accompanied by any complicated mathematics, multicriterion decision aid suffers from a lack of credibility. At the aggregation stage, however, multicriterion methods can largely make up for any hitherto absent mathematical content, and some of them have a ball! They are too complicated and too hermetic. A decision maker who may perfectly well accept the maximizing of some complicated function whose meaning is quite beyond him will insist on understanding how aggregation is performed as soon as he realizes that the various Pareto optima are not equivalent. Since he will not in general have the time to learn, he will reject the method.

The reality is that most decision makers are uncomfortable with multicriterion decision, and this is the main drawback. The authors first realized this after reading a paper by Kottemann and Davis (1991). It has long been known that decision makers have no confidence in complicated multicriterion methods (see Wallenius (1975), Brockhoff (1985), Buchanan and Daellenbach (1987) and Narasimhan and Vickery (1988)). What decision makers prefer is methods that are as close as possible to the trial and error 'method'. Kottemann and Davis hypothesize that decision makers have no confidence in methods that they name 'compensatory' *i.e.* methods which require the decision maker to arbitrate and compensate between criteria. According to these authors, the more a method shows up conflicts between criteria, the more the decision maker's confidence is undermined.

In fact, we feel that several things must be distinguished. Firstly, questions of the compensatory type (*e.g.* 'How much utility on criterion i would you be prepared to sacrifice in order to gain one unit on criterion *j*?') are indeed complicated, tedious and difficult to answer. The decision maker can be forgiven for worrying about a result that depends on answers he is not sure of. This is legitimate anxiety. This mistrust diminishes in methods with direct questions such as the request for the reference point. Thus the decision maker's confidence and the credibility of the method increase for procedures of the 'evolving focal point' and 'level of satisfaction' type, which, in the eyes of the decision maker, are close to simple trial and error. This type of method is generally well accepted. The first aspect of the problem of confidence, then, is confidence in methods that depend on large numbers of questions on local arbitration between criteria. These are refused by decision makers as not being worthy of confidence; this is quite understandable because it is clear that the responses are hardly more than vague approximations. In any case, the tediousness of these methods is often reason enough for them to be rejected.

But contrary to what Kottemann and Davis suggest, it is not the compensation, which is to blame here, but the local tradeoff which is not credible. Global compensation – in the form of weights, for example – is not rejected. We have already had the occasion to observe that it is easy to get the decision maker to provide weights, unaware as he is of the problems that then arise in cardinality and/or measuring scales. The weighted sum is among those methods, which do inspire confidence as the reader will have noted in chapter 10.

What is fundamentally in question – as Kottemann and Davis (1991) state – is the place in the method of the conflict between criteria, *i.e.* how obvious it is that you can't have your cake and eat it. Kottemann and Davis observe – and this is their starting thesis – *that the more a method requires direct arbitration between criteria, and hence shows the decision maker that there is conflict between equally desirable but more or less incompatible ends, the more reluctant is the decision maker to accept the method.*

We would go further than this, because it suggests to us a reason for the lack of success of the multicriterion decision paradigm with company executives and beyond this, poses certain problems on the role of decision aid. The data cannot be interpreted without involving psychology and desire. It is obvious that, faced with more or less contradictory criteria, the individual is aware that there is a conflict. Now, conflict is a stress factor that can cause discomfort or even actual neurosis if the individual does not manage to resolve it or dissimulate it in various ways. One thing that is certain is that the multicriterion paradigm brings conflicts to the fore rather than hiding them. For the decision maker who is not properly prepared, this is a great cause of stress and psychological pressure which he will not necessarily be able to stand, especially if he is personally highly involved in the decision. We can thus understand just why the optimizer paradigm appears much more preferable, and all the more so since the decision maker is easily able to construct a 'rational' psychological model that will persuade him that the 'apprehensive' decision he desires and intends taking really is the best one from that model. And the decision maker retrieves his mental equilibrium. If justification is deemed necessary, the psychological model can be rationalized into a mathematical model leading to the right decision; this is not a particularly difficult task. Our interpretation also explains why the people who simply have to prepare the data and who are not personally involved in the decisions are much more ready than the decision makers themselves to enter into the multicriterion paradigm; they are even openly enthusiastic, the model seeming clearer and more realistic to them in their position safely outside the decision arena. This brings us back to the question: do decision makers like having to arbitrate between criteria? The answer unfortunately is that nobody likes arbitrating. Look at how the mere choice of an automobile or washing machine can cause upset to many people, how relieved they are once they have ordered the goods, and how they post-justify their decisions by rationalization that, to an outside observer, can seem bizarre; you will then understand why the multicriterion paradigm is not spontaneously accepted by large numbers of decision makers to settle questions in which they are highly implicated on a psychological level. For many managers, the decisions are often strategic ones where their personal responsibility is indeed engaged.

A tendency to avoid conflicts has been noted in other domains, and it may well be that this reluctance to enter into conflicts of interest is one manifestation of the avoidance of confrontation which seems to be inherent in human behavior, and which was shown by Crozier (1963) to be present in bureaucratic organizations. Another aspect of this tendency to minimize conflicts appears in our field; it has long been observed that *unaided decision makers tend to limit their investigations to a very small number of alternatives*, to such a point that, in the literature on DSS, widening the search for alternatives is considered to be one of the main and most desirable functions, and we would agree with this (see Jarke and Rademacher, 1988, Lévine-Pomerol, 1989). The increase in the number of alternatives examined by the aided decision maker is one of the measures of the quality of an DSS (*cf.* Sharda *et al.,* 1988). The empirical results obtained in this domain are not very easy to interpret, but it does appear that decision makers only moderately appreciate this enlargement facility, even though it carries with it a better quality of decision.

Two studies, Cats-Baril and Huber (1987) and Abualsamh *et al.* (1990), show that the increase in the number of possible alternatives causes a drop in confidence in the decision maker. Everything is as though increasing the number of alternatives created unease in the decision maker by inexorably confronting him with a much more complex situation with decision made much more difficult than he had spontaneously envisaged. Paradoxically, the study by Cats-Baril and Huber show both an improvement in the performance of decision makers aided by the increase in number and quality of the alternatives produced (career plans in this case), and a lowering of confidence in comparison to unaided decision makers. We think this loss of confidence is directly linked to the increased number of alternatives examined and the correlated increase in uncertainty. Abualsamh *et al.* (1990) confirm this increase in the feeling of uncertainty which also seems to be related to the time spent applying a method of heuristic searching. They also show how the enlargement of the choice set and the correlated exploration time have negative effects on the decision maker's satisfaction. On the other hand, this decision maker 'destabilization' is considered by some experts in decision science to be a positive aspect (Cats-Baril and Huber, 1987, p. 368). Opening the decision maker's eyes and making him less naive is part and parcel of decision aid; but there is a price to pay: *the less limited decision maker is less sure of himself.*

Another interesting theory put forward by Abualsamh *et al.* is that aided decision makers dislike the reduction in ambiguity resulting from the modeling of the problem. This is an example of the effects of introducing undesired, and sometimes undesirable, objectivity – a phenomenon commonly met in the design of expert systems (Pomerol, 1990). Here – though the theory deserves to be studied in more depth – the decision maker challenges the reduction in his freedom, essentially his freedom to change his mind. An interesting result found by Abualsamh *et al.* is that the type of heuristics used (top-down or bottom-up) does not really affect the performance of the decision maker. Extrapolating this finding to our domain, it means that the method of aggregation chosen is finally of little importance provided that there is a method and that the decision maker devotes time to formalizing his decision. This theory is borne out by our experience that *studies of modeling and decision (alternatives + criteria + utilities) constitute the greatest benefits of multicriterion analysis, above aggregation and choice.*

The above results provide some food for thought; decision aid does increase complexity, putting the decision maker into a psychologically uneasy state. Decision aid doubtlessly still has some progress to make, since any thinking about, or study of, decision is likely to increase the psychological choice load and to plunge the decision maker into a state of indecision! Decision aid itself is guilty of the sin of ignorance in this domain.

Conflict will always be psychologically difficult to handle with for anybody who is not prepared or trained. There are main two ways of dealing with it: dissimulate it or talk about it. Optimization comes from the first of these ways, multicriterion decision the second, and apparently it is less spontaneously resorted to than the first. Therefore what is needed to persuade decision makers is more preparation and explanation. *It is the job of the multicriterion decision analysts and modelers to help decision makers to overcome their doubts by showing them the advantages of resolving conflicts rather than dissimulating them.* We should not delude ourselves, a propensity to prefer dissimulation is probably a human psychological trait, and the multicriterion paradigm is an attempt to achieve higher rationalization than optimization. As such, it is demanding and more fragile. As we said at the beginning of this section, multicriterion decision is destabilizing, and in many situations, upsetting. People are so naturally ready to think that they have made the best possible decision that anything which undermines their child-like confidence is unwelcome. The best decision relative to one criterion is not generally the best relative to another, equally desirable, criterion; this is annoying, upsetting and in some cases highly disagreeable, but that's life. Happy are they who make their decisions away from this undeniable reality!

11.6 Applications

In spite of all the obstacles we have seen, multicriterion decision has a considerable number of applications to its credit. Because of these obstacles, many of the applications are in the field of the *preparation of public or semi-public dossiers*. In this type of application, multicriterion decision is used in preparing a decision, which will be used elsewhere and differently. The individual who is preparing the dossier is not directly involved in the decision, which will be political in nature, and will not show the reluctance we talked about in the previous section. The introduction of several criteria shows that all points of view have been taken into consideration, within a framework of negotiation involving various pressure groups.

The second domain of application is resource management. These applications are generally at the technical and operational level, where there is not the emotional content of decisions involving the decision maker personally. The more technical and operational the decision, the easier it is to introduce multicriterion analysis; aggregation also becomes more acceptable. In some technical jobs, automatic multicriterion decision can be considered. It is symptomatic that among the multicriterion DSSs listed by Eom (1989), only 30% were aimed at the strategic level, the rest being designed for routine management operations. To show just how wide the field is, we give below a list of some of the applications classified into various fields. The list is not exhaustive but does show fields in which multicriterion decision is most firmly established. In an appendix to this chapter the reader will find a list of the main publications with significant space given over to applications.

11.6.1 Public investment

In this section is everything concerning choice of locality and construction of public works (roads, subways, airports, hospitals etc.).

Subway line extension: Roy and Huggonard (1982), Roy, Présent and Silhol (1986).
Airport locality: Keeney and Raiffa (1976), Martel and Aouni (1991).
Locality of nuclear and other waste storage sites: Briggs *et al.* (1990), Moskowitz *et al.* (1991), Erkut and Moran (1991), Caruso *et al.* (1993), Perny and Vanderpooten (1998), Rogers *et al.* (2000).
Locality of nuclear and conventional electric power stations: Hobbs (1979), Lugassi *et al.* (1985), Brans and Vincke (1985), Wenstop and Carlsen (1992), Keeney and McDaniels (1992), Rietveld and Ouwersloot (1992).
Choice of freeway, railroad and public transport routes: Cook and Seiford (1984), Khorramshahgol and Moustakis (1988), Pearman *et al.* (1989), Siskos and

Assimakopoulos (1989), Won (1990), Montero *et al.* (1995), Barba-Romero and Mokotoff (1997).
Choice of electric power line routes and network extensions: Massam and Skelton (1986), Bayad and Pomerol (1992).
Choice of research and development projects: Ramanujan and Saaty (1981), Mehrez *et al.* (1982), Feinberg and Smith (1989), Danila (1990), Stewart (1991), Barba-Romero (1994).
Choice of mining land redevelopment projects: Goicoechea *et al.* (1982), Kaden (1984).

11.6.2 Resource allocation and management

Land management and water resource management: Keeney and Wood (1977), Gershon *et al.* (1980), Dauer and Kruegger (1980), Pfaff and Duckstein (1981), Gershon and Duckstein (1983b), Mehrez and Sinuany-Stern (1983), Guariso *et al.* (1985), Ridgley (1989), Barba-Romero and Pérez (1994,1997), Rogers *et al.* (2000).
Forest resource management: Bell (1977), Steuer and Schuler (1978), Duckstein and Gershon (1981), Hallefjord *et al.* (1986), Bare and Mendoza (1988), Davis and Liu (1991), Stam *et al.* (1992).
Human resource management and recruitment: Moscarola (1978), Passy and Levanon (1980).
Production management: Abulbhan and Tabucanon (1979), Wu and Tabucanon (1990).
Public borrowing management: Labat and Futtersack (1992).
Loan allocations : Mareschal and Brans (1991), Bohanec *et al.* (1998).
Choice of computer equipment: Beck and Lin (1981), Fichefet (1985), Brooks and Kirkwood (1988), Tosthzar (1988), Barba-Romero (1994), Paschetta and Tsoukias (1999).

11.6.3 Strategic decision

Choice of a legislative law: Lévine and Pomerol (1989, p.249 *et seq.*).
Choice of commercial distributors: Siskos (1986).

11.6.4 List of works devoted partly or wholly to applications

BANA E COSTA, C.A. (Ed.), 1990, *Readings in multiple criteria decision making*, Springer.
BELL, D.E., KEENEY, R.L. and RAIFFA, H. (Eds.), 1977, *Conflicting objectives in decisions*, Wiley.
CLIMACO J. Ed., 1997, *Multicriteria Analysis*, Proceedings of XIth International Conference on MCDM, Coimbra, Springer.
COCHRANE, J.L. and ZELENY, M. (Eds.), 1973, *Multiple criteria decision making*, Proceedings South Carolina (1972), University of South Carolina Press, Columbia.
COHON, J.L., 1978, *Multiobjective programming and planning*, Academic Press.
COLSON, G. and De BRUIN, C. (Eds.), 1989, *Models and methods in multiple criteria decision making*, Pergamon Press.
DESPONTIN, M., NIJKAMP, P. and SPRONK, J. (Eds.), 1984, *Macroeconomic planning with conflicting goals*, Proceedings Brussels(1982), Springer.

FANDEL, G. and GAL, T. (Eds.), 1980, *Multiple criteria decision making theory and applications*, Springer.

FANDEL, G. and SPRONK, J. (Eds.), 1985, *Multiple criteria decision methods and applications* (Selected readings of the First International Summer School), Springer.

FANDEL, G., GRAUER, M., KURZHANSKI, A. and WIERZBICKI, A.P. (Eds.), 1986, *Large-scale modelling and interactive decision analysis*, Springer.

FRENCH, S., HARTLEY, R., THOMAS, L.C. and WHITE, D.J. (Eds.), 1983, *Multi-objective decision making*, Academic-Press.

GOICOECHEA, A., HANSEN, D.R. and DUCKSTEIN, L., 1982, *Multiobjective decision analysis with engineering and business applications*, Wiley.

GRAUER, M. and WIERZBICKI, A. P. (Eds.), 1984, *Interactive decision analysis* (Proceedings, Laxenburg, Austria, 1983), Springer.

HAIMES, Y.Y. and CHANKONG, V. (Eds.), 1985, *Decision making with multiple objectives*, Springer.

HANSEN, P. (Ed.), 1983, *Essays and surveys on multiple criteria decision making*, Springer.

HWANG, C. and MASUD, A.S.M., 1979, *Multiple objective decision making. Methods and applications survey*, Springer.

HWANG, C. and YOON, K., 1981, *Multiple attribute decision making. Methods and applications survey*, Springer.

JACQUET-LAGREZE, E. and SISKOS, J., 1983, *Méthodes de décision multicritère*, Monographies de L'AFCET, France.

KARPAK, B. and ZIONTS, S. (Eds.), 1989, *Multiple criteria decision making and risk analysis using computers*, Springer.

KEENEY, R.L. and RAIFFA, H., 1976, *Decisions with multiple objectives: preferences and value tradeoffs*, Wiley.

KORHONEN, P., LEWANDOWSKI, A. and WALLENIUS, J. (Eds.), 1991, *Multiple criteria decision support*, Springer.

LARICHEV O. I. and MOSHKOVICH H. M., 1997, *Verbal Decision Analysis for Unstructured problems*, Kluwer.

LOCKETT, A.G. and ISLEI, G., (Eds.), 1989, *Improving decision making in organizations*, Proceedings of the Eight International Conference on MCDM, Manchester, Springer.

MAYSTRE L.Y., PICTET J., SIMOS J., 1994, *Méthodes Multicritères ELECTRE*, Presses Polytechniques et Universitaires Romandes, Lausanne.

MAYSTRE L.Y., BOLLINGER D., 1999, *Aide à la negociation Multicritère: pratique et conseils*, Presses Polytechniques et Universitaires Romandes, Lausanne.

MORSE, J.N. (Ed.), 1981, *Organizations: multiple agents with multiple criteria*, Springer.

NIJKAMP, P. and van DELFT, A., 1977, *Multicriteria analysis and regional planning*, Martinus Nijhoff, The Netherlands.

NIJKAMP, P. and SPRONK, J., 1981, *Multiple criteria analysis: operational methods*, Gower Press, London.

OLSON D. L., 1996, *Decision Aids for Selection Problems*, Springer, New-York.

OSYCZKA, A., 1984, *Multicriterion optimization in engineering with Fortran programs*, Ellis Horwood, London.

RIETVELD, P., 1980, *Multiple objective decision methods and regional planning*, North Holland, Amsterdam.

ROY, B., 1985, *Méthodologie multicritère d'aide a la décision*, Economica, Paris.

ROY B., BOUYSSOU D., 1993a, *Aide multicritère à la décision: méthodes et cas*, Economica, Paris.

SAATY, T.L., 1980, *The analytic hierarchy process*, McGrawHill.

SAATY, T.L. and VARGAS, L.G., 1982, *The logic of priorities*, Kluwer-Nijhoff, Boston.

SAWARAGI, Y., INOUE, K. and NAKAYAMA, H. (Eds.), 1987, *Toward interactive and intelligent decision support systems*, Springer.

SCHÄRLIG A., 1985, *Décider sur plusieurs critères*, Presses Polytechniques et Universitaires Romandes, Lausanne.

SCHÄRLIG A., 1996, *Pratiquer* ELECTRE *et* PROMETHEE, Presses Polytechniques et Universitaires Romandes, Lausanne.

SEO, F. and SAKAWA, M., 1988, *Multiple criteria decision analysis in regional planning: concepts, methods and applications*, Reidel.

SISKOS, J., WASCHER, G. and WINKELS, H-M., 1983, A bibliography on outranking approaches, Cahier du LAMSADE #45, Université de Paris-Dauphine.

SPRONK, J., 1981, *Interactive multiple goal programming: applications to financial planning*, Martinus Nijhoff, Boston.

STADLER, W. (Ed.), 1988, *Multicriteria optimization in engineering and in the sciences*, Plenum Press.

STARR, M. K. and ZELENY, M. (Eds.), 1977, *Multiple criteria decision making*, North-Holland, Amsterdam.

STEUER, R.E., 1986, *Multiple criteria optimization: theory, computation and application*, Wiley.

SZIDAROVSZKY, F., GERSHON, M.E. et DUCKSTEIN, L., 1986, *Techniques for multiobjective decision making in systems management*, Elsevier.

TABUCANON, M.T., 1988, *Multiple criteria decision making in industry*, Elsevier.

TABUCANON, M.T. and CHANKONG, V. (Eds.), 1989, *Multiple criteria decision making: applications in industry and service*, Asian Institute of Technology, Bangkok.

THIRIEZ, H. and ZIONTS, S. (Eds.), 1976, *Multiple criteria decision making*, Proceedings Jouy-en-Josas 1975, Springer.

ZAHEDI, F., 1986, "The analytic hierarchy process. A survey of the method and its applications," *Interfaces*, vol.16, pp.96-108.

ZIONTS, S. (Ed.), 1978, *Multiple criteria problem solving*, Springer

11.7 Conclusion

The modest sum total of actual applications of multicriterion decision belies the strength of the paradigm. Multicriterion modeling is clearer, more attached to the truth and to the real world than monocriterion modeling and optimization. Multicriterion decision takes account of complexity of choice and multiplicity of options and is consequently disturbing and often does disturb for reasons that are sometimes operational and nearly always personal (see 11.5). This explains why, up to now, it has been widely ignored by 'apprehensive man', even though multiplicity of criteria is inherent in human behavior. Everything is as though the multicriterion paradigm represented a leap in rationality in the efforts of man to improve the quality of decisions – and of the processes leading up to those decisions – in an organization. Multicriterion decision has not yet been around for long enough to be able to say that it gives on average a better return on investment than 'optimizing' decision, particularly since there are technical domains where optimization is justified and where 'all things being equal' type measures are difficult to implement in reality. What we can reaffirm, however, is that from an organizational rather than an individual point of view, multicriterion decision is almost always perceived as better, and hence more effective, since it responds to all the hopes and aspirations of the individuals concerned.

12 MULTICRITERION METHODS: FEATURES AND COMPARISONS

12.1 Introduction

Multicriterion decision is a vast area, as the reader will have appreciated by now. There are the fundamental theoretical aspects involving discrete mathematics, such as Arrow's theorem and the conditions for additivity, and then the pragmatic side where we have to recognize that reality is always a question of multiple criteria, and practical solutions are needed before theorems. As Keen (1977) pointed out, "between the constraints and the criteria, there are no solutions"; yet the decision maker (*i.e.* each and every one of us) somehow has to muddle along.

Arrow's theorem proves that an aggregation having all the right properties is impossible to find; but what exactly are the properties that a multicriterion aggregation procedure must satisfy? This is what we shall be looking into in the first part of this chapter, and we shall be doing this by examining the desirable properties of choice functions in axiomatic form. In the second part of the chapter, we shall, from a more practical point of view, be setting out the strengths and weaknesses of multicriterion aggregation procedures.

12.2 A theoretical framework for analyzing the desirable properties of choice functions

12.2.1 Justification and goal

The aim of this section is to define a theoretical framework that will enable us to compare multicriterion aggregation methods. This framework actually exists and is sufficiently well developed provided we restrict ourselves for the moment to ordinal methods (chapter 5); as we have seen, ordinal methods are essentially analogous to methods for voting and group choice, on which there is a vast range of literature with numerous results.

Much of what follows is based on the work of Pérez (1994), whose stated goal was to "provide the elements necessary to carry out a systematic analysis of ordinal aggregation and a comparison of its rules. We have provided a framework for this and have endeavored to structure the properties of internal coherence, external coherence, monotonicity and Pareto optimality." Many of these properties (of which we give precise definitions below) are well known in the context of group decision; their use in multicriterion analysis enables us to judge the quality of aggregation methods. This information is vital to the decision maker when choosing a suitable method for his particular problem.

When choosing from his choice set $\mathcal{A}$, the decision maker faces two main problems: selection and ordering (section 2.6.2). If for the moment we restrict ourselves to the ordinal case, the decision maker will want to make his choice with only the knowledge of the ordinal preferences on each criterion. Questions relating to strategic behavior, though highly important in voting, are irrelevant here since the preferences within each criterion are assumed to have been expressed previously. From ordinal information alone, the decision maker would like to be able to characterize aggregation functions and study their properties. He can do this within the formal framework that we shall now introduce.

12.2.2 Terminology and concepts

Let us remind (definitions 2.7 and 2.14) some notation. An order $\succ$ is a complete, irreflexive therefore asymmetric and transitive relation, and a weak order $\succcurlyeq$ means a complete, reflexive and transitive relation.

Let $C = \{C_1, C_2,...,C_n\}$ be the set of criteria, and $\mathcal{A} = \{a_1, a_2,..,a_m\}$ is the set of alternatives, where m is any integer greater than or equal to 3. Given any subset of alternatives $\mathcal{B}$ of $\mathcal{A}$, with two or more elements, any matrix B expressing a set of preference orderings over $\mathcal{B}$ will be called the decision matrix over $\mathcal{B}$. (We can imagine that, as usual $B = (b_{ij})$ where $b_{ij} = U_j(a_i)$ and U_j is an utility function, although only the preorders $\succcurlyeq_j$ of each criterion j are considered in the two next paragraphs.) Decision matrices over the overall set $\mathcal{A}$ will be denoted A.

The objects to be compared are those aggregation rules f, called *order rules*, which assign a nonempty set of orders $\succ_s$ ($s = 1, 2, ...,q$) on $\mathcal{A}$, denoted $f(A)$, to every decision matrix A over $\mathcal{A}$. Because every weak order $\succcurlyeq_s$ can be assimilated to a set of orders (all those that are compatible with the asymmetric part ($a \succcurlyeq_s b$ and not($b \succcurlyeq_s a$) of that weak order), the above definition of an order rule can be considered wider than that given in sub-section 5.4 of a *social preorder* (function f from $\mathcal{PR}^n = \mathcal{PR} \times \mathcal{PR} \times \ ... \times \mathcal{PR}$ in $\mathcal{PR}$, where the image is a unique preorder of $\mathcal{PR}$, the set of all preorders on $\mathcal{A}$.)

For a given decision matrix A over $\mathcal{A}$, $A/\mathcal{B}$ means the restriction of A to the alternatives in $\mathcal{B}$ in the natural way, that is to say for every criterion, the order on $\mathcal{B}$ will be the result of deleting the alternatives not included in $\mathcal{B}$ from the corresponding order on $\mathcal{A}$. So $A/\mathcal{B}$ is a decision matrix over $\mathcal{B}$ and each of its orders is the restriction to $\mathcal{B}$ of the corresponding order on $\mathcal{A}$. Therefore, $f(A/\mathcal{B})$ is a set of orders on $\mathcal{B}$ determined by the aggregation rule f and the decision matrix $A/\mathcal{B}$. Likewise, $f(A)/\mathcal{B}$ will denote the set of orders on $\mathcal{B}$, each of them being the restriction to $\mathcal{B}$ of one order of $f(A)$.

Attached to every order rule f, a *choice rule* f^* will be defined as follows:

$$f^*(A,\mathcal{B}) = \{ \ b \in \mathcal{B} \ \text{such that } b \text{ is ranked first in some order of } f(A/\mathcal{B}) \ \}.$$

The function f^* says that b is a possible good choice in $\mathcal{B}$ resulting from the aggregation f.

12.2.3 Definition of the rules studied

In order to illustrate the possibilities of this analysis methodology, four ordinal multicriteria aggregation rules are thoroughly tested in Pérez (1994). Formal definitions follow:

Let B be a decision matrix over $\mathcal{B} = \{b_1, b_2, ..., b_p\}$, subset of $\mathcal{A}$, and $h_1, h_2, ..., h_n$ orders or criteria on $\mathcal{B}$. The $p \times p$ matrices $P_B = (p_{ij})$ and $R_B = (r_{ij})$ are called *pairwise preference matrix* and *given rank matrix*, respectively, and are defined as follows:

p_{ij} = number of criteria in which x_i is strictly preferred to x_j
r_{ij} = number of criteria in which x_i is ranked in position j

When we consider orders, the pairwise preference matrix is similar to the pairwise number of votes matrix defined for preorders (notation 5.6), in the framework of outranking it is also the concordance matrix with equal weights.

Definition 12.1 Blin's linear assignment rule (f_{BL})

An order $h \in f_{BL}(B)$ iff $\Sigma_{i,j=1..p} \operatorname{per}(h)_{ij} r_{ij} \geq \Sigma_{i,j=1..p} \operatorname{per}(h')_{ij} r_{ij}$ for every order h' on $\mathcal{B}$, where $R_B = (r_{ij})$ is the given rank matrix of B, and $\operatorname{per}(h)$ is the permutation matrix of any order h, so that: $\operatorname{per}(h)_{ij} = 1$ if b_i is ranked j in h and 0 otherwise

In intuitive terms, $h \in f_{BL}(B)$ means that reordering the rows of $R_B = (r_{ij})$ according to h maximizes the sum of the main diagonal elements, see Blin (1976).

Definition 12.2 Köhler's rule ($f_{K\ddot{o}}$)

An order $h \in f_{K\ddot{o}}(B)$ iff ($b_i \succ_h b_j$ implies $\operatorname{Min}_k \{p_{ik} / k \neq i$ and $b_i \succ_h b_k\} \geq \operatorname{Min}_k \{p_{jk} / k \neq j$ and $b_i \succ_h b_k\}$) where $P_B = (p_{ij})$ is the pairwise preference matrix.

In other words, an order h such that $b_1 \succ_h b_2 \succ_h ... \succ_h b_p$ belongs to $f_{K\ddot{o}}(B)$ when a row in (p_{ij}) whose minimal term, excluding the main diagonal, is maximal, corresponds to b_1; a row whose minimal term, excluding the main diagonal and the column of b_1, is maximal, corresponds to b_2; and so on. Actually, this is a maximin rule on matrix (p_{ij}) with sequential elimination of already chosen alternatives, see Arrow and Raynaud (1986).

Definition 12.3 Kemeny's rule (f_{KE})

An order $h \in f_{KE}(B)$ iff $\Sigma_{i,j=1..p} h_{ij} p_{ij} \geq \Sigma_{i,j=1..p} h'_{ij} p_{ij}$ for every order h' on $\mathcal{B}$, where $P_B = (p_{ij})$ is the pairwise preference matrix of B, and $h'_{ij} = 1$ if $b_i \succ' b_j$ and 0 otherwise.

That is to say, $h \in f_{KE}(B)$ means that re-ordering rows and columns of $P_B = (p_{ij})$ according to h maximizes the sum of terms of the upper triangular matrix, see Kemeny and Snell (1962) and Fishburn (1977). This rule is used as the basic mechanism of aggregation in the QUALIFLEX method, defined in Paelinck (1976).

Definition 12.4 Borda's rule (f_{BO})

An order $h \in f_{BO}(B)$ iff $b_i \succ_h b_j$ implies $\Sigma_{k=1..p} k \, r_{ik} \leq \Sigma_{k=1..p} k \, r_{jk}$, where $R_B = (r_{ij})$ is the given rank matrix of B.

In other words, h is an optimal order when it includes the asymmetrical part of the preorder determined by classical Borda ranking with Borda coefficients $1 < 2 < \ldots < p$ (sub-section 5.2.1). That is to say, when k alternatives are tied, $k!$ optimal orders are produced, one for each permutation of these alternatives. This is a straightforward adaptation of the Borda ranking (sub-section 5.2.1) as an order rule.

Example 12.5

Let B be the following decision matrix over $\mathcal{B} = \{b_1, b_2, b_3, b_4\}$

$$
\begin{aligned}
C_1 = C_2 = C_3 = C_4 : &\quad b_1 \succ b_4 \succ b_2 \succ b_3 \\
C_5 = C_6 = C_7 : &\quad b_2 \succ b_3 \succ b_4 \succ b_1 \\
C_8 = C_9 : &\quad b_3 \succ b_1 \succ b_4 \succ b_2 \\
C_{10} = C_{11} : &\quad b_3 \succ b_4 \succ b_2 \succ b_1 \\
C_{12} : &\quad b_4 \succ b_2 \succ b_1 \succ b_3 \\
C_{13} : &\quad b_2 \succ b_1 \succ b_3 \succ b_4 \\
C_{14} : &\quad b_3 \succ b_4 \succ b_1 \succ b_2 \\
C_{15} : &\quad b_4 \succ b_1 \succ b_2 \succ b_3
\end{aligned}
$$

The corresponding pairwise preference matrix P_B and given rank matrix R_B are:

$P_B =$	b_1	b_2	b_3	b_4
b_1		8	7	7
b_2	7		10	4
b_3	8	5		9
b_4	8	11	6	

$R_B =$	1°	2°	3°	4°
b_1	4	4	2	5
b_2	4	1	7	3
b_3	5	3	1	6
b_4	2	7	5	1

The optimal orders on $\mathcal{B}$ and the chosen alternatives on $\mathcal{B}$ are:

BLIN $f_{BL}(B) = \{b_1 \succ b_4 \succ b_2 \succ b_3, \; b_3 \succ b_4 \succ b_2 \succ b_1\}$; $f^*_{BL}(\mathcal{B}) = \{b_1, b_3\}$

KÖHLER $f_{K\ddot{O}}(B) = \{b_1 \succ b_4 \succ b_2 \succ b_3\}$; $f^*_{K\ddot{O}}(\mathcal{B}) = \{b_1\}$

KEMENY $f_{KE}(B) = \{b_4 \succ b_1 \succ b_2 \succ b_3, \; b_4 \succ b_2 \succ b_3 \succ b_1\}$; $f^*_{KE}(\mathcal{B}) = \{b_4\}$

BORDA $f_{BO}(B) = \{b_4 \succ b_1 \succ b_3 \succ b_2, \; b_4 \succ b_3 \succ b_1 \succ b_2\}$; $f^*_{BO}(\mathcal{B}) = \{b_4\}$

Although Pérez (1994) confines himself to these four rules for reasons of space and because they are widely used in a multicriterion context, there are other rules which could also be analyzed within this methodological framework. These include the Copeland rule and the Bowman and Colantoni rule, variants of the Condorcet rule (or majority rule), as well as the Cook and Seiford rule, proposed in Cook and Seiford (1978), which can be considered as an improvement of the Blin rule. See chapter 5 for the definition of the first two rules, and Pérez (1991) for an analysis of the third. Other choice rules analyzed in Fishburn (1977), such as Dogdson's, Schwartz's and Fishburn's rules could also be adapted and analyzed in this context.

12.2.4 Theoretical properties

There are many different theoretical properties which can be studied (Arrow and Raynaud, 1986; Vincke,1992). Three broad families of properties are analyzed in Pérez (1994): internal consistency, which means consistency in the ranking or the choice of alternatives when the set of alternatives is modified; external consistency, which means consistency in the ranking or choice of alternatives when groups of individual preferences – i.e. the corresponding decision matrices – are combined; and monotonicity, which means that a better ranking for one alternative with respect to another does not result into a worse rank in the social aggregation.

a) Internal consistency properties.

In Figure 12.1 some internal consistency properties are defined. *Let $\mathcal{A}$ be any set* (being fixed) of alternatives, A is any decision matrix on $\mathcal{A}$ and $\mathcal{B}$ is any subset of $\mathcal{A}$.

Internal consistency of orders (ICO)	$f(A) / \mathcal{B} = f(A / \mathcal{B})$ for every $\mathcal{B}$. *(Consistency of optimal orders when $\mathcal{B}$ is changed.)*
Contraction Consistency	Contraction Consistency
α	$b_i \in f^*(A, \mathcal{B})$ implies $b_i \in f^*(A, \mathcal{B}')$ for every subset $\mathcal{B}'$ of $\mathcal{B}$ including b_i. *(If $b_i \in \mathcal{B}' \subset \mathcal{B}$ is chosen in $\mathcal{B}$, it is chosen in $\mathcal{B}'$, consistency of choices for contractions of $\mathcal{B}$.)*
A5	$b_i \in f^*(A,\mathcal{B})$ implies $b_i \in f^*(A,\{b_i, b_j\})$ for some $b_j \in \mathcal{B} - \{b_i\}$. *(If $b_i \in \mathcal{B}$ is defeated by every alternative in $\mathcal{B} - \{b_i\}$, it is not chosen over $\mathcal{B}$.)*
Expansion Consistency	Expansion Consistency
ß⁺	If $b_i \in f^*(A,\mathcal{B})$ and $b_j \in \mathcal{B}' \subset \mathcal{B}$, then $b_j \in f^*(A,\mathcal{B})$ implies $b_i \in f^*(A,\mathcal{B})$. *(If b_i is chosen in $\mathcal{B}' \subset \mathcal{B}$ in the presence of b_j, b_j cannot be chosen in $\mathcal{B}$ if b_i is not also chosen, consistency of choices for expansions of $\mathcal{B}'$.)*
γ2	$\{ \mathcal{B} = \{b_1, b_2, ..., b_p\}$ and $b_i \in f^*(A,\{b_i,b_j\})$ for every $j=1, 2, ..., p \}$ implies $b_i \in f^*(A,\mathcal{B})$. *(If $b_i \in \mathcal{B}$ is not defeated by any alternative in $\mathcal{B}$, it is chosen in $\mathcal{B}$.)*
A weak property	A weak property
μ	$\exists\, b_i, b_j \in \mathcal{B}$ such that $(\forall b_k \in \mathcal{B} - \{b_i\},\ f^*(A,\{b_i, b_k\}) = \{b_i\})$ and $(\forall b_k \in Y - \{b_j\},\ f^*(A,\{b_j,b_k\}) = \{b_k\})$ implies $f^*(A,\mathcal{B}) \neq \{b_j\}$. *(If $b_i \in \mathcal{B}$ defeats every other in $\mathcal{B}$ and $b_j \in \mathcal{B}$ is defeated by every other in $\mathcal{B}$, b_j cannot be the only alternative chosen over $\mathcal{B}$)*

Figure 12.1 Definitions of some internal consistency properties

The logical relations between the properties are shown in Figure 12.2.

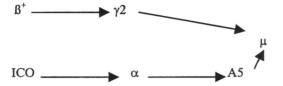

Figure 12.2 Relations among internal consistency properties

Pérez (1994) proves the implications showed in Figure 12.2, and also proves that the Blin rule does not satisfy any of these properties, the Köhler rule only satisfies $\gamma 2$ and μ, the Kemeny rule only satisfies $\gamma 2$, A5 and μ while the Borda rule only satisfies $A5$ and μ.

b) Monotonicity-Pareto Properties.

Monotonicity and Pareto properties can take several not-equivalent forms. The strongest monotonicity property, which implies the others, is the classical and well-known *strong monotonicity property*: if b is chosen on A and is improved in some criterion, everything else remaining the same, b will become the only one chosen by f or f'. The weakest property, implied by each of the others, is the equally classical and well-known *conservative Pareto property*: if b is preferred to c for any criterion and b is not chosen, c will not be chosen by f'. This property implies the unanimity axiom (sub-section 5.4). Pérez (1994) shown that: 1) only Borda's rule satisfies the strongest monotonicity property; 2) Kemeny's rule satisfies many weaker properties; 3) Köhler's rule satisfies only the weakest properties he examined and 4) Blin's rule does not satisfy any.

c) External consistency properties.

The classical *Young consistency property* for choice rules: if some alternative is chosen by a set of criteria C_1 and by C_2, those and only those alternatives chosen by both C_1 and C_2 are chosen when C_1 and C_2 are combined (*i.e.* the group of criteria $C_1 \cup C_2$ is considered) and the *Young and Levenglick consistency property* (1978) for order rules: if some order is optimal for a group of criteria C_1 and for C_2, those and only those orders optimal for C_1 and on C_2 are optimal when C_1 and C_2 are combined, can be analyzed, along with some weaker properties. Only Borda's rule satisfied the six properties considered by Pérez (1994), while the remaining rules have a similar performance.

Review of the results

Pérez' analysis of the above four ordinal aggregation methods tells us that:

    a) none of the methods satisfies all the weak properties of internal consistency,

    b) Blin's rule satisfies none of the weak properties of internal consistency, nor any of the properties of monotonicity-Pareto,

c) Kemeny's rule appears the most satisfactory as it respects all the properties of internal consistency and behaves quite well for the other properties; for these other properties, Borda's rule is the best. Köhler's method lies in an intermediate position when all properties are considered.

The methodological framework proposed and used by Pérez (1994) can be used either alone or combined with other methodologies to analyze and compare other ordinal methods (*Cf.* section 12.3). As a support for the decision maker in his choice of the right properties in multicriterion methods, it is both rigorous and promising. Along the same lines, but limited to one specific case (choice of team members for the U.S. Defense Department), there is the work of Lansdowe (1996), who analyzes methods other than those studied by Pérez. Extending this methodology for comparison to non-ordinal methods appears far from a simple matter, given the complexity that cardinality of data adds to the study of formal properties. This ground is open to research, and the first steps were taken by Vincke (1992, 1999).

Let us also mention the simpler formalism of choice function (*i.e.* a function which associates to each set of alternatives a subset of selected alternatives). Choice functions are thoroughly studied in Makarov *et al.* (1987). Another formalism consists of, given the profiles, associating a score to each alternative, this is extensively studied by Chebotarev and Shanis (1999).

12.3 Empirical comparison of practical properties

In this section we compare various multicriterion methods through the way they actually behave in terms of certain desirable properties. The method consists of using simulation techniques based on the random generation of problems carrying whatever features we wish to study. Some works of this type has already been carried out and more and more is being done.

Pérez and Barba-Romero (1995) carried out a study of this type as a complement to the method of section 12.2. They studied the behavior of the four methods analyzed by Pérez (1994) in relation to the properties that we now introduce.

-The *degree of internal inconsistency*, which is assessed by means of four indicators related to the consequences of eliminating one alternative: the percentage of instances when the ICO, α and β^+ properties (defined in Figure 12.1) fail, and the degree of perturbation ensuing for the resulting aggregation.

-The *degree of discriminability*, which is assessed by means of three indicators: the average number of alternatives chosen, the average number of optimal orders chosen, and the percentage of instances with a single alternative chosen.

-The *computational complexity of the algorithms* which is directly analyzed.

As a result of each of these analyses, the four methods are ordered for each aspect. Combining this with the theoretical criteria of comparison studied in Pérez (1994) and evoked in the previous sub-section, a decision matrix is formed in which the alternatives are the four methods. The resulting decision (data) matrix is analyzed by means of the four aggregation rules. A main result of the paper is that Borda's rule performs the best, followed closely by Kemeny's rule, and that Blin's rule performs very badly, while Köhler's rule occupy an intermediate position.

A similar though less systematic analysis is to be found in Triantaphyllou and Mann (1989), where four cardinal multicriterion methods are compared: weighted sum, weighted product, AHP and the revised version of AHP by Belton and Gear (1983). The comparison criterion is based on the measurement of the non-achievement of two types of property related to independence of irrelevant alternatives, *i.e.* related to the behavior of the choice function on removing or adding alternatives (internal consistency). In some cases, adding an alternative results in the inversion of the ranking of two other alternatives, this phenomenon is known as 'rank reversal'. In the final (multicriterion) comparison, the revised AHP method came out the strongest. In later work, Triantaphyllou and Mann (1994) extend the comparison between AHP and revised AHP through simulations similar to those we just mentioned. Also, a comparison between AHP and DEX is preformed by Bohanec *et al.* (1998).

Buede and Maxwell (1995) also use simulation of random decision matrices to study the discrepancies between the results obtained by the three ordinal multicriterion methods (AHP, TOPSIS, P%) that are subject to the 'rank reversal' phenomenon and compare them with MAUT, where this phenomenon is absent. For MAUT and AHP, they define a utility function with the same parameters (for TOPSIS and P%, there is no reason for such a definition) The results obtained lead the authors to conclude that certain features of the problem (the structure and shape of the criteria, evaluation of the weights) are more important than the choice of one or another method of aggregation).

A later article by Schoner *et al.* (1997) completes the above work by adding two variants of AHP to the methods considered by Buede and Maxwell (1995). Schoner *et al.* (1997) show that these variants give the same result as MAUT.

12.4 Comparison of multicriterion aggregation procedures: the various factors to be considered

We have shown in this chapter that there are various theoretical considerations to be taken into account together with more or less desirable properties that satisfy or should satisfy multicriterion aggregation methods. The properties will also depend on the goal: the requirements will differ according to the way the decision maker handles the problem (*cf.* 2.6.2). Selection of a subset of 'satisficing' alternatives does not involve the same questions as obtaining a complete preorder.

What we have also learned from the above sections is that from Arrow's theorem stems the fact that there is no such thing as a perfect aggregation and that it is impossible to get a complete preorder (and particularly transitivity) with independence of irrelevant alternatives. This is brought out in the opposition between additive or multiplicative methods which produce a complete preorder, and outranking methods which produce partial preorders, priority being given to the problem of selection.

A second opposition, and a very strong one from the point of view of the amount of information to be fed in and the reliability of this information, exists between cardinal and ordinal methods. We shall be reviewing the advantages and disadvantages of the various methods, or rather families of methods, with the same approach that we have used throughout the book: that of basing our judgement on scientifically indisputable properties and concepts. Only then shall we advance several rather more equivocal considerations by introducing the user's feelings. Our analysis then naturally falls into two parts; the next section (12.5) will be from the point of view of the user, and the following one (12.6) from that of the. multicriterion analysis specialist.

From the user's point of view, the ideas that are most frequently put forward (Buchanan and Daellenbach, 1987; Larichev *et al.*, 1993; Buchanan, 1994; Olson, 1996; Olson *et al.*, 1998) are on the ease of use, the decision maker's confidence in the method used and the understandability of the method. These will of course depend on the kind of user involved, and an experienced user will often have a different point of view than a novice. As Belton and Hodgkin (1999) note, the smartness of an interactive system support system ought to vary according to the type of user. The time, and especially, the effort, required of a decision maker for him to reach a conclusion can also be taken into account. Larichev and Nikiforov (1987) tried to formalize the latter idea as 'cognitive effort'. In addition to these classical criteria we could add the quality of exploration and how stimulating the method is. From an entirely constructivist point of view (*cf.* 9.3.1), we can judge how well a method is capable of making a decision maker understand the tradeoffs involved and telling him how much he has learned about the problem and about his real preferences. More than one author has observed – and we share this view – that "the learning process is as much value to the decision maker as the final choice" (Buchanan, 1987). This idea has gradually spread among designers of interactive methods and DSS (*e.g.* Iz and Gardiner, 1992; Belton and Hodgkin, 1999).

It would of course be nonsense to conclude from the above quotation (Buchanan) that the final result of the interaction does not actually matter much; what should be clear is that the result is directly related to the quality of the exploration and the interaction. This 'pedagogical' quality of the software, allowing it more or less successfully to facilitate exploration and learning, is connected with the notion of sensitivity or 'what if' studies. The stimulating nature of 'what if' is widely recognized both in the world of DSS and of multicriterion analysis (*cf.* Rios-Insua and French, 1991).

If we now pass from the user to the specialist, one of the first points to consider is the coherence between the information asked of the user and entered into the system, and the recommendations obtained as output. Stewart (1992) states that "the inputs required from the decision maker should be operationally meaningful"; this is clearly the least one can ask. Knowing, for example, that $m(m-1)/2$ pairwise comparisons are sufficient to rank m alternatives, we might well ask what added advantage is provided by a software application that asks the decision maker to make these comparisons before presenting him with a selection of alternatives.

The quantity and quality of the available information are of central importance and we shall examine their role in each type of method, making a clear distinction between weighting problems and other problems. Next in consideration are the decision maker's terms of reference and the nature of the problem in terms of number of alternatives and criteria. Objective comparisons are based on how well the method suits the decision maker's task (Olson, 1996; Petrovic, 1998).

These aspects, though objective, depend on the task and/or the available information; we must add the properties that we saw in the two previous sections; these too must be studied in relation to the problem. For example, transitivity of the complete preorder is not a fundamental property if we wish to select one or two good alternatives, and it matters little whether the rest of the ranking is transitive or not if the two alternatives selected sufficiently dominate all the others. Likewise, it is really of no importance to know that a method fails to satisfy the axiom of irrelevant alternatives or one of its variants, if the alternatives are fixed, known and intangible and if 'team racing' is not possible (*cf.* chapter 5). On the other hand, some other properties mentioned in the previous section, such as monotonicity, are desirable in most cases.

12.5 The specialist's point of view: choice of a method in terms of information available and terms of reference.

To facilitate our discussion on the advantages and drawbacks of various methods we shall classify them into the main families. We shall distinguish between additive and multiplicative methods which, from the simplest to the most complicated, are based on the idea of the cardinal weighted sum. In these methods, the cardinality of utility and weights strongly influences the result (*cf.* chapters 4 and 6). Among these methods we must distinguish simple weighting SMART, AHP and REMBRANDT (which is essentially a multiplicative AHP method) and methods of the MAUT type (*e.g.* LOGICAL DECISIONS and UTA). We shall deal with the ordinal methods described in chapter 5, distinguishing those based on the sum of ranks (Borda) from those based on the comparison of alternatives (Condorcet). Finally, we shall examine methods of the outranking type, distinguishing between the ELECTRE type and the PROMETHEE type, which are rather different in their philosophy.

For the specialist there are various parts of a multicriterion choice method to be distinguished. The two main ones are the weighting (or the determination of the important factors in the criteria) and the aggregation. These two phases are often wrongly considered to be independent. We do examine them separately, but for the sake of simplification; we then look at the links between them.

12.5.1 Evaluation of weights

The weights are often highly important in the aggregation, and it is easy to see that their value will influence the result considerably (Olson *et al.*, 1995). Turning the problem back to front, a glance at section 4.2 should suffice to show that any point on the Pareto boundary of a convex set is the maximum of a weighted sum of criteria provided the 'right' weights have been chosen. In other words, for any convex-admissible alternative A (definition 8.1), there exist weights such that A achieves the maximum for the weighted sum. As a result, in additive methods it is the weights that determine the solution.

Given the importance of weights, how are they to be determined? The various methods were reviewed in chapter 4. We shall divide our examination of weighting methods into two parts: those methods in which the weights are estimated apart from the aggregation, and those where the two operations are conjoint. (essentially MAUT and UTA). For methods of determining weights alone, we summarize the situation in a table (Figure 12.3). In this table we have indicated the type of evaluation, the section in the book in which the method of estimation is described, and when it is sufficiently specific, the name of the multicriterion method associated with that type of evaluation. The first column indicates whether the amount of information required for the evaluation is linearly dependent on the number of criteria (+), or whether the information requirement varies with the square of the number of criteria (++). The second column indicates the same thing using the same conventions, but applied to the number of alternatives. ('No' means that the number of alternatives does not play a significant role).

The third column indicates whether the method includes any way of checking the consistency of the transitivity of the weight comparisons and/or the consistency of successive choices in case of progressive information. It should be clearly understood that no method will ensure the full quality of weight cardinality; at best, consistency will apply to the plausibility and transitivity of the weight ordering and the coherence of the cardinality with the pairwise comparisons. In the fourth column are estimates of the quantity and difficulty of information asked of the decision maker. The final column is undoubtedly the hardest to fill out as it concerns the reliability of the information gathered. Here, some idea can be had from the difficulty of the questions and the more or less pronounced tendency of decision makers to give a biased response. Thus, it is well known that, although the methods of direct evaluation are considered to be easy by decision makers (Olson *et al.*, 1995), they are highly sensitive to certain types of bias (*cf.* Weber *et al.*, 1988; Weber and Borcherding, 1993; Mousseau, 1992b and the comments in section 4.8.3). Ordinal data on weights, on the other hand, have been found to be fairly stable (Weber and Borcherding, 1993), though some experimental evidence (Olson *et al.*, 1995) shows that caution is called for. Methods requiring ordinal information will thus be considered to give more robust results than the others provided we stick to this kind of information; simple ordinal evaluation, for example, is highly reliable but becomes much less so if the order is transformed into cardinal information. Again, pairwise comparisons are generally considered to be easy (Buchanan, 1994; Olson *et al.*, 1998), and therefore the decision maker's responses can be considered reliable over a small number of responses; consistency problems and in particular, transitivity problems arise when there are too many questions. In addition, when pairwise comparisons with cardinality are required, the information on the differences is very fragile. This is why eigenvalue methods, in spite of the good control of consistency that they afford, must be thought of as only of average reliability.

The problem of the overall coherence of several pairwise comparisons leads Larichev and Nikiforov (1987) to qualify the operation as complex. As for the methods by adjustment, these are highly sensitive to the choice of alternatives presented to the decision maker. To sum up, we can say that methods that provide ordinal information on the weights are relatively reliable as long as no attempt is made to transform the ordering into cardinal information – assuming that the evaluation is followed by ordinal aggregation, which is unfortunately not generally the case.

Method		Depend-ence/ number of criteria	Depend-ence/ number of alterna-tives	Consis-tency control	Quantity and difficulty of inform-ation required	Reliability of inform-ation obtained
Direct	Simple ordinal Evaluation, 4.5.1	+	no	no	low	good*
	Simple cardinal Evaluation, 4.5.2	+	no	no	low	poor
	Ratio, 4.5.2	+	no	no	medium	poor
	Revised Church-man and Ackoff, 4.5.3	++	no	+	high	medium*
Comparison of criteria	Eigenvalues (AHP), 4.6	++	no	++	high	medium
	Adjustment by Regression, 4.6 (REMBRANDT)	++	no	no	high	poor
Comparison of alternatives	Compensation, 4.7	+	no	no	medium	medium*
	Swing, 4.7	+	no	no	medium	medium*
	Price, 4.7	+	no	no	high	poor
	Swing (ZAPROS), 4.7	++	no	+	high	poor
	Adjustment by Linear programs (MACBETH), 2.4.4	+	+	+	high	medium (with cardina-lity)
	Adjustment by Regression (LINMAP), 8.3.3	+	+	no	variable (high)	poor
	Successive adjustments by Comparisons (Zionts), 8.2.1	+	+	+	variable (high)	poor

(* conditional on remaining with ordinal information)

Figure 12.3 Comparison of methods of evaluation of utilities and weights

The AHP method is the only one capable of incorporating weight hierarchization with relative ease. This is a definite advantage, both practically and theoretically, even though the weights so obtained are dependent on the way the criteria are split (Weber *et al.*, 1988; Weber and Borcherding, 1993).

12.5.2 Utilities

At this stage, we are in a position to look at methods of evaluating the utilities U_j which are to be used to evaluate the alternatives for each criterion. Most aggregation methods, except those described in chapter 5, use cardinal information on the alternatives. Several of the most common evaluation methods are described briefly in section 2.4.4. It would be extremely unwise to place too much faith in these evaluations (see section 2.4.5). What the specialist should state is that, whatever the method used, cardinal evaluations of alternatives are highly sensitive to presentation, to alternatives, to scales and to the order in which the alternatives are presented. For safety's sake, some uncertainty factor on the evaluation of the alternatives must be included, even if in general the order of preferences between alternatives, at least if the number of criteria is relatively small, is fairly stable.

It is only the order in which the alternatives are ranked for each criterion that is considered in ordinal methods. This order is usually obtained through pairwise comparisons. When there are many alternatives, transitivity problems arise and a large number of questions is needed to order the alternatives for each criterion. Just as we did for criteria, we can proceed from pairwise comparisons of the alternatives to deduce 'objective' cardinal utilities. Thus in the AHP method, one possibility is to use an alternatives comparison matrix and take the eigenvalue as the cardinal utility (*cf.* section 4.6, where the matrix A is of comparisons between alternatives instead of between criteria). In MACBETH an effort is made to obtain an interval cardinal utility function consistent with the evaluations of the decision maker. In fact, all the evaluation methods used for weights can be applied to alternatives for each criterion, with the same qualities and faults as for weights, so there is no need to repeat what we have just said for weights.

Method	Depend-ence/ number of criteria	Depend-ence/ number of alterna-tives	Consis-tency control	Quantity and difficulty of inform-ation required	Reliability of inform-ation obtained
MAUT Solvability, 6.3.9	*	no	++	medium	Very good
MAUT equidistant point, 6.4.2	+	no	+	high	good
UTA, 6.7	++	+	no	variable (low)	variable **

(* only works easily for 2 criteria)
(** medium to good depending on number of alternatives in A_r)

Figure 12.4 Comparison of methods of joint evaluation of utilities and weights

We do have, on the other hand, a little more to say about methods where we have conjoint evaluation of weights and utilities. This conjoint evaluation is the only correct procedure in methods of aggregation by weighted sum; in chapter 6 the theoretical reasons for this were given. Among these methods we shall look at MAUT, with the two variants, by solvability and by midpoint, and UTA. SMART will not be considered here because its weights and utilities are often evaluated separately. We can use the same table as in figure 12.3. Here, consistency control involves the fundamental assumption of preference independence (*cf.* chapter 6).

In UTA the quality of the result depends on the size of the reference set $\mathcal{A}r$. The preferences in $\mathcal{A}r$ are assumed to be transitive, this being a condition for minimum consistency, but the assumption of independence of preferences has not been tested. Broadly speaking, UTA gives fairly reliable information with a good well classified selection set $\mathcal{A}r$, and this does not require a large amount of information from the decision maker. The MAUT method by midpoint is very reliable provided a large number of questions is asked (to verify independence of preferences), including (see the example in section 6.4.4). The weakness of the method lies in the highly sophisticated nature of the questions to which the decision maker has to respond, and if there are many criteria the consistency of his responses may be in doubt. Decision makers generally consider this method to be difficult to use (*cf.* Olson *et al.*, 1995; Bard, 1993).

12.5.3 Ordinal aggregation methods

Method	Dependence/ number of criteria	Dependence/ number of alternatives	Quantity and difficulty of information required	Index of confidence	Problems
Condorcet, 5.3	+	++	high	medium	no transitivity
Borda, 5.2	+	+	low	high	can be manipulated by 'team racing' or by entering new alternatives
Bowman and Colantoni, 5.5	+	++	high	medium	- many ties if there are circuits - arbitrary of the distance
Lexicographic, 5.6	+	+	low	high	not very discriminatory
QUALIFLEX, 8.4	+	++	high	high	many ties if there are circuits
ZAPROS, 4.7	+	++	high	low	no theoretical justification, a much more complicated version of Borda

Figure 12.5 Comparison of ordinal methods of aggregation.

First we shall deal with ordinal aggregation methods which are in principle insensitive to the cardinality of the evaluation of each alternative, but which are nevertheless dependent on the cardinality of the weights. In table 12.5, the fourth evaluation is no longer on the consistency controls that can be introduced to verify the basic assumptions such as independence of preferences or transitivity of evaluations in relation to each criterion. We have replaced this column with an index showing the confidence that we may have in the final result. The less sensitive the method is to weight cardinality, the more the method takes into account the information supplied, the less arbitrariness there is in the method (*e.g.* in choice of distances) and the more justification there is of aggregation proper, the higher the index. In other words, a high index means that the method gives accurate information with no information that is false with regard to the information fed in.

12.5.4 Methods of aggregation by weighted sum

Method	Depend-ence/ number of criteria	Depend-ence/ number of alterna-tives	Quantity and difficulty of informa-tion required	Index of con-fidence	Problems
AHP, 4.6	+	++	high	medium	-many questions for the user -utility normalization problems
REMBRANDT, 4.6	+	++	high	low	Arbitrariness of adjust-ment by regression
SMART, 4.5.2	+	+	low	medium	dependent on the cardinality and the normalization of the alternatives
Zionts, 8.2.1	++	+	high	low	complicated questions if there are many criteria
TOPSIS, 8.3.2	+	+	low	low	highly dependent on evaluations and choice of metrics
LINMAP , 8.3..3	+	+	(variable) medium to high	low	Arbitrariness of adjustment by regression

Figure 12.6 Aggregation in the form of weighted sum

We are now entering the vast domain of cardinal methods more or less based on sums (or products) of evaluations of each alternative. It is hardly necessary to re-state that the methods are highly dependent on the evaluations of the weights, on the cardinality of the alternatives and on the normalization procedure. This is their greatest weakness. They have the advantage as we shall see of seeming 'natural' to naive decision makers. Among the methods, we have already studied (table 12.2) the MAUT and UTA methods, in which evaluation is concomitant with aggregation, as required by theory. This gives them a definite advantage over the methods we now look at. Among these we have selected AHP, REMBRANDT, SMART, Zionts, TOPSIS and LINMAP. They all use the weighted sum and assume preference independence; when this assumption is not satisfied, the results of regression methods are unreal and certainly less good than those of methods of adjustment by rules (Chung and Silver, 1992).

12.5.5 Outranking methods

We shall be studying interactive methods in the next section, and it remains to examine the outranking methods. Among these methods we shall be looking at ELECTRE, PROMETHEE and TACTIC. Note that the immediate design goal of these methods was not to provide a complete ranking; they deal with the selection problem. The first two methods considered both have numerous variants, but we shall stick to the basic principles.

Method	Depend-ence/ number of criteria	Depend-ence/ number of alterna-tives	Quantity and difficulty of informa-tion required	Index of con-fidence	Problems
ELECTRE, 7.3	+	++	high	medium	sensitive to evaluations for discordance
PROMETHEE, 7.4 and 10.3	+	++	low or very low*	medium	sensitive to the form of the criteria, difficult to follow the ranking rationale
TACTIC, 7.5.2	+	++	very high	high	requires a lot of information

(* depends on the type of criterion used)

Figure 12.7 Comparison of outranking methods

We trust that the reader is not expecting us now to cite the multicriterion choice method that dominates all the others. The choice of method is itself a problem of the multicriterion type and we cannot make the same demands on a simple method requiring little information as we can on a method requiring a considerable number of pairwise comparisons. We can however draw a few tentative conclusions – conclusions which depend on the type of problem we are dealing with. We will repeat here the various types of problem that a decision maker can face (section 2.6.2), summarizing them by the following notation:

P_α = selection of one or more satisficing alternatives
P_β = assignment to predefined classes or sorting
P_γ = complete preorder over all the alternatives

To these three basic problem types we will add P_n negotiation in collective decision-making or progressive elicitation of the decision maker's preferences (*cf.* 2.6.4 and 11.2.1).

For the problem type P_γ of complete classification, weighting methods have their followers. From a theoretical point of view, MAUT with the midpoint method or UTA if the number of criteria is not too large and the decision maker is happy with pairwise comparisons of alternatives (which is usually the case provided there are not too many, *e.g.* Olson *et al.*, 1995), represent two good choices, though MAUT is more difficult to implement.

For those who have no time for theory, the simple weighted sum method of the SMART type with a 'swing' method of determining the weights is easier to use , but highly dependent on the cardinality of the evaluations. Much less dependent on the evaluations and more robust is the Borda method, again with a swing evaluation of weights.

For the more sophisticated user who wishes to have at least a minimum degree of coherence control, AHP or QUALIFLEX, with ordinal information requirements, could be adopted. But bear in mind always that any method based on pairwise comparisons will become information-hungry and heavy on calculation when the number of alternatives is too large and problems arise of overall consistency within the numerous comparisons.

Problems of selection are best handled by outranking methods; they have a greater potential for interaction and negotiation. As they are based on pairwise comparisons they become demanding on information when the number of alternatives starts to grow. They are also dependent on the weight evaluation. From a cardinal point of view they are less dependent on the evaluation of the alternatives (only the ordering of the alternatives is involved) than weighting methods. ELECTRE and TACTIC have the advantage of introducing the notion of the veto, which is important in the field of human decision.

12.6 Choice of method: the user's point of view

Since the work of Wallenius (1975) it has been recognized that ease of use and understandability by the user are important factors in having the user accept a method. These two criteria feature in Wallenius (1975), Buchanan and Daellenback (1987), Larichev *et al.* (1993), Buchanan (1994), Olson *et al.* (1995), Olson (1996), Al Shemmeri *et al.* (1997), Olson *et al.* (1998), Srinivasa Raju and Pillai, (1999) and many others. These two criteria, both related to user appreciation, must be distinguished from the concept of cognitive effort introduced by Larichev and

Nikiforov (1987). This cognitive effort measures the difficulty of the questions and information asked of the decision maker independently of his feelings.

The user's own views on cognitive effort are not always very coherent. Thus, while it is accepted that pairwise comparisons of alternatives pose no problems for the user (Buchanan and Daellenbach, 1987), comparing two alternatives that have more than seven attributes is a very stiff task that becomes quite impossible once there are more than ten attributes. Moreover, when many comparisons have to be made, as there are in alternatives comparison methods, no unaided user is able to maintain overall consistency, *e.g.* transitivity (*cf.* Larichev and Nikiforov, 1987).

Whatever the real cognitive effort, it is not always felt as such by the user; despite all the associated problems of scale, context and reference point, it has always been found easy to obtain weights from the decision maker. What may be difficult for the specialist is easy for the naive user. Among the questions carrying a high cognitive effort, there are those related to tradeoff rates. It is always difficult to obtain reasonable, independent responses to questions on preference such as: how much would you be prepared to sacrifice on such and such a criterion in order to gain x on another one? Yet this type of question is central in MAUT methods, and here lies doubtless the main weakness of this type of method.

Coming back to the first two criteria of ease of use and understandability: two methods stand out, and users always express confidence in them: exploration by trial and error, and weighted sum. The trial and error method in particular enjoys an undeniable popularity among decision makers (Wallenius, 1975; Buchanan and Daellenbach, 1987; Corner and Buchanan, 1997; Olson *et al.*, 1998). We shall be returning to the trial and error method and interactive methods favored by decision makers. The weighted sum method has always been perceived by users as the most natural, provided they have not been warned about questions of scale and normalization. Stewart (1992) also points out that it is a method that is easy to explain to the general public. This argument is a little specious, because the public can also be shown that the final result of the method depends on the normalization chosen. Such difficulty is explicitly recognized and managed in the package SMC (Barba-Romero and Mokotoff, 1998).

To return to interactive methods, starting with the trial and error method: what is the purpose of interactive methods? We shall give two answers to this question. The first concerns construction (section 9.3.1). The decision maker has to elicit his own (aggregated) choice function; thus he becomes aware little by little that to gain on one criterion he must lose on another and be prepared to arbitrate. The decision maker will feed information into the system in the form of aspiration levels, or acceptance or refusal of solutions proposed by the system, the latter thus learning about the decision maker's preferences as the decision maker realizes them himself (*cf.* chapter 9).

The second purpose of interactive methods is during negotiations, to allow the various people involved in a multicriterion decision to reveal their preferences and reach a compromise (*cf.* 2.6.4 and 11.2).

The real purpose of interactive methods is in fact to allow an individual or collective learning process to take place; the progressive revelation of the decision maker's preferences and the collective search for a compromise are two aspects of this learning process. This idea has received quite a lot of attention recently. Buchanan (1994) writes: "The learning process is of as much value to the decision maker as the final choice". This learning is both by the method and the software. Belton and Hodgkin (1998) speak of the stimulus component of the software package (method + program). One way of stimulating the decision maker and facilitating his exploration is by the use of 'what if?' scenarios, or through sensitivity analysis: "sensitivity is a means of stimulating" (Rios Insua and French, 1991). Some software based on methods studied in the preceding section offers very elaborate facilities for analysis and sensitivity, for example PROMETHEE/GAIA and EXPERT CHOICE.

Many of the methods we have mentioned elsewhere can easily and fruitfully be made interactive, specially distance methods (*cf.* chapter 8) and outranking methods. Here we shall concentrate on methods that are intrinsically interactive, of which there are three categories (see chapter 9): exploration through aspiration levels, of which the prototype is MULTIDECISION based on the PRIAM algorithm; the distance method using either aspiration levels or scalarizing (temporary aggregation) functions; and the methods with domain elimination through pairwise comparisons.

We shall review a few representatives of these various types of method and then evaluate them according to the relevant criteria: quantity and difficulty of the information to be supplied; ease of learning, stimulation features (which may or may not be linked to sensitivity analysis). We must also ask: does a given method lead to a Pareto-optimal result? This is the theoretical yardstick for including a method here; distance methods do not necessarily lead to a non-dominated alternative (*cf.* 9.4.2).

Method and type	Pareto optimal result	Quantity and difficulty of information required	Ease of learning	Stimulation of exploration
MULTIDECISION Free Exploration from aspiration levels 9.3.3	yes	low	medium	high
AIM, Distance aspiration level 8.3.4 and 9.4.2	depends on the distance	medium	high	high
TRIPLE C, Distance to the ideal 8.3.4	no	medium	high	high
VIMDA, Distance and elimination 9.4.2	yes	medium to high	high	medium
Koksalan *et al.* (1984) Elimination of domains, 8.2.2	yes	high	low	medium

Figure 12.8 Comparison of interactive methods

Table 12.8 only contains methods that were not studied in the previous section. The outranking methods, especially ELECTRE and PROMETHEE, also belong in this realm since they allow learning and are stimulating to exploration through their facilities for sensitivity analyses due to the interactive fixing of the various decision parameters. Here, ELECTRE appears more interactive than PROMETHEE since the building up of the various orderings can be followed through the variations in concordance and discordance thresholds. In a comparative study, Salminem *et al.* (1998) show that the progressive distillation procedure in ELECTRE leads to choices that are more readily accepted by actual decision makers than those from PROMETHEE which, according to these authors, are close to those obtained from SMART.

12.7 Conclusion

Optimization of a function having a real value (a scalarization function), is often defended against multicriterion choice. Provided we forget that the function to be optimized is arbitrary, or at best reflects the preferences of a hidden decision maker, partisans of optimization are in a comfortable position. If however we choose to face up to complexity and try to analyze what is arbitrary, what is personal and what can be shared, multicriterion analysis is incomparably superior to optimization.

However, multicriterion analysis is not a ready-made recipe that can be applied to any situation at all using any method at all. Methods will be chosen according to the value and quantity of information available, the nature of the task and the expectations of the various people involved in the decision. Nevertheless, we should like to give a few recommendations.

If there is only certain information on ranking, ordinal methods will be preferred. Borda is one of the most robust, and one of the simplest in this case. Apart from the weakness of 'irrelevant alternatives' dealt with in chapter 5, it also has a strong theoretical basis (section 12.3). If weights or coefficients of importance are to be introduced, they must be determined with care, preferably by the revised swing method of Churchman and Ackoff, or by eigenvalues if there is sufficient time, information and at most ten or so criteria. If the weaknesses of Borda seem prohibitive, QUALIFLEX, which is a sort of improved Condorcet, could be used.

If there is a lot of cardinal information and if utility functions are to be used on each attribute, weighting methods should be favored. As independent evaluation of weights and attributes is out of the question, there then remain MAUT and UTA, according to whether utilities are to be constructed on each attribute or comparisons of alternatives are to be relied upon (which assumes a small number of criteria (<10)). But in any case weighting methods presuppose a small number of criteria in order to respect preference independence.

Where simple methods are required that are readily accepted by decision makers and have a pedagogical role of exposing preferences and tradeoffs, interactive outranking methods should be used. From the simplest such as MULTIDECISION through the more elaborate such as TRIPLE C and AIM, there is a range of methods which, according to how well designed the accompanying software is, promote rapid learning in the individual decision maker. An interactive ELECTRE method requiring more information than the aforementioned ones will also contribute to the training of the decision maker. In many cases multicriterion decision is institutional and the preponderant philosophy is of negotiation and consensus; it should nevertheless be possible, interactively, to construct, explain and 'sell' the decision. Here we feel that only outranking methods (essentially ELECTRE and PROMETHEE) and AHP (for proponents of the weighted sum) can meet the minimum criteria of reliability, robustness and enable discussion and explanation. These three methods should of course be used interactively with numerous discussions on the various parameters. They make ample use of pairwise comparisons, limiting their field of application to a reasonable number of alternatives (≤ 20). When there are many alternatives, individual decision methods such as MULTIDECISION, AIM and TRIPLE C are still very effective. In an institutional context a simplistic weighted sum method such as SMART could be used, while bearing in mind its theoretical and practical weaknesses; in any case the resulting choice will always be worth more than a purely arbitrary one!

Having completed this survey, it is clear that if we reason in terms of type of method and of requirements, the decision maker's choice is fairly limited. According to whether an individual or an institutional decision is involved, to whether there is a lot of cardinal information or relatively little information, and mainly ordinal at that, the choice will be limited to one or two methods. Thus we are a long way from the tangled forest that the detractors of multicriterion analysis like to refer to. Proponents of multicriterion analysis choice methods are becoming more and more numerous, using above all the methods featured in our own 'best of' list (AHP, AIM, BORDA, ELECTRE, MAUT, MULTIDECISION, PROMETHEE, QUALIFLEX, SMART, TRIPLE C, UTA). All users of these methods praise the merits of the multicriterion analysis process inherent in them, and though the final decision may have a certain arbitrariness when the alternatives are very close to each other, the emergent choice is otherwise clear. And the value of the decision process and its great institutional usefulness are always being recognized. Finally, as is sometimes suggested (Olson *et al.*, 1995), a wise analyst may be tempted to use two methods based on different methodologies, such as outranking and weighting, to expand the dialogue with the decision maker and have a better appreciation of the evaluations and the final decision.

REFERENCES

ABUALSAMH R., CARLIN B., McDANIEL Jr. R.R., 1990, «Problem Structuring Heuristics in Strategic Decision Making», *Organizational Behavior and Human Decision Processes vol. 45*, p. 159-174.

ABULBHAN P., TABUCANON M.T., 1979, «A biobjective model for production planning in a cement factory», *International Journal of Computers and Industrial Engineering, vol. 3*, p. 41-51.

ALLAIS M., 1953a, «Fondements d'une théorie positive des choix comportant un risque et critiques des postulats et axiomes de l'école américaine», in Econométrie, *Colloques internationaux du CNRS*, n° 40, p. 257-332.

ALLAIS M., 1953b, «Le comportement de l'homme rationnel devant le risque: critique des postulats et axiomes de l'école américaine», *Econometrica, vol. 21*, p. 503-546.

ALLAIS M., 1979, «The so-called Allais paradox and rational decisions under uncertainty», in ALLAIS M., HAGEN O., Eds., *Expected Utility Hypotheses and the Allais Paradox*, Reidel, Dordrecht, p. 437-681.

AL-SHEMMERI T., AL-KLOUB B. and PEARMAN A., 1997, «Model choice in multicriteria decision aid», *European Journal of Operational Research* 97, 550-560.

ANCOT J.-P., 1988, *Micro-Qualiflex*, Kluwer.

ANCOT J.-P., PAELINCK J.H.P., 1982, «Recent experiences with the QUALIFLEX multicriteria method», in *Qualitative and quantitative mathematical economics*, PAELINCK J.H.P., Ed., 1982, Martinus Nijhoff, The Hague, p. 217-266.

ANGEHRN A.A., 1989, «TRIPLE C, Overview and How to Use it», *User's Guide, INSEAD.*

ANGEHRN A.A., 1990, «Supporting multicriteria decision making: new perspectives and new systems», *Proceedings of the First ISDSS Conference*, Austin, USA, p. 521- 535.

ANGEHRN A.A., 1991a, «Designing humanized systems for multiple criteria decision making», *Human Systems Management, vol. 10*, p. 221-231.

ANGEHRN A.A., 1991b, «Modeling by example: a link between users, models and methods in Decision Support Systems», *European Journal of Operational Research, vol. 55*, p. 293-305.

ANGEHRN A.A., DUTTA S., 1992, «Integrating Case-Based Reasoning in Multicriteria Decision Support Systems», in JELASSI T., KLEINM.R., MAYON-WHITE W.M., Eds., *Decision support systems: Experiences and expectations*, North Holland, Amsterdam, p. 133-150.

ANGEHRN A A., LÜTHI H.J., 1990, «Intelligent Decision Support Systems: a visual Interactive Approach», *Interfaces, vol. 20, n° 6*, p. 17-28.

ARROW K.J., 1951, *Social Choice and Individual Values*, Wiley, New-York.

ARROW K.J., RAYNAUD H., 1986, *Social Choice and Multicriterion Decision-Making*, MIT Press, Cambridge Ma.

BADRAN Y., 1988, «Preference ranking of discrete multiattribute instances», *European Journal of Operational Research, vol. 34,* p. 308-315.

BALESTRA G., TSOUKIÀS A., 1990, «Multicriteria Analysis Represented by Artificial Intelligence Techniques», *Journal of the Operational Research Society 41, n° 5,* p. 419-430.

BANA E COSTA C.A., 1986, «A multicriteria decision aid methodology to deal with conflicting situations on the weights», *European Journal of Operational Research, vol. 26,* p. 22-34.

BANA E COSTA C.A., 1988, «A methodology for sensitivity analysis in three-criteria problems: a case study in municipal management», *European Journal of Operational Research, vol. 3,* p. 159-173.

BANA E COSTA C. A., Ed., 1990, *Readings in multiple criteria decision making,* Springer.

BANA E COSTA C. A., 1991, «MCDS under poor weighting information: the outweight approach», in *Multiple Criteria Decision Support,* KORHONEN P., LEWANDOWSKI A, WALLENIUS J., Eds., *Lecture Notes in Economics and Mathematical Systems 356,* p. 87-93.

BANA E COSTA C A, VANSNICK J-C., 1997, «Applications of the MACBETH approach in the framework of an additive aggregation model», *Journal of Multicriteria Decision Analysis, vol. 6, n° 2,* p. 107-114.

BANA E COSTA C A, VANSNICK J-C., 1998, «The MACBETH approach: basic ideas, software and an application», in *Advances in Decision Analysis,* MERKENS N., ROUBENS M., Eds., Kluwer.

BARBA-ROMERO S., 1987, «Panorámica actual de la decisión multicriterio discreta», *Investigaciones Económicas, vol. 11, n° 2,* p. 279-308.

BARBA-ROMERO S., 1988, «Un sistema de soporte a las decisiones multicriterio discretas», *Novática, vol. 14, n° 72,* p. 15-22.

BARBA-ROMERO S., 1990, «A comparative review of discrete MCDM software», presented at OR'90, *International Conference on Operations Research,* Viena, August 1990.

BARBA-ROMERO S., 1993, «Redes neuronales artificiales y optimización», in *Redes Neuronales Artificiales. Fundamentos y Aplicaciones,* OLMEDA I., BARBA-ROMERO S., Eds., Universidad de Alcalá, p. 165-182.

BARBA-ROMERO S., 1994, «Evaluación multicriterio de proyectos», in *Ciencia, Tecnología y Desarrollo: Interrelaciones Teóricas y Metodológicas,* MARTÍNEZ E., Ed., Nueva Sociedad, Caracas, p. 455-507.

BARBA-ROMERO S., 1998, «Conceptos y soportes informáticos de la decisión multicriterio discreta», in *Evaluación Multicriterio: Reflexiones Básicas y Experiencias en América Latina,* MARTÍNEZ E., ESCUDEY M., Eds., Cap. 3, p. 47-68, Ed. Universidad de Santiago de Chile.

BARBA-ROMERO S., 1999, «A discrete multicriteria decision support system for the selection of hardware and software acquisitions by Spanish public administration», *Interfaces,* to appear.

BARBA-ROMERO S., PÉREZ J., 1994, «La decisión multicriterio en el análisis y la gestión de los recursos naturales», in *Análisis Económico y Gestión de Recursos Naturales,* AZQUETA D., FERREIRO A., Eds., Alianza Editorial, Madrid, p. 137-162.

BARBA-ROMERO S., MOKOTOFF, E., 1997, «Multicriteria comparison of alternate investments plans on roads at the Spanish secretary of public works. The CMC decision support system», *proceedings of the International Conference on Methods and Applications of Multicriteria Decision Making,* Mons (Belgium).

BARBA-ROMERO S., MOKOTOFF, E., 1998, «A system to support discrete multicriteria evaluations and decisions: The SMC package», *proceedings of CESA'98 IMACS Multiconference Computational Engineering in Systems Applications,* Hammamet (Tunisia).

BARBA-ROMERO S., PÉREZ J., 1997, «la metodologia multicriterio en el analisis y la planificación territorial», *Estudios Territoriales XXIX n° 112,* p. 323-333.

BARD J. F., 1993, «A comparison of the analytic hierarchy process with multiattribute utility theory : a case study», *Operations research/ Management Science 33,* p. 689-691.

BARE B.B., MENDOZA G., 1988, «Multiple objective forest land and management planning: an illustration», *European Journal of Operational Research, vol. 34*, p. 44-55.

BARTHÉLEMY J.-P., MONJARDET B., 1981, «The median procedure in cluster analysis and social choice theory», *Mathematical Social Sciences, vol. 1*, p. 235-267.

BARTHÉLEMY J.-P., MULLET E., 1986, «Choice basis: a model for multiattribute preference», *British Journal of Mathematical and Statistical Psychology, vol. 39*, p. 106-124.

BARTHÉLEMY J.-P., MULLET E., 1989, «Choice basis: a model for multiattribute preferences: some further evidence», in *Mathematical Psychology in Progress*, ROSKAM E.E., Ed., Springer, p. 179-196.

BARTHÉLEMY J.-P., MULLET E., 1992, «A model of selection by aspects», *Acta Psychologica, vol. 79*, p. 1-19.

BARZILAÏ J., and GOLANY B., 1990, «Deriving weights from parwise comparison matrices: the additive case», *Operations Research Letters 9 n°6*, 407-410.

BARZILAÏ J., COOK W. and GOLANY B., 1987, "Consistent weights for judgements matrices of the relative importance for alternatives", *Operations Research Letters 6 n°3*, p. 131-134.

BATANOVIC V., 1989, «Multicriteria evaluation of an urban traffic control system: Belgrade case study», in *Models and methods in multiple criteria decision making*, COLSON G., DE BRUIN C., Eds.,1989, Pergamon Press, p. 1411-1417.

BAYAD R., POMEROL J.-Ch., 1992, «An «intelligent» DSS for the reinforcement of urban electrical power networks», in *Decision Support Systems: Experiences and Expectations*, JELASSI T., KLEIN M.R., MAYON-WHITE W.M., Eds., Elsevier, Amsterdam, p. 153-165.

BECK M.P., LIN B.W., 1981, «Selection of automated office systems: a case study», *OMEGA, vol. 9, n° 2*, p. 169-176.

BELL D.E., 1977, «A decision analysis of objectives for a forest pest problem», in *Conflicting objectives in decisions*, BELL D.E., KEENEY R.L., RAIFFA H., Eds., Wiley, p. 389-421.

BELL D.E., KEENEY R.L., RAIFFA H., Eds., 1977, *Conflicting objectives in decisions*, Wiley.

BELTON V., 1985, «The use of a simple multiple-criteria model to assist in selection from a shortlist», *Journal of the Operational Research Society, vol. 36, n° 4*, p. 265-274.

BELTON V., GEAR T., 1983, «On a shortcoming of Saaty's method of analytic hierarchies», *Omega, vol.11, n°3*, p. 228-230.

BELTON V. and HODGKIN J., 1999, «Facilitators Decision Makers, D. I. Y. Users : is intelligent Multicriteria Decision Support for all feasible or desirable?», to appear in *Multicriteria Decision Support and Artificial Intelligence*, C. ZOUPOUNIDIS and J. SISKOS Eds., *European journal of Operational research 113*, p. 247-260.

BELTON V., VICKERS S., 1990, «Use of a simple multiattribute value function incorporating visual interactive sensitivity analysis for multiple criteria decision making», in *Readings in multiple criteria decision making*, BANA E COSTA C.A., Ed., Springer, p. 319-334.

BENAYOUN R., DE MONTGOLFIER J., TERGNY J., LARICHEV O.I., 1971, «Linear programming with multiple objective functions: Step method STEM», *Mathematical Programming, vol. 1*, p. 366-375.

BENAYOUN R., TERGNY J., 1969, «Critères multiples en programmation mathématique: une solution dans le cas linéaire», *RIRO, vol. 3, n° 2*, p. 31-56.

BENCHIMOL G., LÉVINE P., POMEROL J.-Ch., 1990, *Systèmes experts dans l'entreprise*, augmented version, Hermes-Science, Paris. English translation: *Developing Expert Systems for Business*, North Oxford Academic, 1987.

BERGE C., 1970, *Graphes et hypergraphes*, Dunod, Paris.

BERNARDO J.J., BLIN J.M., 1977, «A programming model of consumer choice among multiattributed brands», *Journal of Consumer Research, vol. 4, n° 2*, p. 111-118.

BEVAN R.G., 1980, «Measurement for evaluation», *Omega, vol. 8, n° 3*, p. 311-321.

BLACK D., 1948a, «On the rationale of group decision making», *Journal of Political Economy, vol. 56*, p. 23-24.

BLACK D., 1948b, «The decisions of a committee using a special majority», *Econometrica, vol. 16*, p. 245-261.

BLACK D., 1948c, «The elasticity of committee decisions with an altering size of majority», *Econometrica, vol. 16*, p. 262-270.

BLACK D., 1958, *The theory of committees and elections*, Kluwer, Boston.

BLAIR D., POLLAK R., 1983, «La logique du choix collectif», *Pour la science*, octobre 1983, p. 104-111.

BLIN J.M., 1976, «A linear assignment formulation of the multiattribute decision problem», *revue Française d'automatique, d'informatique et de recherche operationnelle*, vol. 10, n° 6, p. 21-32.

BLIN J.M., DODSON Jr. J.A., 1978, «A multiple criteria model for repeated choice situations», in ZIONTS S., Ed., 1978, *Multiple criteria problem solving*, Springer, p. 8-22.

BOHANEC M., BRATKO I., RAJKOVIC V., 1983, «An expert system for decision making», in *Processes and Tools for Decision Support*, SOL H.G., Ed., North Holland, p. 235-248.

BOHANEC M., CESTNIK B., RAJKOVIC V., 1998, «Evaluation models for housing loan allocation in the context of floats», in *Context-sensitive Decision Support Systems*, D. BERKELEY, G. WIDMEYER, P. BRÉZILLON and V. RAJCOVIC Eds., Chapman and Hall, p. 174-189.

BOHANEC M., RAJKOVIC V., 1987, «An expert system approach to multiattribute decision making», in *IASTED International Conference on Expert Systems*, HAMZA M.H., Ed., Acta Press, Genève.

BOHANEC M., RAJKOVIC V., 1988, «Knowledge acquisition and explanation for multi- attribute decision making», *8ᵉ Journées internationales sur les systèmes experts et leurs applications, vol. 1*, Avignon, p. 59-78.

BOHANEC M., RAJKOVIC V., 1990, «DEX: an Expert System Shell for Decision Support», *Sistémica 1, n° 1*, p. 145-158.

BOHANEC M., URH B., RAJKOVIC V., 1991, «Evaluating Options by Combined Qualitative and Quantitative Methods», *13th Conference on Subjective Probability, Utility and Decision Making*, Fribourg, 1991.

BORCHERDING K., EPPEL T., VON WINTERFELDT D., 1991, «Comparison of weighting judgments in multiattribute utility measurement», *Management Science, vol. 37, n° 12*, p. 1603-1619.

BORDES G., TIDEMAN N., 1991, «Independence of irrelevant alternatives in the theory of voting», *Theory and Decision, vol. 30*, p. 163-186.

BOSCARINO M., GIANOGLIO G. and OSTANELLO A., 1993, «A neural network designed to support decision making with a complex multicriteria problem», *Journal of Decision Systems .2*, p. 149-171.

BOUCHON-MEUNIER B., 1993, *La logique floue, que sais-je ?*, Presses Universitaires de France, Paris.

BOUYSSOU D., 1984, «Decision-aid and expected utility theory: a critical survey», *in Progress in utility and risk theory*, HAGEN O. and WENSTOP F. Eds., Reidel, Dordrecht, p. 181-216.

BOUYSSOU D., 1986, «Some remarks on the notion of compensation in MCDM», *European Journal of Operational Research, vol. 26*, p. 150-160.

BOUYSSOU D., 1989, «Problèmes de construction de critères», *Cahier du LAMSADE, n° 91*, Université de Paris-Dauphine, Paris.

BOUYSSOU D., PERNY P., 1992, «Ranking methods for valued preferences relations: a characterization of a method based on leaving and entering flows», *European Journal of Operational Research, vol. 61*, p. 186-194.

BOUYSSOU D., PIRLOT M., 1997, «Non Transitive Decomposable Conjoint Measurement: Non additive Representations of Non Transitive Preferences on Product Sets», Working Paper, ESSEC.

BOUYSSOU D., PIRLOT M., VINCKE Ph., 1997, «A General Model of Preference Aggregation», in *Essays in Decision Making, A volume in honour of Stanley Zionts*, M. H. KARWAN, J. SPRONK and J. WALLENIUS Eds., Springer, p. 120-134.

BOUYSSOU D., VANSNICK J.-C., 1986, «Noncompensory and generalized noncompensatory preference structures», *Theory and Decision, vol. 21*, p. 251-266.

BOUYSSOU D., VANSNICK J.-C., 1990, «Utilité cardinale dans le certain et choix dans le risque», *Revue économique, vol. 41, n° 6*, p. 979-1000.

BOWMAN V.J., COLANTONI C.S., 1973, «Majority rule under transitivity constraints», *Management Science, vol. 19, n° 9*, p. 1029-1041.

BRANS J.P., MARESCHAL B., 1989, «The Promethee methods for MCDM ; the PROMCALC, GAIA and BANKADVISER software», *Working report ST00/224*, Vrije Universiteit, Brussel, January 1989.

BRANS J.P., MARESCHAL B., 1992, «PROMETHEE V: MCDM problems with segmentation constraints», *Information Systems and Operations Research vol. 30, n° 2* , p. 85-96.

BRANS J.P., MARESCHAL B., 1991, «The PROMCALC and GAIA Decision Support System for Multicriteria Decision Aid», *IFORS SPC1*, Bruges, mars 1991.

BRANS J.P., MARESCHAL B., 1994, «The PROMCALC and GAIA Decision Support System for Multicriteria Decision Aid», *Decision Support Systems, vol. 12*, p. 297-310.

BRANS J.P., MARESCHAL B., VINCKE Ph., 1984, «PROMETHEE: a new family of outranking methods in multicriteria analysis», in *Operational Research'84*, BRANS J.P., Ed., North-Holland, p. 408-421.

BRANS J.P., VINCKE Ph., 1985, «A preference ranking organization method, the PROMETHEE method», *Management Science, vol. 31*, p. 647-656.

BRANS J.P., VINCKE Ph., MARESCHAL B., 1986, «How to select and how to rank projects: The PROMETHEE method», *European Journal of Operational Research, vol. 24*, p. 228-138.

BRIGGS Th., KUNSCH P.L., MARESCHAL B., 1990, «Nuclear waste management: an application of the multicriteria PROMETHEE methods», *European Journal of Operational Research, vol. 44, n° 1*, p. 1-10.

BROCKHOFF K., 1985, «Experimental test of MCDM algorithms in a modular approach», *European Journal of Operational Research, vol. 22*, p. 159-166.

BROOKS D.G., KIRKWOOD C.W., 1988, «Decision analysis to select a microcomputer networking strategy: a procedure and a case study», *Journal of the Operational Research Society, vol. 39, n° 1*, p. 23-32.

BROWN C.A., STINSON D.P., GRANT R.W., 1986, *Multi-Attribute Tradeoff System: Personal computer version user's manual*, Bureau of Reclamation, U.S. Depart. of the Interior.

BUEDE D. M., 1996, «Second overview of the MCDA software market», *Journal of Multicriteria Decision Analysis, vol. 5, n° 6* , p. 312-316.

BUEDE D. M., MAXWELL D. T. 1995, «Rank disagreement: A comparison of multi-criteria methodologies», *Journal of MultiCriteria Decision Analysis, vol. 4, n° 1*, p. 1-21.

BUCHANAN J. T., 1994, «An experimental Evaluation of Interactive MCDM Methods and the Decision Making Process», *Journal of Operational Research Society 45*, p. 1050-1059.

BUCHANAN J.T., DAELLENBACH H.G., 1987, «A comparative evaluation of interactive solution methods for multiple objective decision models», *European Journal of Operational Research, vol. 29*, p. 353-359.

CALPINE H.C., GOLDING A., 1976, «Some properties of Pareto-optimal choices in decision problems», *OMEGA, vol. 4, n° 2*, p. 141-147.

CAMACHO A., 1982, *Societies and Social Decision Functions, a Model with Focus on the Information Problem*, Reidel, Dordrecht.

CAMACHO A., 1983, «Cardinal utility and decision making under uncertainity», in *Foundations of utility and risk theory with applications*, STIGUM B.P., WENSTOP F., Eds., Reidel, p. 347-370.

CARMONE F. J., SCHAFFER C.M., 1995, «Conjoint LINMAP», *Journal of Marketing Research, February*, p. 113.

CARPENTER G.A., CROSSBERG S., 1987, «A massively parallel architecture for a self-organizing pattern recognition machine», *Computer Vision, Graphics and Image processing 37*, p. 54-115.

CARUSO C., COLORNI A., PARUCCINI M., 1993, «The regional urban solid waste management system: a modelling approach», *European Journal of Operational Research, vol. 70, n° 1*, p. 16-30.

CATS-BARIL W.L., HUBER G.P., 1987, «Decision support systems for ill-structured problems: an empirical study», *Decision Science, vol. 18*, p. 350-372.

CHANKONG V., HAIMES Y.Y., 1983, *Multiobjective decision making: theory and methodology*, North-Holland, New-York.

CHANKONG V., HAIMES Y.Y., THADATHIL J., ZIONTS S., 1985, «Multiple criteria optimization: a state of the art review», in *Decision making with multiple objectives*, HAIMES Y.Y., CHANKONG V., Eds., Springer, p. 36-90.

CHARNES A., COOPER W., 1961, *Management Models and Industrial Applications of Linear Programming*, John Wiley and Sons.

CHEBOTAREV P.Y., SHAMIS E., 1998, in *Special issue on Preference Modelling, Annals of Operations Research, 80*, p. 299-332.

CHEUNG Y.L., 1991, «Correspondence analysis as an aid to multicriteria decision making», *OMEGA, vol. 19, n° 2/3*, p. 149-155.

CHEUNG Y.L., 1992, «The application of correspondance analysis to multiple-criteria decision making», *Journal of Multiple-Criteria Decision Analysis, vol. 1, n°3*, p. 155-163.

CHIPMAN J.S., 1971, «On the lexicographic representation of preference orderings», in *Preferences, utility and demand*, CHIPMAN J.S., HURWICZ L., RICHTER M.K., SONNESCHEIN H.F., Eds., 1971, Harcourt Brace, New-York, p. 276-288.

CHUNG H. M., SILVER M. S., 1992, «Rule-Base Expert Systems and Linear Models: An empirical Comparison of Learning-by-Examples methods», *Decision Sciences 23*, p. 687-707.

CHURCHMAN C.W., ACKOFF R.L., 1954, «An aproximate measure of value», *Journal of the Operational Research Society of America, vol. 2, n° 2*, p. 172-187.

CHVATAL V., LOVASZ L., 1974, «Every directed graph has a semikernel», in *Hypergraph Seminar*, BERGE C., RAY-CHAUDURI D., Eds., Heidelberg, Springer.

CLAESSENS M.N.A.J., LOOTSMA F.A., VOOGD F.J., 1991, «An elementary proof of Paelinck's theorem on the convex hull of ranked criterion weights», *European Journal of Operational Research, vol. 52, n° 2*, p. 255-258.

CLIMACO J., ANTUNES C.H., 1989, «Implementation of a user-friendly software package a guided tour of TRIMAP», *Mathematical and Computer Modeling, vol. 12*, p. 1299-1309.

CLIMACO J. Ed., 1997, *Multicriteria Analysis*, Proceedings of XIth International Conference on MCDM, Coimbra, Springer.

COCHRANE J.L., ZELENY M., Eds., 1973, «Multiple criteria decision making», *Proceedings South Carolina*, (1972), University of South Carolina Press, Columbia (USA).

COHEN M., 1992, «Security level, potential level, expected utility: a three-criteria decision model under risk», *Theory and Decision, vol. 33*, p. 101-134.

COHEN M., JAFFRAY J.-Y., SAÏD T., 1987, «Experimental comparison of individual behavior under risk and under uncertainty for gains and for losses», *Organizational Behavior and Human Decision Processes, vol. 39*, p. 1-22.

COHON J.L., 1978, *Multiobjective programming and planning*, Academic Press.

COLSON G., 1989, «MARS: a multiattribute utility ranking support for risk situations, with a P Q I R relational system of preferences», in *Models and methods in multiple criteria decision making*, COLSON G., DE BRUIN C., Eds., Pergamon Press, p. 1269-1297.

COLSON G., DE BRUIN C., Eds., 1989, *Models and methods in multiple criteria decision making*, Pergamon Press.

COLSON G., ZELENY M., 1980, «Multicriterion concept of risk under incomplete information», in *Computers and Operations Research: Special Issue on Mathematical Programming with Multiple Objectives*, ZELENY M., Ed., *vol. 7, n° 1/2*, p. 125-141.

CONDORCET J.-M., MARQUIS DE, 1785, *Essai sur l'application de l'analyse à la probabilité des décisions rendues à la pluralité des voix*, Imprimerie Royale, Paris. Réédité par Chelsea, New-York, 1972.

COOK W.D., GOLAN I., KAZAROV A., KRESS M., 1988, «A case study of a non- compensatory approach to ranking transportation projects», *Journal of the Operational Research Society, vol. 39, n° 10*, p. 901-910.

COOK W.D., KRESS M., 1988, «Deriving weights from pairwise comparison ratio matrices: an axiomatic approach», *European Journal of Operational Research, vol. 37, n° 3*, p. 355-362.

COOK W.D., KRESS M., 1991, «A multiple criteria decision model with ordinal preference data», *European Journal of Operational Research, vol. 54*, p. 191-198.

COOK W.D., SEIFORD L.M., 1978, «Priority ranking and consensus formation», *Management Science, vol. 24, n° 16*, p. 1721-1732.

COOK W.D., SEIFORD L.M., 1984, «An ordinal ranking model for the highway corridor selection problem», *Computers, Environment and Urban Systems, vol. 9, n° 4*, p. 271-276.

COOMBS C.H., 1958, «On the use of inconsistency of preferences in psychological measurement», *Journal of Experimental Psychology, vol. 55*, p. 1-7.

COPELAND A.H., 1951, A «reasonable» social welfare function», University of Michigan Seminar on Applications of Mathematics to the Social Sciences, mimeo.

CORNER J. L. and BUCHANAN J. T., 1997, «Capturing decision maker preference: Experimental comparison of decision analysis and MCDM techniques», *European journal of Operational Research 98*, p. 85-97.

CRAMA Y., HANSEN P., 1983, «An introduction to the ELECTRE research programme», in *Essays and surveys on multiple criteria decision making*, HANSEN P., Ed., 1983, Springer, p. 31-42.

CROZIER M., 1963, *Le phénomène bureaucratique*, Le Seuil, Paris.

DANEV B., SLAVOV G., METTEV B., 1987, «Multicriteria comparative analysis of discrete alternatives», in *Towards interactive and intelligent decision support systems*, SAWARAGI Y., INOUE K., NAKAYAMA H., Eds., Springer, p. 57-64.

DANILA N.V., 1990, «The Dante model: Dynamic appraisal of network technologies and equipment», in *Selection and evaluation of advanced manufacturing technologies*, LIBERATORE M.J., Ed., 1990, Springer, p. 163-185.

DASARATHY B.V., 1976, «Smart: Similarity measured anchored ranking technique for the analysis of multidimensional data», *IEEE Trans. on Systems, Man and Cybernetics, vol. SMC-6, n° 10*, p. 708-711.

DAUER J. P., KRUEGGER R.J., 1980, «A multiobjective optimization model for water resources planning», *Applied Mathematical Modelling*, vol. 4, n° 3, p. 171-175.

DAVID H.A., 1988, *The method of paired comparisons*, Oxford University Press.

D'AVIGNON G.R., VINCKE Ph., 1988, «An outranking method under uncertainty», *European Journal of Operational Research, vol. 36, n° 3*, p. 311-321.

DAVIS L.S., LIU G., 1991, «Integrated forest planning across multiple ownerships and decision makers», *Forest Science, vol. 37, n° 1*, p. 200-226.

DEBREU G., 1954, «Representation of a preference ordering by a numerical function», reprinted in Readings in *mathematical economics*, NEWMAN P., Ed., 1968, John Hopkins University Press, Baltimore, USA, p. 257-263.

DEBREU G., 1960, «Topological methods in cardinal utility theory», in *Mathematical Methods in the Social Sciences*, ARROW K.J., KARLIN S., SUPPES P., Eds, Stanford University Press, p. 16-26.

DE GRAAN J.G., 1980, «Extensions of the multiple criteria analysis method of T.L. Saaty», presented at Euro IV, Cambridge, UK, July 1980.

DESPONTIN M., LEHERT F., ROUBENS M., 1986, «Multi-attribute decision making by consumers associations», *European Journal of Operational Research, vol. 23*, p. 194-201.

DESPONTIN M, MOSCAROLA J., SPRONK J., 1983, «A user-oriented listing of multiple criteria decision methods», *Revue Belge d'Informatique et de Recherche Opérationnelle, vol. 23, n° 4*, p. 3-110.

DESPONTIN M., NIJKAMP P., SPRONK J., Eds., 1984, *Macroeconomic planning with conflicting goals*, Proceedings Brussels (1982), Springer.

DIAKOULAKI D., KOUMOUTSOS N., 1991, «Cardinal ranking of alternative actions: Extensions of the Promethee method», *European Journal of Operational Research, vol. 53*, p. 337-347.

DIAKOULAKI D., MAVROTAS G., PAPAYANNAKIS L., 1992, «Objective weights of criteria for interfirm comparisons», *36ᵉ Journées du groupe européen Aide Multicritère à la Décision*, Luxembourg.

DINKELBACH W., 1971, «Über einen Lösungsanstz zum Vektormaximumproblem», in *Unternehmensforschung-Heute*, BECKMANN M., Ed, Sringer, Berlin, p. 1-13.

DINKELBACH W., ISERMANN H., 1973, «On decision making under multiple criteria and under incomplete information», in *Multiple Criteria Decision Making*, COCHRANE J.L., ZELENY M., Eds., University of south Carolina Press, Columbia, USA, p. 302-312.

DUBOIS D., KONING J.-L., 1991, «Social choice axioms for fuzzy set aggregation», *Fuzzy Sets and Systems, vol. 43*, p. 257-274.

DUBOIS D., PRADE H., 1980, *Fuzzy sets and systems, Theory and Applications*, Academic Press, New-York.

DUBOIS D., PRADE H., 1987, *Théorie des possibilités, applications à la représentation des connaissances en informatique*, Masson, 2ᵉ édition, Paris. English translation: *Possibility Theory, an approach to the computer uncertainty processing*, Plenum Press, New-York, 1988.

DU BOIS Ph., BRANS J.-P., CANTRAINE F., MARESCHAL B., 1989, «MEDICIS: an Expert System for Computer-Aided Diagnosis using the PROMETHEE Multicriteria Method», *European Journal of Operational Research, vol. 39*, p. 284-292.

DUCKSTEIN L., GERSHON M., 1981, «Multiobjective analysis of a vegetation management problem using ELECTRE II», *Working Paper, n° 11*, University of Arizona, Tucson.

DUCKSTEIN L., KEMPF J., 1981, «Multicriteria Q-Analysis for Plan Evaluation», in *Multiple Criteria Analysis: Operational Methods*, NIJKAMP P., SPRONK J., Eds., Grower, Londres.

DUPUY J. P., 1982, *Ordres et Désordres, Enquête sur un nouveau paradigme*, Le Seuil, Paris.

DYER J.S., 1990, «Remarks on the analytic hierarchy process», *Management Sciences, vol.36, n°3*, p.249-258.

DYER J.S., FISHBURN P.C., STEUER R.E., WALLENIUS J., ZIONTS S., 1992, «Multiple criteria decision making, multiattribute utility theory: the next ten years», *Management Science, vol. 38, n° 5*, p. 645-654.

DYER R.F., FORMAN E.H., 1992, «Group decision with the analytic hierarchy process», *Decision Support Systems, vol. 8*, p. 99-124.

ECKENRODE R.T., 1965, «Weighting multiple criteria», *Management Science, vol. 12, n° 3*, p. 180-192.

ECKER J.G., KOUADA I.A., 1978, «Finding efficient points for linear multiple objective programs», *Mathematical Programming, vol. 8, n° 3*, p. 249-261.

EDWARDS W., 1977, «Use of multiattribute utility measurement for social decision making», in *Conflicting objectives in decisions*, BELL D.E., KEENEY R.L., RAIFFA H., Eds., 1977, Wiley, p. 247-276.

EFSTATHIOU J., RAJKOVIC V., BOHANEC M., 1986, «Expert systems and rules based decision support systems», in *Computer assisted decision making*, MITRA G., Ed., Elsevier, p. 165-174

EKELAND I., 1979, *Eléments d'économie mathématique*, Hermann, Paris.

EOM B.H., 1989, «The Current State of Multiple Criteria Decision Support Systems», *Human Systems Management, vol. 8*, p. 113-118.

ERKUT E., MORAN S.R., 1991, «Locating obnoxious facilities in the Public Sector: an application of the Analytic Hierarchy Process to municipal landfill siting decisions», *Socio-Economic Planning Sciences, vol.25, n°2*, p. 89-102.

ESSID S., 1990, Aide à la décision dans le risque: modèle et logiciel, thèse de l'université P. et M. Curie, Paris.

EVANS J., STEUER R., 1973, «Generating efficient extreme points in linear multiple objective programming», in Multiple Criteria Decision Making, J. COCHRANE and M. ZELENY Eds., University of South Carolina press, Columbia.

FANDEL G., GAL T., Eds., 1980, *Multiple criteria decision making theory and applications*, Springer.

FANDEL G., GRAUER M., KURZHANSKI A., WIERZBICKI A.P., Eds., 1986, *Large-scale modelling and interactive decision analysis*, Springer.

FANDEL G., SPRONK J., Eds., 1985, *Multiple criteria decision methods and applications*, Springer.

FANG L., HIPEL K.W., KILGOUR D.M., 1988, «The graph model approach to environmental conflict resolution», *Journal of Environmental Management, vol. 27*, p. 195-212.

FARQUHAR P.H., 1984, «Utility Assessment Methods», *Management Science, vol. 30, n° 11*, p. 1283-1300.

FARQUHAR P.H., 1987, «Applications of utility theory in artificial intelligence research», in *Towards Interactive and Intelligent Decision Support Systems*, SAWARAGI Y., INOUE K., NAKAYAMA H., Eds., Springer, New-York, p. 155-161.

FEINBERG A., SMITH J.H., 1989, «Lessons from six applications of multicriterion decision anlysis to rank advanced technology alternatives», in *Multiple criteria decision making: Applications in industry and service*, TABUCANON M.T., CHANKONG V., Eds., 1989, *Proceedings, International Conference on MCDM*, Asian Institute of Technology, Bangkok, p. 661-675.

FERBER J., GHALLAB M., 1988, «Problématiques des univers multiagents intelligents», *Proceedings of the PRC-GRECO Conference*, Teknea, Toulouse, p. 295-320.

FERNANDEZ G.M., 1991, Extensión a los métodos Promethee de nuevas estructuras de preferencia para la toma de decisiones multicriterio discretas, Tesis Doctoral, Universidad de Alcala, Madrid.

FIALA P., 1991, Problem Solving Methods in Multicriteria Analysis, Diskussionsbeitrag Fernuniversität Hagen, n° 181.

FIALA P., 1992, «Artificial intelligence methods in multicriteria analysis», in *Multicriteria Decision Making, Methods, Algorithms, Applications*, CERNY M., GLÜCKAUFOVA D., LOULA D., Eds, Institute of Economics, Czechoslovak Academy of Sciences, Prague, p. 58-65.

FICHEFET J., 1985, «Computer selection and multicriteria decision aid», in *Multiple criteria decision methods and applications*, FANDEL G., SPRONK J., Eds., 1985, Selected Readings of the First International Summer School, Springer, p. 337-346.

FINKE G., BURKARD R.E., RENDL F., 1987, «Quadratic assignment problems», in *Surveys in combinatorial optimization*, MARTELLO S., LAPORTE G., MINOUX M., RIBEIRO C., Eds., 1987, North-Holland, p. 61-82.

FISCHHOFF B., 1980, «Clinical decision analysis», *Operations Research, vol. 28*, p. 28-43.

FISHBURN P.C., 1965, «Independance in utility theory with whole product sets», *Operations Research, vol. 13*, p. 28-45.

FISHBURN P.C., 1967, «Methods of estimating additive utilities», *Management Science, vol. 13, n° 7*, p. 435-453.

FISHBURN P.C., 1970, Utility for Decision Making, *Publications in Operations Research, n° 18*, Wiley, New-York.

FISHBURN P.C., 1973, *The theory of social choice*, Princeton University Press, USA.

FISHBURN P.C., 1974, «Lexicographic orders, utilities and decision rules: a survey», *Management Science, vol. 20, n° 11*, p. 1442-1471.

FISHBURN P.C., 1976, «Noncompensatory preferences», *Synthese, vol. 33*, p. 393-403.

FISHBURN P.C., 1977, «Condorcet social choice functions», *Siam Journal of Applied Mathematics, vol. 33, n°. 3*, p. 469-489.

FISHBURN P. C., 1991, «Non-transitive preferences in Decision Theory», *Journal of Risk and Uncertainty 4*, p. 113-134.

FISHBURN P.C., 1992a, «Multiattribute signed orders», *Journal of Multi-Criteria Decision Analysis, vol. 1*, p. 3-16.

FISHBURN P.C., 1992b, «Additive differences and simple preference comparisons», *Journal of Mathematical Psychology, vol. 36*, p. 21-31.

FISHER R., URY W., 1982, *Comment Réussir une négociation*, Le Seuil, Paris.

FRENCH S., HARTLEY R., THOMAS L.C., WHITE D.J., Eds., 1983, *Multiobjective decision making*, Academic-Press.

GABRIEL K.R., 1971, «The biplot graphic display of matrices with application to principal component analysis», *Biometrika, vol. 58, n° 3*, p. 453-467.

GARDINER L.R., VANDERPOOTEN D., 1997, «Interactive Multicriteria Procedures: some reflections», in *Multicriteria analysis, Proceedings of the XIth Conference on MCDM*, Climaco J. Ed., p. 290-301.

GEHRLEIN W.V., 1983, «Condorcet's paradox», *Theory and Decision, vol. 15*, p. 161-197.

GEOFFRION A.M., 1965, «A parametric programming solution to the vector maximum problem with applications to decisions under uncertainty», *Operations Research Program, Stanford University Technichal Report, n° 11*.

GEOFFRION A.M., DYER A.M., FEINBERG A., 1972, «An interactive approach for multicriterion optimization with an application to the operation of an academic department», *Management Science, vol. 19*, p. 357-368.

GEORGESCU-ROEGEN N., 1954, «Choice, expectations and measurability», *Quarterly Journal of Economics, vol. 68, n° 4*, p. 503-541.

GERSHON M., DUCKSTEIN L., 1983a, «An algorithm for choosing a multiobjective technique», in *Essays and surveys on multiple criteria decision making*, HANSEN P., Ed., 1983, Springer, p. 53-62.

GERSHON M., DUCKSTEIN L., 1983b, «Multiobjective approach to river basin planing», *Journal of Water resources planing and management, vol. 109*, p. 13-28.

GERSHON M., DUCKSTEIN L., 1984, «A procedure for the selection of a multiobjective technique with application to water and mineral resources», *Applied Mathematics and Computation, vol. 14, n° 3*, p. 245-271.

GERSHON M., McANIFF R., DUCKSTEIN L., 1980, «Multiobjective river basin planning with qualitative criteria», *American Water Resources Association Proceedings*, Las Vegas, p. 41-50.

GIBBARD A., 1973, «Manipulation of voting schemes: a general result», *Econometrica, vol. 41*, p. 587-601.

GIBBARD A., 1974, «A Pareto-consistent libertarian claim», *Journal of Economic Theory, vol. 7*, p. 338-330.

GLEIZES M.-P., GLIZE P., 1990, *Les systèmes multi-experts*, Hermès, Paris.

GLOVER F., 1975, «Improved linear integer programming formulations of nonlinear integer problems», *Management Science, vol. 22*, p. 455-460.

GOICOECHEA A., HANSEN D.R., DUCKSTEIN L., 1982, *Multiobjective decision analysis with engineering and business applications*, Wiley.

GOICOECHEA A., LI F., 1994, «Evaluating alternative systems with ARIADNE », presented at the XIth International Conference on Multicriteria Decision Making, Coimbra (Portugal).

GOLDSTEIN W.M., 1990, «Judgments of relative importance in decision making: global vs. local interpretations of subjective weights», *Organizational Behavior and Human Decision Processes, vol. 47*, p. 313-336.

GOMES L.F.A.M., LIMA M.M.P.P., 1992, «From modeling individual preferences to multicriteria ranking of discrete alternatives: a look at prospect theory and the additive difference model», *Foundations of Computing and Decision Sciences, vol.17, n°3*, p. 171-184.

GONZALEZ C., 1996, «Additive utilities when some components are solvable and others are not», *J. of Mathematical psychology 40 (2)*, p. 141-151.

GRABISCH M. and ROUBENS M., 1999, «An axiomatic approach to the concept of interaction among players in cooperative games», *Int. Journal of Game theory*, to appear.

GRAUER M., LEWANDOWSKI A., WIERZBICKI A.P., 1984, «DIDASS, Theory, Implemen- tation and Experiences», in *Interactive Decision Analysis*, GRAUER M., WIERZBICKI A. Eds., Lecture Notes in Economics and Mathematical Systems 229, Springer, p. 22-30.

GRAUER M., WIERZBICKI A.P., Eds., 1984, *Interactive decision analysis*, Proceedings, Laxenburg, Austria, 1983, Springer.

GRECO S., MATARAZZO B., SLOWINSKI R., 1999, «The use of rough sets and fuzzy sets in MCDM», in *Multicriteria Decision making, Advances in MCDM models, algorithms, theory and applications*, T. GAL, T.J. STEWART and T. HANNE Eds., Kluwer, p.14-1, 14-59.

GUARISO G., RINALDI S., SONCINI-SESSA R., 1985, «Decision support systems for water management: The Lake Como case study», *European Journal of Operational Research, vol. 21, n° 3*, p. 295-306.

GUIGOU J.L., 1973, *Analyse des données et choix à critères multiples*, Dunod.

HABENICHT W., 1982, «Quad Trees, a Datastructure for discrete vector optimization problems», *Lecture notes in Economics and Mathematical Science n°209*, Springer p. 136-145.

HABENICHT W., 1992, «ENUQUAD: a DSS for discrete vector optimization problems», In *Multicriteria Decision Making Methods, Algorithms and Applications*, M. CERNY, D. GLÜCKAUFOVA and D. LOULA Eds., Czechoslovak Academy of Sciences, p. 66-74.

HADZINAKOS I., YANNACOPOULOS D., FALTSETAS C., ZIOURKAS K., 1991, «Application of the MINORA decision support system to the evaluation of landslide favourability in Greece», *European Journal of Operational Research, vol. 50, n° 1*, p. 61-75.

HAIMES Y.Y., CHANKONG V., Eds., 1985, *Decision making with multiple objectives*, Springer.

HALLEFJORD A., JÖRNSTEN K., ERIKSSON O., 1986, «A long range forestry planning problem with multiple objectives», *European Journal of Operational Research, vol. 26*, p. 123-133.

HANSEN P., Ed., 1983, *Essays and surveys on multiple criteria decision making*, Springer.

HANSEN P., ANCIAUX M., VINCKE P., 1976, «Quasi-kernels of outranking relations», in *Multiple criteria decision making*, THIRIEZ H., ZIONTS S., Eds., Jouy-en-Josas, Springer, p. 53-63.

HARKER P.T., 1987, «Incomplete pairwise comparisons in the analytic hierarchy process», *Mathematical Modelling, vol. 9, n° 11*, p. 837-848.

HARKER P.T., VARGAS L.G., 1987, «The theory of ratio scale estimation: Saaty's analytic hierarchy process», *Management Science, vol. 33, n° 11*, p. 1383-1403.

HENS L., PASTIJN H., STRUYS W., 1992, «Multicriteria analysis of the burden sharing in the European Community», *European Journal of Operational Research, vol. 59, n° 2*, p. 248-261.

HERSHEY J.C., KUNREUTHER H.C., SCHOEMAKER P.J.H., 1982, «Sources of bias in assessment procedures for utility functions», *Management Science, vol. 28, n° 8*, p. 936-954.

HERWIJNEN VAN M., JANSSEN R., 1989, «DEFINITE: a support system for decisions on a finite set of alternatives», in *Improving decision making in organizations*, LOCKETT A.G., ISLEY G., Eds., 1989, *Proceedings of the Eight International Conference on MCDM*, Manchester, August 1988, Springer, p. 534-543.

HINLOOPEN E., NIJKAMP P., RIETVELD P., 1983, «The REGIME method: a new multicriteria technique», in *Essays and surveys on multiple criteria decision making*, HANSEN P., Ed., Springer, p. 146-155.

HIPEL K.W., FRASER N.M., COOPER A.F., 1990, «Conflict analysis of the trade in services dispute», *Information and Decision Technologies, vol. 16, n° 4*, p. 347-360.

HITCH C.J., 1953, «Suboptimization in operations problems», *Operations Research, vol. 1*, p. 89.

HOBBS B.F., 1979, «Analytical Multiobjective decision methods for power plant sitting: a review of theory and applications», *Working Paper*, Brookhaven National Laboratory, Upton.

HOLDER R.D., 1990, «Some comments on the analtic hierarchy process», *Journal of the Operational Research Society, vol. 41, n° 11*, p. 1073-1076.

HONDA N., MIMAKI T., 1986, «Multiobjective decision method using heuristic rules», in *Artificial Intelligence in Economics and Management*, PAU L.F., Ed., p. 157-165.

HONG H.B., VOGEL D.R., 1991, «Data and Model Management in a generalized MCDM- DSS», *Decision Sciences vol. 22*, p. 1-25.

HORSKY D., RAO M.R., 1984, «Estimation of attribute weights from preference comparisons», *Management Science, vol. 30, n° 7*, p. 801-822.

HUGONNARD J.-C., ROY B., 1983, «Le plan d'extension du métro en banlieue parisienne», in *Méthodes de décision multicritère*, JACQUET-LAGRÈZE E., SISKOS J., Eds., 1983, Monographies de l'AFCET, France, p. 39-65.

HWANG C., MASUD A.S.M., 1979, *Multiple objective decision making. Methods and applications survey*, Springer.

HWANG C., YOON K., 1981, *Multiple attribute decision making.Methods and applications survey*, Springer.

IGNIZIO J. P., 1976, *Goal Programming and Extensions*, Heath, Lexington.

ISERMANN H., 1977, «The enumeration of the set of all efficient solutions for a multiple objective linear program», *Operations Research Quaterly, vol. 28*, p. 711-725.

ISERMANN H., 1978, «Duality in MOLP», in *Multicriteria Problem Solving*, ZIONTS S. Ed., Lecture Notes in Economics and Mathematical Systems 155, Springer, Heidelberg, p. 274-285.

IZ P. H. and GARDINER L. R., 1992, «Analysis of Multiple Criteria Decision Support Systems for Cooperative Groups», Working Paper.

JACQUET-LAGRÈZE E., 1969, «L'aggrégation des opinions individuelles», *Informatique et Sciences Humaines, n° 4*.

JACQUET-LAGRÈZE E., 1978, «De la logique d'agrégation de critères à une logique d'agrégation-désagrégation de préférences et de jugements», *Cahiers de l'ISMEA, tome 13*, p. 839-859.

JACQUET-LAGRÈZE E., 1982, «A behavioral model of the decision process and an application for designing a decision support system», Nato Research Institute on *Understanding and aiding Human Decision Making*, Williamsburgh.

JACQUET-LAGRÈZE E., 1983, «Concepts et modèles en analyse multicritère», in JACQUET-LAGRÈZE E., SISKOS J., Eds., 1983, *Méthodes de décision multicritère*, Monographies de l'AFCET, France, p. 7-37.

JACQUET-LAGRÈZE E., 1990, «Interactive assessment of preferences using holistic judgements: the PREFCALC system», in *Readings in multiple criteria decision making*, BANA E COSTA C.A., Ed., 1990, Springer, p. 335-350.

JACQUET-LAGRÈZE E., MEZIANI R., SLOWINSKI R., 1987, «MOLP with an interactive assesment of a piecewise linear utility function», *European Journal of Operational Research, vol. 31*, p. 350-357.

JACQUET-LAGRÈZE E., SHAKUN M.F., 1984, «Decision support systems for semi-structured buying decisions», *European Journal of Operational Research, vol. 16, n° 1*, p. 48-58.

JACQUET-LAGRÈZE E., SISKOS J., 1982, «Assessing a set of additive utility functions for multicriteria decision-making ; the UTA method», *European Journal of Operational Research, vol. 10*, p. 151-164.

JACQUET-LAGRÈZE E., SISKOS J., Eds., 1983, *Méthodes de décision multicritère*, Monographies de L'AFCET, France.

JAFFRAY J.-Y., 1974, «On the extension of additive utilities to infinite sets», *Journal of Mathematical Psychology, vol. 11, n° 4*, p. 431-452.

JAFFRAY J.-Y., 1982, *Cours de troisième cycle*, Polycopié de l'université P. et M. Curie, Paris.

JAFFRAY J.-Y., 1988, «Choice under risk and the security factor: an axiomatic model», *Theory and Decision, vol. 24*, p. 169-200.

JAFFRAY J.-Y., POMEROL J.-Ch., 1989, «A direct proof of the Kuhn-Tucker necessary optimality theorem for convex and affine inequalities», *SIAM Review, vol. 31, n° 4*, p. 671-674.

JAHN J., 1999, «Theory of vectormaximisation: various concepts of efficient solutions», in *Multicriteria Decision making, Advances in MCDM models, algorithms, theory and applications*, T. GAL, T.J. STEWART and T. HANNE Eds., Kluwer, p.2-1,2-32.

JANSSEN R., 1992, *Multiobjective decision support for environmental management*, Kluwer.

JANSSEN R., HERWIJNEN VAN M., 1991, «Graphical decision support applied to decisions changing the use of agricultural land», in *Multiple Criteria Decision Support*, KORHONEN P., LEWANDOWSKI A., WALLENIUS J., Eds., Lecture Notes in Economics and Mathematical Systems 356, p. 293-302.

JANSSEN R., NIJKAMP P., RIETVELD P., 1990, «Qualitative multicriteria methods in the Netherlands», in BANA E COSTA C.A., Ed., 1990, *Readings in multiple criteria decision making*, Springer, p. 383-409.

JARKE M., JELASSI T., STOHR E.A., 1984, «A data-driven user interface generator for a generalized multiple criteria decision support system», *Working Paper, n° 84-72*, New-York University.

JARKE M., RADERMACHER F.J., 1988, «The AI-Potential of model management and its central role in decision support», *Decision Support Systems, vol. 4, n° 4*, p. 387-404.

JELASSI T., 1986, «MCDM: «From «Stand-Alone», Methods to Integrated and Intelligent DSS», in *Toward Interactive and Intelligent DSS*, SAWARAGI Y., INOUE K., NAKAYAMA H., Eds, Lecture Notes in Economics and Mathematical Systems 286, Springer, Berlin, p. 90-99.

JELASSI T, JARKE M., STOHR E.A., 1985, «Designing a Generalized Multiple Criteria Decision Support System», *Journal of Management Information Systems, vol. 1, n° 4*, p. 24-43.

KADEN S., 1984, «Analysis of regional water policies in open cast mining areas. A multicriteria approach», in *Interactive decision analysis*, GRAUER M., WIERZBICKI A.P., Eds., 1984, Proceedings, Laxenburg, Austria, 1983, Springer, p. 218-226.

KAHNEMAN D., TVERSKY A., 1979, «Prospect theory: an analysis of decision under risk», *Econometrica, vol. 47*, p. 263-292.

KALLIO M., LEWANDOWSKI A., ORCHARD-HAYS W., 1980, «An implementation of the reference point approach for multiobjective optimization», *Working Paper, n° 35*, IASA, Laxenburg, p. 1-29.

KANT J.-D., 1995, «CATEG-ART: a neural network for automatic extraction of human categorization rules», in Proceedings of the international conference on artificial neural networks, Paris, p.479-484.

KANT J.-D., LEVINE D.S., 1997, «A neural network for decision making under the influence of reinforcement», Proceedings of the international conference on artificial neural networks, Houston.

KARKAZIS J., 1989, «Facilities location in a competitive environment: A Promethee based multiple criteria analysis», *European Journal of Operational Research, vol. 42, n° 3*, p. 294-304.

KARPAK B., ZIONTS S., Eds., 1989, *Multiple criteria decision making and risk analysis using computers*, Springer.

KARWAN M.H., ZIONTS S., VILLAREAL B., RAMESH R., 1985, «An improved interactive multicriteria integer programming algorithm», in *Decision making with multiple objectives*, HAIMES Y.Y., CHANKONG V., Eds., 1985, Springer, p. 261-271.

KASANEN E., ÖSTERMARK R., ZELENY M., 1991, «Gestalt system of holistic graphics: new management support view of MCDM», *Computers and Operations Research, vol. 18, n° 2*, p. 233-239.

KEEN P.G.W., 1977, «The evolving concept of optimality», in *Multiple Criteria Decision Making*, STARR M.K., ZELENY M., Eds., TIMS Study in Management Sciences 6, orth Holland, p. 31-57.

KEEN P.G.W., 1987, «Decision support systems: the next decade», *Decision Support Systems, vol. 3*, p. 253-265.

KEEN P.G.W., SCOTT MORTON M.S., 1978, *Decision Support Systems*, Addison Wesley, Reading.

KEENEY R.L., 1988, «Value-driven Expert Systems for Decision Support», *Decision Support Systems vol. 4, n° 4*, p. 405-412.

KEENEY R. L., 1992, *Value-Focused thinking*, Harvard University Press.

KEENEY R.L., MCDANIELS T.L., 1992, «Value-focused thinking about strategic decisions at BC Hydro», *Interfaces, vol. 22, n° 6*, p. 94-109.

KEENEY R.L., RAIFFA H., 1976, *Decisions with multiple objectives: Preferences and value tradeoffs*, Wiley.

KEENEY R.L., WOOD E.F., 1977, «An illustrative example of the use of multiattribute utility theory for water resources planning», *Water Resources Research, vol. 13, n° 4*, p. 705-712.

KELLY J.S., 1978, *Arrow impossibility theorems*, Academic Press.

KEMENY J.G., SNELL J.L., 1962, *Mathematical models in the social sciences*, MIT Press, Cambridge, USA.

KENDALL M., 1970, *Rank correlation methods*, Charles Griffin, London, 4ᵉ édition

KEPNER C. H. and TREGOE B. B., 1965, *The rational Manager : A systematic Approach to Problem Solving and Decision Making*, Mc Graw-Hill, New-York.

KEPNER C.H., TREGOE B.B.., 1981, *The new rational manager*, Charles Kepner Associates, Princeton, USA.

KERNI R., SANCHEZ P. and RAO TUMMALA V. M., 1990, «A comparative study of multiattribute decision making methodologies », *Theory and Decision 29*, p. 203-222.

KHAIRULLAH Z.Y., 1982, A study of algorithms for multicriteria decision making, Doctoral Thesis, School of Management, State University of New-York at Buffalo.

KHAIRULLAH Z.Y., ZIONTS S., 1981, «An empirical evaluation of some multiple criteria methods for discrete alternatives», in *Organizations: Multiple agents with multiple criteria*, MORSE J.N., Ed., 1981, Springer, p. 171.

KHAIRULLAH Z.Y., ZIONTS S., 1987, «An approach for preference ranking of alternatives», *European Journal of Operational Research, vol. 28*, p. 329-342.

KHORRAMSHAHGOL R., MOUSTAKIS V.S., 1988, «Delphic hierarchy process, DHP: a methodology for priority setting derived from the Delphic method and analytical hierarchy process», *European Journal of Operational Research, vol. 37, n° 3*, p. 347-354.

KIRKWOOD C.W., VAN DER FELTZ L.C., 1986, *Personal Computer Programs for Decision Analysis: User's manual*, Arizona State University, Technical Report DIS-86/87-4.

KISS L.N., MARTEL J.M., NADEAU R., 1992, «ELECCALC. An interactive software for modeling the decision maker's preference», *submitted to Decision Support Systems*.

KLAHR C.N., 1958, «Multiples objectives in mathematical programming», *Operations Research, vol. 6*, p. 849-855.

KLEE A.J., 1971, «The role of decision models in the evaluation of competing environmental health alternatives», *Management Science, vol. 18, n° 2*, p. B52-B67.

KNOLL A.L., ENGELBERG A., 1978, «Weighting multiple objectives. The Churchman-Ackoff technique revisited», *Computers and Operations Research, vol. 5*, p. 165-177.

KOKSALAN M., 1984, *Discrete multiple criteria decision*, Unpublished Doctoral Dissertation, Dpt. Industrial Engineering, State University of New-York at Buffalo.

KOKSALAN M., 1989, «Identifying and ranking a most preferred subset of alternatives in the presence of multiple criteria», *Naval Research Logistics, vol. 36, n° 4*, p. 359-372

KOKSALAN M., KARWAN M.H., ZIONTS S., 1984, «An improved method for solving multiple criteria problems involving discrete alternatives», *IEEE Trans. on Systems, Man and Cybernetics, vol. 14, n° 1*, p. 24-34.

KOKSALAN M., KARWAN M.H., ZIONTS S., 1986, «Approaches for discrete alternative multiple criteria problems for different types of criteria», *IIE Transactions, vol. 18, n° 3*, p. 262-270.

KOKSALAN M., TANER O.V., 1992, «An approach for finding the most preferred alternative in the presence of multiple criteria», *European Journal of Operational Research, vol. 60, n° 1*, p. 52-60.

KOOPMANS T.C., 1951, «Activity Analysis of Production and Allocation», *Cowles Commission Monograph, n° 13*, John Wiley, New-York.

KORHONEN P., 1986, «A hierarchical interactive method for ranking alternatives with multiple qualitative criteria», *European Journal of Operational Research, vol. 24*, p. 265-276.

KORHONEN P., 1988, «A visual reference direction approach to solving discrete multiple criteria problems», *European Journal of Operational Research, vol. 34, n° 2*, p. 152-159.

KORHONEN P., 1991, «Using harmonious houses for visual pairwise comparison of multiple criteria alternatives», *Decision Support Systems, vol. 17*, p. 47-54.

KORHONEN P., KARAIVANOVA J., 1999, «An algorithm for Projecting a Reference direction onto the Nondominated Set of given Points», *IEEE SMC 29-5*, p. 429-435.

KORHONEN P., LAAKSO J., 1986, «A visual interactive method for solving the multiple criteria problem», *European Journal of Operational Research, vol. 24*, p. 277-287.

KORHONEN P., LEWANDOWSKI A., WALLENIUS J., Eds., 1991, *Multiple criteria decision support*, Springer.

KORHONEN P., MOSKOWITZ H., WALLENIUS J., 1986, «A progressive algorithm for modelling and solving multiple criteria decision problems», *Operations Research, vol. 34*, p. 726-731.

KORHONEN P., NARULA S.C., WALLENIUS J., 1989, «An evolutionary approach to decision-making, with an application to media selection», *Mathematical Computation and Modelling, vol. 12, n° 10/11*, p. 1239-1244.

KORHONEN P., SOISMAA M., 1981, «An interactive criteria approach to ranking alternatives», *Journal of the Operational Research Society, vol. 32*, p. 577-585.

KORHONEN P., SOISMAA M., 1988, «A multiple criteria model for pricing alcoholic beverages», *European Journal of Operational Research, vol. 37, n° 2*, p. 165-175.

KORHONEN P., WALLENIUS J., 1986, «Some theory and an approach to solving sequential multiple criteria decision problems», *Journal of the Operational Research Society, vol. 37*, p. 501-508.

KORHONEN P., WALLENIUS J., ZIONTS S., 1984, «Solving the discrete multiple criteria problem using convex cones», *Management Science, vol. 30, n° 11*, p. 1336-1345.

KORNBLUTH J.S.H., 1978, «Ranking with multiple objectives», in *Multiple criteria problem solving*, ZIONTS S., Ed., 1978, Springer, p. 345-361.

KOTTEMANN J.E., DAVIS D.R., 1991, «Decisional conflict and user acceptance of multicriteria decision-making aids», *Decision Sciences, vol. 22*, p. 918-926.

KRANTZ D.H., LUCE R.D., SUPPES P., TVERSKY A., 1971, *Foundations of Measurement, vol. 1*, Academic Press, New-York.

KRUSKAL J.B., WISH M., 1978, *Multidimensional scaling*, Sage publications, USA.

KRZYSZTOFOWICZ R., KOCH J.B., 1989, «Estimation of cardinal utility based on a nonlinear theory», *Annals of Operations research, vol. 19*, p. 181-204.

KUHN H.W., TUCKER A. W., 1951, «NonLinear Programming», *Proceedings of the second Berkeley Symposium on Mathematical Statistics and Probability*, University of California Press, Berkeley.

KUNG H.I, LUCIO F., PREPARATA F.P., 1975, «On finding the maxima of a set of vectors», *Journal of the ACM, vol. 25*, p. 469-478.

LABAT J.-M., FUTTERSACK M., 1992, «Analyse multicritère et système expert – un SIAD pour l'affectation des aides en faveur des chômeurs de longue durée», *Journal of Decision Systems, vol 1, n° 1*, p. 79-92.

LAI Y.J., LIU T-Y., HWANG C-L., 1994, «TOPSIS for MODM», *European Journal of Operational Research, vol. 76, n° 3*, p. 486-500.

LANCASTER K., 1971, *Consumer demand, a new approach*, Columbia University Press, New-York.

LANGLOTZ C.P., SHORTLIFFE E.H., 1989, «Logical and Decision-Theoretic Methods for Planning Uncertainty», *AI Magazine*, Spring 89, p. 39-47.

LANSDOWNE Z.F. 1996, «Ordinal ranking methods for multicriterion decision making», *Naval Research Logistics, vol. 43*, p. 613-627.

LARICHEV O.I., NIKIFOROV A.D., 1987, «Analytical survey of procedures for solving multicriteria mathematical programming problem (MMPP)», in *Decision Support Systems and MCDM*, Kyoto proceedings, p. 400-414.

LARICHEV O.I., POLYAKOV O.A., NIKIFOROV A.D., 1987, «Multicriterion linear programming problems (Analytical Survey)», *Journal of Economic Psychology, vol. 8*, p. 389-407.

LARICHEV O. I. and MOSHKOVICH H. M., 1991, «ZAPROS : a Method and System for ordering multiattribute alternatives on the base of a decision maker's preferences», preprint, Moscou.

LARICHEV O. I. and MOSHKOVICH H. M., 1997, *Verbal Decision Analysis for Unstructured problems*, Kluwer.

LARICHEV O. I. and MOSHKOVICH H. M., MECHITOV A. I. and OLSON D. L. , 1993, « Experiments Comparing Qualitative Approches to Rank Ordering of Multiattribute Alternatives », *J. of Multicriteria Decision Analysis 2*, p. 5-26.

LAVIALLE O., VIDAL C., 1992, «Recherche d'un quasi ordre à éloignement faible d'un profil de relations binaires quelconques», *36ᵉ Journées du groupe européen «Aide Multicritère à la Décision»*, Luxembourg.

LECLERCQ J.-P., 1984, «Propositions d'extension de la notion de dominance en présence des relations d'ordre sur les pseudo-critères: La méthode MELCHIOR», *Revue Belge de Statistique, d'Informatique et de Recherche Opérationnelle, vol. 24, n° 1*, p. 32-46.

LEONTIEF W.W., 1947, «Introduction to a theory of the internal structure of functional relationships», *Econometrica, vol. 51*, p. 361-373.

LEVEN S.J., LEVINE D.S., 1996, «Multiattribute decision making in context: a dynamic neural network methodology», Cognitive science 20, p. 271-299.

LÉVINE P., MAILLARD J.-Ch., POMEROL J.-Ch., 1987, «An intelligent support system for strategic decisions», in *Expert Systems and Artificial Intelligence in Decision Support Systems*, SOL H.G., TAKKENBERG C.A. Th., DE VRIES ROBBÉ P.F., Eds., Reidel, p. 247-255.

LÉVINE P., POMEROL J.-Ch., 1983, «An interactive program for the discrete multiobjective problem (extended abstract)», *Methods of Operations Research, vol. 45*, p. 479-480.

LÉVINE P., POMEROL J.-Ch., 1986, «PRIAM, an interactive program for choosing among multiple attribute alternatives», *European Journal of Operational Research, vol. 25*, p. 272-280.

LÉVINE P., POMEROL J.-Ch., 1989, *SIAD et systèmes experts*, Hermès, Paris.

LÉVINE P., POMEROL J.-Ch., 1991, «Negotiation support systems: an overview and some knowledge based examples», in *Defense Decision Making*, AVENHAUS R., KARKAR H., RUDNIANSKI M., Eds., Springer, Berlin, p. 241-256.

LÉVINE P., POMEROL J.-Ch., SANEH R., 1986, «Bridging an Expert System and a Multicriteria Decision Support System», *Presented at EUROVIII*, Lisbon.

LÉVINE P., POMEROL J.-Ch., 1995, «The role of the decision maker in DSSs and representation levels», in *Proceedings of the 28th Hawaii International Conference on System Science*, NUNAMAKER J.F. and SPRAGUE R.H., Eds., IEEE, vol. 3, p. 42-51.

LÉVINE P., POMEROL M.-J., SANEH R., 1990, «Rules integrate Data in a Multicriteria Decision Support System», *IEEE Transactions on Systems Man and Cybernetics, vol. 20, n° 3*, p. 678-686.

LEWANDOWSKI A., GRANAT J., 1991, «Dynamic BIPLOT as the interaction interface for aspiration based decision support systems», in *Multiple Criteria Decision Support*, KORHONEN P., LEWANDOWSKI A., WALLENIUS J., Eds., Lecture Notes in Economics and Mathmatical Systems 356, p. 229-241.

LIANG G.-S., 1999, «Fuzzy MCDM based on ideal and anti-ideal concepts», *European Journal of Operational Research 112*, p. 682-691.

LINDBLOM C.E., 1959, «The Science of Muddling Through», *Public Administration Review, vol. 19*, p. 79-88.

LINDEN J. van der, STIJNEN H., 1995, *QUALIFLEX version 2.3. A software package for multi-criteria analysis*, Kluwer.

LIROV Y., 1991, «Algorithmic multiobjective heuristics construction in the A* search», *Decision Support Systems 7*, p. 159-167.

LOCKETT A.G., ISLEY G., Eds., 1989, *Improving decision making in organizations*, Proceedings of the Eight International Conference on MCDM, Manchester, August 1988, Springer.

LOOTSMA F.A., 1987, «Modélisation du jugement humain dans l'analyse multicritère au moyen de comparaisons par paires», *RAIRO/Recherche Opérationnelle, vol. 21*, p. 241-257.

LOOTSMA F.A., 1992 a, «The REMBRANDT system for multicriteria decision analysis via pairwise comparisons or direct rating», TU DELFT Report, n° 92-05, Delft University of Technology.

LOOTSMA F.A., 1992 b, «Category scaling within a given context as a common basis for the AHP, SMART and ELECTRE», TU DELFT Report, n° 92-22, Delft University of Technology.

LOOTSMA F. A., 1994, «The relative importance of the criteria in the multiplicative AHP and SMART», Report 94-07, University of Michigan.

LOOTSMA F. A. and SCHUIJT, 1997, «The multiplicative AHP, SMART, and ELECTRE in a common context», *Journal of Multi-Criteria Decision Analysis 6*, p. 185-196.

LOTFI V., STEWART T.J., ZIONTS S., 1992, An Aspiration-level Interactive Model for multiple criteria decision making, *Computers and Operations Research, vol. 19, n° 7*, p. 671-681.

LOTFI V., TEICH J.E., 1991, «Multicriteria decision making using personal computers», in *Multiple criteria decision support*, KORHONEN P., LEWANDOWSKI A., WALLENIUS J., Eds., 1991, Springer, p.152-158.

LOTFI V., ZIONTS S., 1990, «AIM, Aspiration-level Interactive Method for Multiple Criteria Decision Making», *User's Guide*, State University of New-York at Buffalo.

LUCE R.D., 1956, «Semiorders and a theory of utility discrimination», *Econometrica, vol. 24*, p. 178-191.

LUCE R.D., RAIFFA H., 1957, *Games and Decisions*, John Wiley, New-York.

LUCE R.D., TUKEY J.W., 1964, «Simultaneous conjoint measurement», *Journal of Mathematical Psychology, vol. 4*, p. 1-27.

LUGASSI Y., MEHREZ A., SINUANY-STERN Z., 1985, «Nuclear power plant site selection: a case study», *Nuclear Technologies, vol. 69*, p. 7-13.

MACCRIMMON K.R., 1973, «An overview of multiple objective decision making», in *Multiple criteria decision making*, COCHRANE J.L., ZELENY M., Eds., University of South Carolina Press, Columbia, USA, p. 18-44.

MAJCHRZAK J., 1985, «DISCRET — A package for multicriteria optimization and decision problems with discrete alternatives», in *Plural rationality and interactive decision process*, GRAUER M., THOMPSON M., WIERZBICKI A.P., Eds., 1985, Proceedings, Sopron, Hungary, 1984, Springer, p. 319-324.

MAJCHRZAK J., 1989, «A methodological guide to the decision support system DISCRET for discrete alternatives problems», in *Aspiration based decision support systems*, LEWANDOWSKY A., WIERZBICKI A., Eds., 1989, Springer, p. 193-212.

MAKAROV I.M., VINOGRADSKAYA T.M.., RUBCHINSKY A.A. and SOKOLOV V., 1987, *The Theory of Choice and Decision Making*, MIR, Moscou.

MALAKOOTI B., 1988, «A decision support system and a heuristic interactive appraoch for solving discrete multiple criteria problems», *IEEE Transactions on Systems Man and Cybernetics, vol. 18, n° 2*, p. 273-284.

MALAKOOTI B., ZHOU Y.Q., 1992, «Feedforward artificial neural networks for solving discrete MCDM problems», *Management Science 40*, p. 1542-1561.

MALMBORG C.J., JONES J.E., JONES M.S., 1986, «IMAP II: a microcomputer based decision support system for multiattribute value function assessment», *Computers and Industrial Engineering, vol. 11, n° 1-4*, p. 232-235.

MANDIC N.J., MAMDANI E.H., 1984, «A multi-attribute decision- making model with fuzzy rule-based modification of priorites», in *Fuzzy sets and decision analysis*, ZIMMERMANN H.J., ZADEH L.A., GAINES B.R., Eds., 1984, North-Holland, p. 285-306.

MANHEIM M.L., HALL F., 1967, «Abstract representation of goals: A method for making decisions in complex problems», in *Proceedings of Transportation: a service*, New York Academy of Sciences, New-York.

MARCOTORCHINO F., MICHAUD P., 1979, *Optimisation en analyse ordinale des données*, Masson, Paris.

MARESCHAL B., 1986, «Stochastic multicriteria decision making and uncertainty», *European Journal of Operational Research, vol. 26*, p. 58-64.

MARESCHAL B., 1988, «Weight stability intervals in multicriteria decison aid», *European Journal of Operational Research, vol. 33*, p. 54-64.

MARESCHAL B., BRANS J.-P., 1988, «Geometrical Representations for MCDA», *European Journal of Operational Research, vol. 34*, p. 69-77.

MARESCHAL B., BRANS J.-P., 1991, «BANKADVISER: an industrial evaluation system», *European Journal of Operational Research, vol. 54, n° 3*, p. 318-324.

MARTEL J.-M., AOUNI B., 1991, «Méthode multicritère de choix d'un emplacement: le cas d'un aéroport dans le nouveau Québec», *Cahier du LAMSADE, n° 104*, Université de Paris-Dauphine.

MASSAM B.H., ASKEW I.D., 1982, «Methods for comparing policies using multiple criteria: an urban example», *OMEGA, vol. 10, n° 2*, p. 195-104.

MASSAM B.H., SKELTON I., 1986, «Application of three plan evaluation procedures to a highway alignment problem», *Transportation Research Record, 1076*, p. 54-58.

MATARAZZO B., 1986, «Multicriterion analysis of preferences by means of pairwise actions and criterion comparisons», *Applied Mathematics and Computations, vol. 18*, p. 119-141.

MATARAZZO B., 1988, «Preference ranking global frequencies in multicriterion analysis, PRAGMA», *European Journal of Operational Research, vol. 36, n° 1*, p. 36-49.

MATARAZZO B., 1990, «A pairwise criterion comparison approach: the MAPPAC and PRAGMA methods», in *Readings in multiple criteria decision making*, BANA E COSTA C.A., Ed., 1990, Springer, p. 253-273.

MATARAZZO B., 1991, «MAPPAC as a compromise between outranking methods and MAUT», *European Journal of Operational Research, vol. 54, n° 1*, p. 48-65.

MAY K.O., 1954, «Intransitivity, Utility, and the agregation of preference patterns», *Econometrica, vol. 22*, p. 1-13.

MAYSTRE L.Y., PICTET J., SIMOS J., 1994, *Méthodes Multicritères ELECTRE*, Presses Polytechniques et Universitaires Romandes, Lausanne.

MAYSTRE L.Y., BOLLINGER D., 1999, *Aide à la negociation Multicritère: pratique et conseils*, Presses Polytechniques et Universitaires Romandes, Lausanne.

MCCORD M., DE NEUFVILLE R., 1983a, «Empirical demonstration that expected utility decision analysis is not operational», in *Foundations of Utility and Risk Theory*, STIGUM B.P., WENSTOP F., Eds., Reidel, Dordrecht.

MCCORD M., DE NEUFVILLE R., 1983b, «Fundamental deficiency of expected utility analysis», in *Multiobjective Decision-Making*, FRENCH S., HARTLEY R., L. THOMAS L.C., WHITE D.J., Eds., Academic Press, Londres, p. 279-305.

MCCORD M., DE NEUFVILLE R., 1986, «Lottery equivalents: reduction of the certainty effect problem in utility assessment», *Management Science, vol. 32*, p. 56-60.

MEHREZ A., MOSSERY S., SINUANY-STERN Z., 1982, «Project selection in small university», *R&D Management, vol. 12*, p. 169-174.

MEHREZ A., SINUANY-STERN Z., 1983a, «An interactive approach for project selection», *Journal of the Operational Research Society, vol. 34*, p. 621-626.

MEHREZ A., SINUANY-STERN Z., 1983b, «Ressource allocation to interrelated risky project using a multiattribute utility function», *Management Science, vol. 29*, p. 430-439.

MILI F., 1984, Réalisation d'un système hiérarchique de décision multicritère: PRIAMH, thèse de 3ᵉ cycle de l'université P. et M. Curie, Paris.

MILLER G.A., 1956, «The magical number seven, plus or minus two: Some limits on our capacity for processing information», *Psychological Review, vol. 63*, p. 81-97.

MINTZBERG H., 1975, «The manager's job : Folklore and facts», *Harward Business Review, July/Aug*, p. 46-61.

MITCHELL T.R. and BEACH L. R., 1990, «Do I love thee? Let me Count..." Toward an understanding of Intuitive and Automatic Decision Making», *Organizational Behavior and Human Decision Processes 47*, p. 1-20.

MLADINEO N., MARGETA J., BRANS J.-P., MARESCHAL B., 1987, «Multicriteria ranking of alternative locations for small scale hydroplants», *European Journal of Operational Research, vol. 31*, p. 215-222.

MOND B., ROSINGER E.E., 1985, «Interactive weight assessment in multiple attribute decision making», *European Journal of Operational Research, vol. 22*, p. 19-25.

MONTERO J., PEARMAN A.D., TEJADA J., 1995, «Fuzzy multicriteria decision support for budget allocation in the transport sector», *TOP, vol. 3, n° 1*, p. 47-68.

MONTGOMERY H., 1983, «Decision rules and the search for a dominance structure: Towards a process model of decision making», in *Analysing and aiding decision processes*, HUMPHREYS P., SVENSON O., VARI A., Eds., 1983, North-Holland, p. 343-369.

MORSE J.N., Ed., 1981, *Organizations: multiple agents with multiple criteria*, Springer.

MOSCAROLA J., 1978, «Multicriteria decision aid: two application in education management», in *Multiple Criteria Problem Solving*, ZIONTS S., Ed., Springer, New- York, p. 402-423.

MOSCAROLA J., SISKOS J., 1983, «Analyse a posteriori d'une étude d'aide à la décision en matière de gestion de réseau de distribution», in *Méthodes de décision multicritère*, JACQUET-LAGRÈZE E., SISKOS J., Eds., 1983, Monographies de l'AFCET, p. 143-167.

MOSKOWITZ H., PRECKEL P.V., YANG A., 1991, «Multiple Criteria Robust Interactive Decision Analysis (MCRID): a tool for multiple criteria decision support», in *Multiple Criteria Decision Support*, KORHONEN P., LEWANDOWSKI A., WALLENIUS J., Eds., Lecture Notes in Economics and Mathematical Systems 356, p. 118-127.

MOSKOWITZ H., PRECKEL P.V., YANG A., 1992, «Multiple Criteria Robust Interactive Decision Analysis (MCRID) for optimizing public policies», *European Journal of Operational Research, vol. 56*, p. 219-236.

MOTE J., OLSON D.L., VENKATARAMAN M.A., 1988, «A comparative Multiobjective Programming study», *Mathematical and Computer Modelling 10*, p. 719-729.

MOUSSEAU V., 1992a, «Analyse et classification de la littérature traitant de l'importance relative des critères en aide multicritère à la décision», *revue Française de recherche opérationnelle, vol. 26, n° 4*, p. 367-389.

MOUSSEAU V., 1992b, «Are judgments about relative importance of criteria dependent or independent of the set of alternatives ? An experimental approach», *Cahier du LAMSADE, n° 111*, université de Paris-Dauphine.

MOUSSEAU V., 1995, «Eliciting information concerning the relative importance of criteria», in *Advances in Multicriteria Decision Aid*, PARDALOS, SISKOS, ZAPOUNIDIS (Eds.), Kluwer.

MOUSSEAU V., SLOWINSKI R., ZIELNIEWICZ P., 1999, «ELECTRE TRI 2.0a, Methodological Guide and User's Manual», *Document du Lamsade # 111*, Université de Paris-Dauphine.

NAGEL S., LONG J., 1986, «P/G% analysis: a decision-aiding program», in *Artificial intelligence in economics and management*, PAU L.F., Ed., North-Holland, p. 137-145.

NARASIMHAN R., VICKERY S.K., 1988, «An experimental evaluation of articulation of preferences in multiple criteria decision making (MCDM) methods», *Decision Sciences, vol. 19*, p. 880-888.

NEUFVILLE DE R., MCCORD M., 1984, «Unreliable measurement of utility: significant problems for decision analysis», in *Operational Research 84*, BRANS J.-P., Ed., Elsevier, p. 354-366.

NEUMANN VON J., MORGENSTERN O., 1944, *Theory of Games and Economic Behavior*, Princeton University Press.

NEWELL A., SIMON H. A., 1972, *Human Problem Solving*, Prentice-Hall, Englewood Cliffs.

NIJKAMP P., 1979, *Multidimensional spatial data and decision analysis*, Wiley, New-York.

NIJKAMP P., VAN DELFT A., 1977, *Multicriteria analysis and regional planning*, Martinus Nijhoff, The Netherlands.

NIJKAMP P., SPRONK J., 1981, *Multiple criteria analysis: operational methods*, Gower Press, London.

NIJKAMP P., VOOGD H., 1985, «An informal introduction to multicriteria evaluation», in *Multiple criteria decision methods and applications*, FANDEL G., SPRONK J., Eds., 1985, Springer, p. 61-84.

NILSSON N.J., 1982, *Principles of Artificial Intelligence*, Springer.

NITZSCH von R., WEBER M., 1993, «The effect of attribute ranges on weights in multiattribute utility measurements», *Management Science*, vol. 39, p. 937-943.

NUTT P.C., 1980, «Comparing methods for weighting decision criteria», *OMEGA, vol. 8, n° 2*, p. 163-172.

O'LEARY D.E., 1986, «Multiple criteria decision making in accounting expert systems», *Comptes rendus des 6ᵉ journées internationales sur les systèmes experts et leurs applications*, Avignon, p. 1017-1035.

OLSON D. L., 1996, *Decision Aids for Selection Problems*, Springer, New-York.

OLSON D. L., MECHITOV A. I. and MOSHKOVITCH H. M., 1998, «Cognitive Effort and Learning Features of Decision Aids : Review of Experiments», *Journal of Decision Systems 7*, p. 129-146.

OLSON D. L., MOSHKOVITCH H. M., SCHELLENBERGER R. and MECHITOV A. I., 1995, «Consistency and Accuracy in Decision Aids : Experiments with four Multiattribute Systems», *Decision Sciences 26*, p. 723-748.

ORAL M., KETTANI O., 1990, «Modeling outranking process as a mathematical program- ming problem», *Computers and Operations Research, vol. 17, n° 4*, p. 411-423.

OSYCZKA A., 1984, *Multicriterion optimization in engineering with Fortran programs*, Ellis Horwood, London.

OZERNOY V.M., 1988, «Multiple criteria decision making in the USSR: a survey», *Naval Research Logistics, vol. 35*, p. 543-566.

OZERNOY V.M., 1991, «Developing an interactive Decision Support for discrete alternative MCDM Method Selection», in *Multiple Criteria Decision Support*, KORHONEN P., LEWANDOWSKI A., WALLENIUS J., Eds., Lecture Notes in Economics and Mathematical Systems 356, p. 242-257.

OZERNOY V. M., 1992, « Choosing the « best » multiple criteria decision-making method », *INFOR 30*, p. 159-171.

PAELINCK J.H.P., 1976, «Qualitative multiple criteria analysis, environmental protection and multiregional development», *Papers of the Regional Science Association, vol. 36*, p. 59-74.

PAELINCK J.H.P., 1977, «Qualitative multiple criteria analysis: An application to airport location», *Environment and Planning, vol. 9*, p. 883-895.

PAELINCK J.H.P., 1983, *Formal spatial economic analysis*, Gower Press, Aldershot.

PARIZEK P., BERTOK I., VASKO T., 1991, «MDS, a decision support software for evaluation of discrete alternatives», in *Multiple criteria decision support*, KORHONEN P., LEWANDOWSKI A., WALLENIUS J., Eds., 1991, Springer, p. 175-182.

PASCHE C., 1991, «EXTRA: An expert system for multicriteria decision making», *European Journal of Operational Research 52*, p. 224-234. ·

PASCHETTA E., TSOUKIAS A., 1999, «A real world MCDA application: evaluating software», *Document LAMSADE # 113*, Unniversité de Paris-Dauphine.

PASSY U., LEVANON Y., 1980, «Manpower allocation with multiple objectives — the min- max approach», in *Multiple Criteria Decision Making: Theory and Applications*, FANDEL G., GAL T., Eds., Springer, New-York, p. 329-344.

PASTIJN H., LEYSEN J., 1989, «Constructing an outranking relation with ORESTE», in *Models and methods in multiple criteria decision making*, COLSON G., DE BRUIN C., Eds., 1989, Pergamon Press, p. 1255-1268.

PEARMAN A.D., MACKIE P.J., MAY A.D., SIMON D., 1989, «The use of multicriteria techniques to rank highway investment proposals», in *Improving Decision Making in Organizations*, Proceedings of the Eight International Conference on MCDM, LOCKETT A.G., ISLEIG., Eds., Manchester, August 1988, Springer, p. 157-165.

PEKELMAN D., SEN S.K., 1974, «Mathematical programming models for the determination of attribute weights», *Management Science, vol. 20*, p. 1217-1229.

PÉREZ J., 1991, *Propiedades de Consistencia en los Métodos de la Decisión Multicriterio Discreta*, Tesis Doctoral, Universidad de Alcalá, Madrid.

PÉREZ J., 1994, «Theoretical elements of comparison among ordinal discrete multicriteria methods», *Journal of Multicriteria Decision Analysis, vol. 3, n° 3*, p. 157-176.

PÉREZ J., 1995, «Some comments on Saaty's AHP», *Management Science, vol. 41, n° 6*, p. 1091-1095.

PÉREZ J., BARBA-ROMERO S., 1995, «Three practical criteria of comparison among ordinal preference aggregating rules», *European Journal of Operational Research, vol. 85, n° 3*, p.473-487.

PERNY P., 1992, Modélisation, agrégation et exploitation de préférences floues dans une problématique de rangement, thèse de l'université de Paris-Dauphine.

PERNY P., 1998 «Multicriteria Filtering Methods based on Concordance and Non-Discordance Principles», in *Special issue on Preference Modelling, Annals of Operations Research, 80*, p. 137-165.

PERNY P. and POMEROL J.-Ch., 1999, «Use of Artificial Intelligence in MCDM», in *Multicriteria Decision making, Advances in MCDM models, algorithms, theory and applications*, T. GAL, T.J. STEWART and T. HANNE Eds., Kluwer, p.15-1,15-32.

PERNY P., ROY B., 1992, «The use of fuzzy outranking relations in preference modelling», *Fuzzy Sets and Systems*, vol. 49, p. 33-53.

PERNY P., VANDERPOOTEN D., 1998, «An interactive Multiobjective Procedure for selecting medium-term countermeasures after nuclear accidents», *Journal of Multi-Criteria Decision Analysis 7*, p. 48-60.

PETROVIC S., 1998, «Case-based Reasoning in a DSS for Multicriteria analysis», in *Decision Support Systems*, T. JELASSI (Ed.), *Journal of Decision Systems*, Special Issue, p. 99-119.

PFAFF R., DUCKSTEIN L., 1981, «Transportation design decisions via multicriterion Q-analysis (MCQA)», *ORSA/TIMS meeting*, Houston.

PHILLIPS L. D., 1990, «Decision analysis for group decision support», in *Tackling strategic problems: the role of group decision support*, EDEN C. F., RADFORD G., Eds., Sage Publications, London.

PIRLOT M., VINCKE Ph., 1992, «Lexicographic agregation of semiorders», *Journal of Multi-Criteria Decision Analysis, vol. 1*, p. 47-58.

PIRLOT M., VINCKE Ph., 1997, *Semi-orders, Properties, Representations, Applications*, Kluwer.

POINCARÉ H., 1902, *La science et l'hypothèse*, Flammarion, Paris.

POMEROL J.-Ch., 1985, *MULTIDECISION, un logiciel d'aide à la décision multicritère*, Manuel de référence, STRATEMS, Paris.

POMEROL J.-Ch., 1988, «MCDM and AI: Heuristics, Expert and Multi-Expert systems», *Presented at EURO IX*, Paris.

POMEROL J.-Ch., 1990, «Systèmes experts et SIAD: enjeux et conséquences pour les organisations», *Technologies de l'Information et Société, vol. 3, n° 1*, p. 37-64.

POMEROL J.-Ch., 1992, «Editorial», *Journal of Decision Systems 1*, p. 11-13.

POMEROL J.-Ch., 1993, «Multicriteria DSSs: State of the art and problems», *Central european Journal for Operations Research and Economics 3*, p. 197-211.

POMEROL J.-Ch., 1998, «Scenario development and practical decision making under uncertainty: robustness and "risk control"», in *Proceedings CESA-98*, P. BORNE, H. KSOURI, A. EL KAMEL Eds., vol. 2, p. 238-242.

POMEROL J.-Ch., 1999, «Scenario development and practical decision making under uncertainty», in *Proceedings ISDSSS-99*, International Society for DSS, p. 1-9.

POMEROL J.-Ch., BARBA-ROMERO S., 1993, *Choix multicritère dans l'entreprise. Principes et pratique*, Hermes, Paris.

POMEROL J.-Ch., BRÉZILLON P., 1997, «Organizational experiences with multicriteria decision support systems», in *Proceedings of the 30th Hawaii conference on System Science*, NUNAMAKER J.F. and SPRAGUE R.H. Eds., vol. 3, p. 90-98.

POMEROL J-Ch., ROY B., ROSENTHAL-SABROUX C., SAAD A., 1995, «An «Intelligent» DSS for the Multicriteria Evaluation of Railway Timetables», *Foundations of Computing and Decision Science 20*, p. 219-238.

POMEROL J.-Ch., TRABELSI T., 1987, «An adaptation of PRIAM to continuous multi- objective linear programming», *European Journal of Operational Research, vol. 31*, p. 335-341.

POULTON E.C., 1977, «Quantitative subjective judgements are almost always biased, sometimes completely misleading», *British Journal of Psychology, vol. 68*, p. 409-425.

QUINLAN J., 1979, «Discovering rules by induction from large collections of examples», in *Expert Systems in the micro-electronic age*, D. MICHIE Ed., Edinburgh University Press.

QUINLAN J., 1983, «Learning efficient classification procedures and their application to chess end-games», in *Machine learning: an AI approach*, R. MICHALSKI, T. MITCHELL and J. CARBONNELL Eds., Tioga Press.

RAJKOVIC V., BOHANEC M., BATAGELV V., 1988, «Knowledge Engineering Techniques for Utility Identification», *Acta Psychologica 68*, p. 271-286.

RAMANUJAN V., SAATY T.L., 1981, «Technological choices in less developing countries», *Technological Forecasting and Social Changes, vol. 19*, p. 81-98.

REYES J., BARBA-ROMERO S., 1986, «Expert systems and multicriteria decision making», presented at *EURO VIII, Eight European Conference on Operational Research*, Lisbon.

RICHARD J.-L., 1983, «Aide à la decision stratégique en P.M.E.», in *Méthodes de décision multicritère*, JACQUET-LAGRÈZE E., SISKOS J., Eds., 1983, Monographies de l'AFCET, p. 119-142.

RIDGLEY M., 1989, «Water and urban land use planning in Cali, Colombia», *Journal of Water Resource Planning and Management, vol. 115, n° 5*, p. 753-774.

RIETVELD P., 1980, *Multiple objective decision methods and regional planning*, North Holland, Amsterdam.

RIETVELD P., 1984, «The use of qualitative information in macroeconomic policy analysis», in *Macroeconomic planning with conflicting goals*, DESPONTIN M., NIJKAMP P., SPRONK J., Eds., 1984, Proceedings Brussels, 1982, Springer, p. 263-280.

RIETVELD P., OUWERSLOOT H., 1992, «Ordinal data in multicriteria decision making, a stochastic dominance approach to sitting nuclear power plants», *European Journal of Operational Research, vol. 56, n° 2*, p. 249-262.

RÍOS INSUA D. and FRENCH S., 1991, «A framework for sensitivity analysis in discrete multi-objective decision making», *European Journal of Operational Research 54*, p. 176-190.

RIVETT B.H.P., 1977, «Multidimensional scaling for multiobjective policies», *OMEGA, vol. 5, n° 4*, p. 367-379.

ROBERTS F.S., 1979, *Measurement theory with applications to decision making, utility and the social sciences*, Addison-Wesley.

ROCKAFELLAR R.T., 1970, *Convex Analysis*, Princeton University Press, New Jersey.

ROGERS M., BRUEN M., MAYSTRE L.-Y., 2000, *ELECTRE and Decision Support, Methods and applications in Engineering and Infrastructure Investment*, Kluwer.

ROMMEL Y., 1989, Apport de l'Intelligence Artificielle à l'Aide à la décision multicritère, thèse de l'université de Paris-Dauphine, Paris.

ROMMEL Y., 1991, «Une procédure interactive reposant sur des techniques d'apprentissage symbolique: bibliographie sur l'apprentissage symbolique par induction à partir d'exemples et présentation de la procédure interactive Plexiglas», *cahier LAMSADE, n° 67*, université de Paris Dauphine, Paris.

ROSINGER E.E., 1991, «Beyond preference information based multiple criteria decision making», *European Journal of Operational Research, vol. 53*, p. 217-227.

ROUBENS M., 1982, «Preference relations on actions and criteria in multicriteria decision making», *European Journal of Operational Research, vol. 10*, p. 51-55.

ROUBENS M., VINCKE Ph., 1984, «On families of semiorders and interval orders imbedded in a valued structure of preference: a survey», *Information Sciences, vol. 34*, p. 187-198.

ROUBENS M., VINCKE Ph., 1985, «Preference Modelling», *Lecture Notes in Economics and Mathematical Systems 250*, Springer, Heidelberg.

ROY B., 1968a, «Classement et choix en présence de points de vue multiples, la méthode ELECTRE», *R.I.R.O., vol. 2, n° 8*, p. 57-75.

ROY B., 1968b, «Il faut désoptimiser la recherche opérationnelle», *Bulletin de l'AFIRO, n° 7*, Paris, p. 1.

ROY B., 1971, «Problems and methods with multiple objective functions», *Mathematical Programming vol. 1*, p. 239-266.

ROY B., 1974, «Management scientifique et aide à la décision», *Direction Scientifique de la SEMA rapport, n° 86*, Paris.

ROY B., 1975, «Vers une méthodologie générale d'aide à la décision», *Metra, vol. 14, n° 3*, p. 459-497.

ROY B., 1976, «Optimisation et aide à la décision», *Journal de la société de statistique de Paris, tome 117, n° 3*, p. 208-215.

ROY B., 1977, «Critique et dépassement de la problématique de l'optimisation», *Cahiers de la Sema, n° 1*, p. 65-79; repris dans *AFCET/INTERFACES, n° 13*, novembre 1983, p. 35-39.

ROY B., 1978, «ELECTRE III: Un algorithme de rangement fondé sur une représentation floue des préférences en présence de critères multiples», *Cahiers du Centre d'études de recherche operationnelle, vol. 20*, p. 3-24.

ROY B., 1981, «The optimisation problem formulation: Criticism and overstepping», *Journal of the Operational Research Society, vol. 32*, p. 427-436.

ROY B., 1985, *Méthodologie Multicritère d'Aide à la Decision*, Economica, Paris. English translation: *Multicriteria Methodology for Decision Aiding*, Kluwer, 1996.

ROY B., 1987a, «Meaning and validity of interactive procedures as tools for decision making», *European Journal of Operational Research, vol. 31, n° 3*, p. 297-303.

ROY B., 1987b, «Des critères multiples en recherche opérationnelle: pourquoi ?», *Cahier du LAMSADE, n° 80*, université de Paris-Dauphine.

ROY B., 1990a, «The outranking approach and the foundations of ELECTRE methods», in *Readings in multiple criteria decision making*, BANA E COSTa C.A., Ed., 1990, Springer, p. 155-183.

ROY B., 1990b, «Decision-aid and decision-making», *European Journal of Operational Research, vol. 45*, p. 324-331.

ROY B., 1990c, «Science de la décision ou science de l'aide à la décision ?», *Cahier du LAMSADE n° 97*, Université de Paris-Dauphine, revised in 1992.

ROY B., 1999, «Decision-Aiding Today: What should we expect», in *Multicriteria Decision making, Advances in MCDM models, algorithms, theory and applications*, T. GAL, T.J. STEWART and T. HANNE Eds., Kluwer, p.1-1, 1-35.

ROY B., BERTIER P., 1973, «La méthode ELECTRE II. Une application au media-planning», in *Operations Research '72*, Dublin 1972, ROSS M., Ed., 1973, North-Holland, p. 291-302.

ROY B., BOUYSSOU D., 1985, «An exemple of comparison of two decision-aid models», in *Multiple Criteria Decision Methods and Applications*, FANDEL G., SPRONK J., Eds., p. 361-381.

ROY B., BOUYSSOU D., 1987, «Aide multicritère à la décision: méthodes et cas», *documents LAMSADE, n° 37-41-42*, Université de Paris-Dauphine. Repris dans ROY B., BOUYSSOU D., 1993.

ROY B., BOUYSSOU D., 1993a, *Aide multicritère à la décision: méthodes et cas*, Economica, Paris.

ROY B. and BOUYSSOU D.,1993b, «Decision-aid: an Elementary Introduction with Emphasis on Multiple Criteria», *Investigacion Operativa 3*, p. 175-190.

ROY B. and FIGUEIRA J., 1998, «Détermination des poids des critères dans les méthodes de type ELECTRE avec la technique de Simos révisée», Cahier du Lamsade n° 109, Paris.

ROY B., HUGONNARD J.-C., 1982, «Ranking of suburban line extension projects on the Paris metro system by a multicriteria method», *Transportation Research, vol. 16A, n° 4*, p. 301-312.

ROY B., PRÉSENT M., SILHOL D., 1986, «A programming method for determining which Paris metro stations should be renovated», *European Journal of Operational Research, vol. 24*, p. 318-334.

ROY B., SKALKA J.-M., 1985, «ELECTRE IS, Aspects méthodologiques et guide d'utilisation», *Cahier du LAMSADE, n° 30*, université de Paris-Dauphine, Paris.

ROY B., VINCKE Ph., 1982, «Relational systems of preference with one or several pseudo-criteria: new concepts and new results», *Cahier du LAMSADE, n° 28 bis*, université de Paris-Dauphine, Paris.

SAAD A., 1989, ARGUMENT: un générateur de systèmes experts basé sur les schémas de dépendance fonctionnelle, thèse de l'université P. et M. Curie, Paris.

SAATY T.L., 1977, «A scaling method for priorities in hierarchical structures», *Journal of Mathematical Psychology, vol. 15, n° 3*, p. 234-281.

SAATY T.L., 1980, *The Analytic Hierarchy Process*, McGrawHill.

SAATY T.L., 1987, «Rank generation, preservation and reversal in the analytic hierarchy process», *Decision Sciences, vol. 18*, p. 157-177.

SAATY T.L., 1990, «An exposition of the AHP in reply to the paper «Remarks on the analytic hierarchy process»», *Management Science, vol.36, n°3*, p. 259-268.

SAATY T.L., 1995, *Decision Making for Leaders: The Analytic Hierarchy Process for Decisions in a Complex World*, RWS Publications, 3'd edition, Pittsburgh, USA.

SAATY T.L., VARGAS L.G., 1982, *The logic of priorities*, Kluwer-Nijhoff, Boston.

SAATY T.L., VARGAS L.G., 1984a, «Inconsistency and rank preservation», *Journal of Mathematical Psychology, vol. 28, n° 2*, p. 205-214.

SAATY T.L., VARGAS L.G., 1984b, «Comparison of eigenvalue, logarithmic least squares and least squares methods in estimating ratios», *Journal of Mathematical Modeling, vol. 5*, p. 309-324.

SAATY T.L., VARGAS L.G., 1984c, «The legitimacy of rank reversal», *OMEGA, vol. 12, n° 5*, p. 513-516.

SADAGOPAN S., RAVINDRAN A., 1986, «Interactive algorithms for multiple criteria nonlinear programming problems», *European Journal of Operational Research, vol. 25*, p. 247-257.

SAGE A.P., WHITE III C.C., 1984, «ARIADNE: A knowledge-based interactive system for planning and decision support», *IEEE Trans.on Systems, Man and Cybernetics, vol. SMC-14, n° 1*, p. 35-47.

SALMINEN P., 1992, «Solving the discrete multiple criteria problem using linear prospect theory», to appear in *European Journal of Operational Research*.

SALMINEN P., HOKKANEN J., LAHDELMA R., 1998, «Comparing multicriteria methods in the context of environmental problems», *European Journal of Operational Research 104*, p. 485-496.

SALMINEN P., KORHONEN P., WALLENIUS J., 1989, «Testing the form of decision-maker's multiattribute value function based on pairwise preference information», *Journal of the Operational Research Society, vol. 40, n° 3*, p. 299-302.

SALO A.A., HAMALAINEN R.P., 1995, «Preference programming through approximate ratio comparisons», *European Journal of Operational Research, vol. 82, n° 3*, p. 458-475.

SAMUELSON P.A., 1938, «The empirical implications of utility analysis», *Econometrica, vol. 6*, p. 344-356.

SARIN R.K., 1977, «Interactive evaluation and bound procedure for selecting multi-attributed alternatives», in *Multiple Criteria Decision Making*, STARR M.K., ZELENY M., Eds., North Holland, Amsterdam, p. 211-224.

SARIN R.K., DYER J.S., NAIR K., 1980, «A comparative evaluation of three approaches for preference function assessment», *Working paper, ORSA/TIMS meeting*, Washington.

SAVAGE L. J., 1954, *The Foundations of Statistics*, John Wiley, New-York.

SAWARAGI Y., INOUE K., NAKAYAMA H., Eds., 1987, *Towards interactive and intelligent decision support systems*, Springer.

SAWARAGI Y., NAKAYAMA H., TANINO T., 1985, *Theory of Multiobjective Optimization*, Academic Press.

SCHÄRLIG A., 1985, *Décider sur plusieurs critères*, Presses Polytechniques et Universitaires Romandes, Lausanne.

SCHÄRLIG A., 1996, *Pratiquer ELECTRE et PROMETHEE*, Presses Polytechniques et Universitaires Romandes, Lausanne.

SCHÄRLIG A., PASCHE D., 1980, «La multiplication des ratios, une méthode multicritère préférable à la somme dans les problèmes de localisation industrielle», *SPUR*, université Catholique de Louvain, WP 8006, Avril 1980.

SCHILLING D.A., REVELLE C., COHON J., 1983, «An approach to the display and analysis of multiobjective problems», *Socio-Economic Planning Sciences, vol. 17, n° 2*, p. 57-63.

SCHLAGER K., 1968, «The rank-based expected value method of plan evaluation», *Highway Research Record, n° 238*, p. 153-158.

SCHOEMAKER P.J.H., WAID C.C., 1982, «An experimental comparison of different approaches to determining weights in additive utility models», *Management Science, vol. 28*, p. 182-196.

SCHONER B., CHOO E. U. WEDLEY W.C., 1997, « A comment on 'Rank disagreement: A comparison of multi-criteria methodologies'», *Journal of MultiCriteria Decision Analysis, vol. 6, n°. 4*, p. 197-200.

SCHONER B., WEDLEY W.C., 1989, «Ambigous Criteria Weights in AHP: consequences and solutions», *Decision Sciences, vol. 20*, p. 462-475.

SCHUIJT, H. 1994, «Comparison Between ELECTRE III and AHP/ REMBRANDT?», Working Paper.

SCOTT D., 1964, «Measurement structures and linear inequalities», *Journal of Mathematical Psychology, vol. 1*, p. 233-247.

SCOTT D., SUPPES P., 1958, «Foundational aspects of theories of measurement», *Journal of Symbolic Logic, vol. 23*, p. 113-128.

SEN A.K., 1970, *Collective Choice and Social Welfare*, Oliver and Boyd, London, réédité North Holland, 1979.

SEN A. K., 1977, «Social choice theory: a reexamination», *Econometrica, vol. 45, n° 1*, p. 53-89.

SEO F., SAKAWA M., 1988, *Multiple criteria decision analysis in regional planning: concepts, methods and applications*, Reidel.

SHANNON C.E., WEAVER W., 1949, *The mathematical theory of communication*, University of Illinois Press, Chicago.

SHAPIRA Z., 1981, «Making trade-offs between job attributes», *Organizational Behavior and Human Decision Making 28*, p. 331-355.

SHARDA R., BARR S.H., MCDONNEL J.C., 1988, «Decision support systems effectiveness: a review and an empirical test», *Management Science, vol. 34*, p. 139-159.

SHORTLIFFE E.H., 1976, *Computer-based medical consultation*, MYCIN, Elsevier, Amsterdam.

SIEGEL S., 1957, «Level of aspiration and decision making», *Psychological Review, vol. 64*, p. 253-262.

SIMON H.A., 1955, «A behavioral model of rational choice», *Quaterly Journal of Economics, vol. 69*, p. 99-118. Repris dans *Models of Man*, Wiley, New-York, 1957, p. 241-260.

SIMON H.A., 1956, «Rational choice and the structure of the environment», *Psychological Review, vol. 63*, p. 129-138.

SIMON H.A., 1977, *The new science of management decision*, revised edition, Prentice- Hall, Englewood Cliffs.

SIMON H.A., 1983, *Reason in Human Affairs*, Basil Blackwell, Oxford.

SIMOS J., 1990, *Evaluer l'impact sur l'environnement : une approche originale par l'analyse multicritère et la négociation*, Presses Polytechniques et Universitaires romandes, Lausanne.

SISKOS J., 1982, «Evaluating a system of furniture retail outlets using an interactive ordinal regression method», *Cahier du LAMSADE, n° 38*, université de Paris- Dauphine.

SISKOS J., 1986, «Evaluating a system of furniture retail outlets using an interactive ordinal regression method», *European Journal of Operational Research, vol. 23*, p. 179-193.

SISKOS J., ASSIMAKOPOULOS N., 1989, «Multicriteria highway planning: a case study», *Mathematical Computer Modeling, vol. 12, n° 10-11*, p. 1401-1410.

SISKOS J., LOMBARD J., OUDIZ A., 1986, «The use of multicriteria outranking methods in the comparison of control options against a chemical pollutant», *Journal of the Operational Research Society, vol. 37, n° 4*, p. 357-371.

SISKOS J., WÄSCHER G., WINKELS H-M., 1983, «A bibliography on outranking approaches», *Cahier du LAMSADE, n° 45*, université de Paris-Dauphine.

SISKOS J., YANNACOPOULOS D., 1985, «UTASTAR: An ordinal regression method for building additive value functions», *Investigaçao Operacional, vol. 5*, p. 39-53.

SISKOS J., ZOPOUNIDIS C., 1987, «The evaluation criteria of the venture capital investment activity: An interactive assessment», *European Journal of Operational Research, vol. 31*, p. 304-313.

SKALKA J.-M., BOUYSSOU D., VALLÉE D., 1992, «ELECTRE III et IV, aspects méthodologiques et guide d'utilisation», *Cahier du LAMSADE, n° 25*, 4ª edition, Université de Paris-Dauphine, Paris.

SKULIMOWSKI A.M.J., 1987, «Theoretical foundations for decision support systems based on reference points», *Cahier du LAMSADE, n° 83*, université de Paris-Dauphine.

SKULIMOWSKI A.M.J., 1996, *Decision Support Systems Based on Reference Sets*, Wydawnictwa AGH, Krakow.

SLATER P., 1961, «Inconsistencies in a schedule of paired comparisons», *Biometrika, vol. 48*, p. 303-312.

SMITH G.F., 1989, «Representational effects on the Solving of an Unstructural Decision Problem», *IEEE Transactions Systems Man and Cybernetics vol.19*, p. 1083-1090.

SOLYMOSI T., DOMBI J., 1986, «A method for determining the weights of criteria: the centralized weights», *European Journal of Operational Research, vol. 26*, p. 35-41.

SPENDL R., BOHANEC M., RAJCOVIC V., 1999, «Hierarchical Decision models: Experimental Comparison of AHP and DEX», in *Decision support for the new millenium*, F. BURSTEIN Ed., Proceedings ISDSSS-99, Melbourne.

SPRONK J., 1981, *Interactive multiple goal programming: applications to financial planning*, Martinus Nijhoff, Boston.

SRINIVASAN V., SHOCKER A.D., 1973a, «Linear programming for multidimensional analysis of preferences», *Psychometrika, vol. 38*, p. 337-369.

SRINIVASAN V., SHOCKER A.D., 1973b, «Estimating the weights for multiple attributes in a composite criterion using pairwise judgements», *Psychometrika, vol. 38*, p. 473-493.

SRINIVASAN V., SHOCKER A.D., 1982, «LINMAP, Version IV», *Journal of Marketing Research, vol. 19*, p. 601-602.

SRINIVASA RAJU K., PILLAI C.R.S., 1999, «Multicriterion decision making in river basin planning and development», *European Journal of Operational Research 112*, p. 249-257.

STADLER W., Ed., 1988, *Multicriteria optimization in engineering and in the sciences*, Plenum Press.

STAM A., KUULA M., CESAR H., 1992, «Transboundary air pollution in Europe: an inter- active multicriteria tradeoff analysis», *European Journal of Operational Research, vol. 56, n° 2*, p. 263-277.

STARR M.K., ZELENY M., Eds., 1977, *Multiple criteria decision making*, TIMS Study in Management Sciences, 6, North-Holland, Amsterdam.

STARR M.K., ZELENY M., 1977, «MCDM-State and future of the arts», in *Multiple Criteria Decision Making*, STARR M.K., ZELENY M., Eds., *TIMS Study in Management Sciences 6*, North Holland, p. 5-29.

STEUER R.E., 1977, «An interactive multiple objective linear programming procedure», in *Multiple Criteria Decision Making*, STARR M.K., ZELENY M., Eds., North Holland, Amsterdam, p. 225-240.

STEUER R.E., 1986, *Multiple Criteria Optimization: Theory, Computation, and Application*, John Wiley, New-York.

STEUER R. E., CHOO E.U., 1983, «An interactive weighted Tchebycheff procedure for multiple objective programming», *Mathematical Programming, vol. 26*, p. 326-344.

STEUER R.E., SCHULER A.T., 1978, «An interactive multiple objective linear programming approach to a problem in forest management», *Operations Research, vol. 25, n° 2*, p. 254-269.

STEWART T.J., 1981, «A descriptive approach to MCDM», *Journal of the Operational Research Society, vol. 32*, p. 45-53.

STEWART T.J., 1987, «Pruning of decision alternatives in multiple criteria decision making, based on the UTA method for estimating utilities», *European Journal of Operational Research, vol. 28*, p. 79-88.

STEWART T.J., 1991, «A Multi-Criteria Decision Support System for R&D Project Selection», *Journal of the Operational Research Society, vol. 42, n° 1*, p. 17-26.

STEWART T. J. ,1992, «A critical survey on the status of Multiple Criteria Decision Making Theory and Practice», *Omega 20*, p. 569-582.

STILLWELL W.G., WINTERFELDT VON D., JOHN R.S., 1987, «Comparing hierarchical and nonhierarchical weighting methods for eliciting multiattribute value models», *Management Science, vol. 33, n° 4*, p. 442-450.

STIMSON D.H., 1969, «Utility measurement in public health decision making», *Management Science, vol. 16*, p. B17-B30.

SUN M. and STEUER R. E., 1996, «InterQuad : an interactive Quad Tree based Procedure for solving discrete alternative multiple criteria Problem», *European Journal of Operational Research.*

SUPPES P., KRANTZ D.H., LUCE R.D., TVERSKY A., 1989, *Foundations of Measurement, vol. 2*, Academic Press, New-York.

SZIDAROVSZKY F., GERSHON M.E., DUCKSTEIN L., 1986, *Techniques for multiobjective decision making in systems management*, Elsevier.

TABUCANON M.T., 1988, *Multiple criteria decision making in industry*, Elsevier.

TABUCANON M.T., CHANKONG V., Eds., 1989, *Multiple criteria decision making: applications in industry and service*, Asian Institute of Technology, Bangkok.

TAKEDA E., COGGER K.O., YU P.L., 1987, «Estimating criterion weights using eigenvectors: a comparative study», *European Journal of Operational Research, vol. 29*, p. 360-369.

TANER O.V., KOKSALAN M.M., 1991, «Experiments and an improved method for solving the discrete alternative multiple criteria problem», *Journal of the Operational Research Society, vol. 42, n° 5*, p. 383-392.

TANNER L., 1991, «Selecting a text-processing system as a qualitative multiple criteria problem», *European Journal of Operational Research, vol. 50, n° 2*, p. 179-187.

TEGHEM J., DELHAYE C., KUNSCH P.L., 1989, «An Interactive Decision Support System (IDSS) for Multicriteria Decision Aid», *Mathematical Computation Modelling, vol.12, n° 10*, p. 1311-1320.

TERSSAC DE G., 1992, *Communication Orale*, Toulouse 7/2/92.

THIRIEZ H., ZIONTS S., Eds., 1976, *Multiple criteria decision making*, Proceedings, Jouy-en-Josas 1975, Springer.

TOSTHZAR M., 1988, «Multicriteria decision making approach to computer software evaluation: Application of the analytical hierarchy process», *Mathematical and Computer Modelling, vol. 11*, p. 276-281.

TRIANTAPHYLLOU, E. MANN S. H., 1989, «An examination of the effectiveness of multi-dimensional decision-making methods: a decision-making paradox», *Decision Support Systems, vol. 5*, p. 303-312.

TRIANTAPHYLLOU, E. MANN S. H., 1994, « A computational evaluation of the original and revised Analytic Hierarchy Process», *Computers and Industrial Engineering, vol. 26, n°. 3*, p. 609-618.

TSOUKIAS A., 1991, «Preference modeling as a reasoning process: a new way to face to uncertainty in multiple criteria decision support systems», *European Journal of Operational Research, vol. 55*, p. 309-318.

TSOUKIAS A., 1992, «Extended preference models in decision aid», Presented at *EURO XII / TIMS XXXI*, Helsinki.

TVERSKY A., 1964, «Finite additive structures», *Michigan Mathematical Psychology Programm, n° 64-6*, University of Michigan.

TVERSKY A., 1967a, «A general theory of polynomial conjoint measurement», *Journal of Mathematical Psychology, vol. 4*, p. 1-20.

TVERSKY A., 1967b, «Additivity, utility and subjective probability», *Journal of Mathematical Psychology, vol. 4*, p. 175-201.

TVERSKY A., 1969, «Intransitivity of preferences», *Psychological Review, vol. 76, n° 1*, p. 31-48.

TVERSKY A., 1972a, «Elimination by aspects», *Psychological Review, vol. 79*, p. 281-299.

TVERSKY A., 1972b, «Choice by elimination», *Journal of Mathematical Psychology, vol. 9*, p. 341-367.

TVERSKY A., KAHNEMAN D., 1974, «Judgement under uncertainty: Heuristics and Biases», *Science, vol. 185*, p. 1124-1131.

TVERSKY A., KAHNEMAN D., 1981, «The framing of decisions and the psychology of choice», *Science, vol. 211*, p. 453-458.

TVERSKY A., KAHNEMAN D., 1988, «Rational choice and the framing of decisions», in *Decision Making*, BELL D.E., RAÏFFA H., TVERSKY A., Eds., Cambridge University Press, p. 167-192.

TVERSKY A. and KAHNEMAN D., 1992, «Loss aversion in riskless choice : A reference dependent model», *Quaterly Journal of Economics 107*, p. 1039-1041.

USHER M., ZACKAY D., 1993, «A neural network model for attribute-based decision processes», *Cognitive Science 17*, p. 349-396.

VANDERPOOTEN D., 1989, «The interactive approach in MCDA: a technical framework and some basic conceptions», *Mathematical Computing Modelling, vol. 12, n° 10/11*, p. 1213-1220.

VANDERPOOTEN D, 1990a, «The construction of prescriptions in outranking methods», in *Readings in multiple criteria decision making*, BANA e COSTA C.A., Ed., 1990, Springer, p. 184-215.

VANDERPOOTEN D., 1990b, L'approche interactive dans l'aide multicritère à la décision, thèse de l'université de Paris-Dauphine, Paris.

VANDERPOOTEN D., VINCKE Ph., 1989, «Description and analysis of some representative interactive multicriteria procedures», *Mathematical Computing Modelling, vol. 12, n° 10/11*, p. 1221-1238.

VANSNICK J.-C., 1984, «Strength of preference: theoretical and practical aspects», in *Operational Research 84*, BRANS J.-P., Ed., Elsevier, p. 367-381.

VANSNICK J-C., 1986, «On the problem of weights in multiple criteria decision making, the noncompensatory approach», *European Journal of Operational Research, vol. 24*, p. 288-294.

VANSNICK J.-C., 1987, «Intensity of preference», in *Towards Interactive and Intelligent decision Support Systems*, SAWARAGI Y., INOUE K., NAKAYAMA H., Lecture Notes in Economics and Mathematical Systems 285, Springer, p. 220-229.

VANSNICK J.-C., 1988, *Principes et applications des méthodes multicritères*, Cours de l'université de Mons.

VARGAS L.G., 1990, «Overview of the analytic hierarchy process and its applications», *European Journal of Operational Research, vol. 48*, p. 2-8.

VETSCHERA R., 1986, «Sensitivity analysis for the ELECTRE multicriteria method», *Zeitschrift Operations Research, vol. 30*, p. B99-B117.

VETSCHERA R., 1988, «An interactive outranking system for multiattribute decision making», *Computers and Operations Research, vol. 15, n° 4*, p. 311-322.

VETSCHERA R., 1989, «The IDEAS interactive outranking system: a user-oriented description», *Ludwig Boltzmann Institut working paper 8912*, Wien.

VETSCHERA R., 1992, «A preference-preserving projection technique for MCDM», *European Journal of Operational Research, vol. 61, n° 1-2*, p. 195-203.

VETSCHERA R., 1994, «McView: An integrated graphical system to support multiattribute decisions», *Decision Support Systems, vol. 11*, p. 363-371.

VIDAL C., YEHIA ALCOUTLABI A., 1990, «Méthode d'aide à la décision sur des évaluations multicritères par plusieurs juges», *Mathématiques informatique et sciences humaines, vol. 28, n° 112,* p. 27-36.

VIELI M., 1984, «Une expérience en matière de pondération des critères», presented at *20ᵉ journées sur l'aide à la décision multicritère,* Paris.

VILLAR A., 1988, «La logica de la eleccion social: una revision de los resultados basicos», *Investigaciones Economicas, vol. 12, n° 1,* p. 3-44.

VILLE J., 1946, «Sur les conditions d'existence d'une ophélimité totale et d'un indice du niveau des prix», *Annales de l'université de Lyon, vol. 9, série A,* p. 32-39.

VINCKE Ph., 1976, «Une méthode interactive en programmation linéaire à plusieurs fonctions économiques», *RAIRO, Recherche opérationnelle, vol. 10, n° 6,* p. 5-20.

VINCKE Ph., 1977, «Quasi-ordres généralisés et modélisation des préférences», *Cahier du LAMSADE, n° 9,* université de Paris-Dauphine, Paris.

VINCKE Ph., 1980, «Vrais, quasi, pseudo et précritères dans un ensemble fini: propriétés et algorithmes», *Cahier du LAMSADE, n° 27,* université de Paris-Dauphine, Paris.

VINCKE Ph., 1986, «Analysis of multicriteria decision aid in Europe», *European Journal of Operational Research, vol. 25,* p. 160-168.

VINCKE Ph., 1989, *L'aide multicritère à la décision,* Editions de l'université de Bruxelles. English translation: *Multicriteria Decision Aid,* J. Wiley and Sons, Chichester, 1992.

VINCKE Ph., 1991, «Exploitation d'une relation non valuée dans une problématique de rangement complet», *Document du LAMSADE n°62,* Université de Paris-Dauphine.

VINCKE Ph., 1992, «Exploitation of a crisp relation in a ranking problem», *Theory and Decision, vol. 32,* p. 221-240.

VINCKE Ph., 1995, «A short note on a methodology for choosing a decision-aid method», in *Advances in Multicriteria Analysis,* P. M. PARDALOS, Y. SISKOS and C. ZOPOUNIDIS Eds., Kluwer, p. 3-6.

VINCKE Ph., 1999a, «Robust and neutral methods for aggregating preferences into an outranking relation», *European Journal of Operational Research, vol. 112,* p. 405-412.

VINCKE Ph., 1999b, «Outranking approach», in *Multicriteria Decision making, Advances in MCDM models, algorithms, theory and applications,* T. GAL, T.J. STEWART and T. HANNE Eds., Kluwer, p. 11-1, 11-29.

VOOGD H., 1983, *Multicriteria evaluation for urban and regional planning,* Pion, London.

VUK D., KOZELJ B., MLADINEO N., 1991, «Application of multicriterional analysis on the selection of the location for disposal of communal waste», *European Journal of Operational Research, vol. 55, n° 2,* p. 211-217.

WAKKER P.P., 1986, «The repetition appraoch to characterize cardinal utility», *Theory and Decision, vol. 17,* p. 33-40.

WAKKER P.P., 1989, *Additive Representations of Preferences,* Kluwer, Dordrecht.

WALLENIUS J., 1975, «Comparative evaluation of some interactive approaches to multicriterion optimization», *Management Science, vol. 21, n° 12,* p. 1387-1396.

WANG J., MALAKOOTI B., 1992, «A feedforward neural network for multiple criteria decision making», *Computer and Operations Research, vol. 19,* p. 151-167.

WEBER M. and BORCHERDING K., 1993, «Behavioral influences on weight judgments in multiattribute decision making», *European Journal of Operational Research 65,* p. 1-12.

WEBER M., EISENFÜHR F., von WINTERFELDT D., 1988, «The effects of splitting attributes on weights in multiattribute utility measurement», *Management Science, vol. 33, n° 4,* p. 431-445.

WENSTOP F.E., CARLSEN A.J., 1988, «Ranking hydroelectric power projects with multicriteria decision analysis», *Interfaces, vol. 18, n° 4,* p. 36-48.

WESSSELING J.A.M., GABOR A., 1994, «Decision modelling with HIPRE3+», *Computational Economics, vol. 7,* p. 147-154.

WHITE C.C. III, 1990, «A survey on the integration of Decision Analysis and Expert Systems for Decision Support», *IEEE Transactions on System Man and Cybernetics, vol SMC 20, n° 2*, p. 358-364.

WHITE C.C. III., SAGE A.P., 1980, «A multiple objective optimization-based approach to choice making», *IEEE Transactions on Systems, Man and Cybernetics, vol. SMC-10, n° 6*, p. 315-326.

WHITE C.C. III, STEWART B.S., CARRAWAY R.L., 1992, «Multiobjective, preference-based search in acyclic OR-graphs», *European Journal of Operational Research 56*, p. 357-363.

WIERZBICKI A., 1980, «The use of reference objectives in multiobjective optimization», in *Multiple Criteria Decision Making: Theory and Applications*, FANDEL G., GAL T., Eds., Lecture Notes in Economics and Mathematical Systems 177, Springer, Berlin, p. 468-486.

WIERZBICKI A., 1983, «A mathematical basis for satisficing decision making», *Mathematical Modeling 3*, p. 391-405.

WIERZBICKI A., 1984, «Interactive decision analysis and interpretative Computer intelligence», in *Interactive decision Analysis*, GRAUER M., WIERZBICKI A., Eds, Lecture Notes in Economics and Mathematical Systems 229, Springer, p. 2-19.

WIERZBICKI A., 1999, «Reference point Approaches», in *Multicriteria Decision making, Advances in MCDM models, algorithms, theory and applications*, T. GAL, T.J. STEWART and T. HANNE Eds., Kluwer, p. 9-1,9-39.

WILSON R.J., 1972, *Introduction to graph theory*, Longman, London.

WINTERFELDT VON D., EDWARDS W., 1986, *Decision Analysis and Behavioral Research*, Cambridge University Press.

WON J., 1990, «Multicriteria evaluation approaches to urban transportation projects», *Urban Studies, vol.27, n°1*, p. 119-138.

WU De-Shih, TABUCANON M.T., 1990, «Multiple Criteria Decision Support System for Production Management», *Engineering Costs and Production Economics, vol. 20*, p. 203-213.

YOON K., 1987, «A reconciliation among discrete compromise solutions», *Journal of the Operational Research Society, vol. 38, n° 3*, p. 277-288.

YOON K., HWANG C., 1985, «Manufacturing plant location analysis by multiple attribute decision making: Part I — single plant strategy», *International Journal of Production Research, vol. 23*, p. 345-359.

YOON Y.S., LOTFI V., ZIONTS S., 1991, «Convergent aspiration-level search, CASE method for multiple objective linear programming: Theory and comparative tests», University of New-York at Buffalo, *Working Paper, n° 754*.

YOUNG H.P., 1974, «An axiomatization of Borda's rule», *Journal of Economic Theory, vol. 9*, p. 43-52.

YOUNG H.P., LEVENLGLICK A., 1978, «A consistent extension of Condorcet election principle», *Siam J. Applied Mathematics 35*, p. 285-300.

YU P.L., 1973, «A class of solutions for group decision problems», *Management Science, vol. 19, n° 8*, p. 936-946.

YU WEI, 1992, Aide multicritère à la décision dans le cadre de la problématique du tri, thèse de l'université de Paris-Dauphine, Paris.

ZAHEDI F., 1986, «The analytic hierarchy process. A survey of the method and its applications», *Interfaces, vol. 16*, p. 96-108.

ZAPATERO E. G., SMITH C.H. , WEISTROFFER H. R., 1997, «Evaluating multiple-attribute decision support systems», *Journal of Multicriteria Decision Analysis, vol. 6*, p. 201-214.

ZELENY M., 1973, «Compromise programming», in *Multiple criteria decision making*, COCHRANE J.-L., ZELENY M., Eds., 1973, *Proceedings South Carolina 1972*, University of South Carolina Press, Columbia, USA, p. 262-301.

ZELENY M., 1982, *Multiple criteria decision making*, McGrawHill, New-York.

ZIONTS S., Ed., 1978, *Multiple Criteria Problem Solving*, Springer.

ZIONTS S., 1981, «A multiple criteria method for choosing among discrete alternatives», *European Journal of Operational Research,vol. 7*, p. 143-147.

ZIONTS S., 1992, «The state of multiple criteria decision making: past, present and future», in *Multiple Criteria Decision Making*, A. GOICOECHEA, L. DUCKSTEIN and S. ZIONTS Eds., Springer, p. 33-43.

ZIONTS S., WALLENIUS J., 1976, «An interactive programming method for solving the multiple criteria problem», *Management Science, vol. 2, n° 6*, p. 652-653.

ZIONTS S., WALLENIUS J., 1983a, «Identifying efficients vectors: some theory and computational results», *Operations Research, vol. 28, n° 3*, p. 788-793.

ZIONTS S., WALLENIUS J., 1983b, «An interactive multiple linear programming method for a class of underlying non-linear utility functions», *Management Science, vol. 29, n° 5*, p. 519-529.

AUTHOR INDEX

SUBJECT INDEX

The first page reference is to the first appearance of the term, followed by pages in which the term is prominent. Computer multicriterion methods and software are in capitals. Multicriterion methods are listed under that heading, while software is listed under its name.